Lecture Notes in Mathematics

Edited by A. Dold and B. Eckmann

T0214800

976

X. Fernique · P.W. Millar
D.W. Stroock · M. Weber

Ecole d'Eté de Probabilités de Saint-Flour XI – 1981

Edité par P.L. Hennequin

Springer-Verlag
Berlin Heidelberg New York 1983

Auteurs

X. Fernique
Université Louis Pasteur
Institut de Recherche Mathématique Avancée
Laboratoire Associé au C.N.R.S.
Rue du Général Zimmer, 67084 Strasbourg Cédex, France

P.W. Millar
Department of Mathematics, University of California
Berkeley, CA 94720, USA

D.W. Stroock
Department of Mathematics, University of Colorado
Campus Box 426, Boulder, Colorado 80309, USA

M. Weber
Université Louis Pasteur
Institut de Recherche Mathématique Avancée
Laboratoire Associé au C.N.R.S.
Rue du Général Zimmer, 67084 Strasbourg Cédex, France

Editeur

P.L. Hennequin
Université de Clermont II, Complexe Scientifique des Cézeaux
Département de Mathématiques Appliquées
B.P. 45, 63170 Aubière, France

AMS Subject Classifications (1980): 60-02, 60G15, 60G17, 60G57, 60H05, 60H15, 60J60, 62-02, 62C20, 62E20, 62F12, 62F35, 62G20

ISBN 3-540-11987-6 Springer-Verlag Berlin Heidelberg New York
ISBN 0-387-11987-6 Springer-Verlag New York Heidelberg Berlin

Printing and binding: Beltz Offsetdruck, Hemsbach/Bergstr.
2146/3140-543210

INTRODUCTION

La Onzième Ecole d'Eté de Calcul des Probabilités de Saint-Flour s'est tenue du 6 au 22 Juillet 1981 et a rassemblé, outre les conférenciers, une quarantaine de participants. Ceux-ci ont apprécié une nouvelle fois la qualité de l'accueil du Foyer des Planchettes.

Les quatre conférenciers, Messieurs Fernique, Millar, Stroock et Weber, ont entièrement repris la rédaction de leurs cours pour en faire un texte de référence, ce qui explique le délai qu'a nécessité leur publication.

En outre, les exposés suivants ont été faits par les participants durant leur séjour à Saint-Flour :

M. AHMED	Modèles probabilistes pour choisir un itinéraire dans un réseau de trafic
A. ACQUAVIVA	Mesures aléatoires t-régulières
A. ACQUAVIVA	Mesures aléatoires localement finies
P. BALDI	Petites perturbations d'un phénomène peano
D. BAKRY	Semimartingales à deux indices
G. BEN AROUS	Equations stochastiques à coefficients analytiques et séries de Taylor stochastiques
L. BIRGE	Tests robustes pour des variables indépendantes et des chaînes de Markov
M. CHALEYAT-MAUREL	Exemples de grossissement gaussien de la filtration brownienne
G. DESLAURIERS	Refroidissement d'un objet convexe
A. EHRHARD	Lois stables et propriétés de Slepian
L. GALLARDO	Comportement asymptotique d'une classe de chaînes de Markov sur $\mathbb{N}$
L. GALLARDO	Une transformation de Cramer sur le dual de SU(2)
J. GLOVER	Markov processes and their last exit distributions
B. MAISONNEUVE	Sur les chaos de Wiener
D. NUALART	Martingales non fortes à variation indépendante du chemin
D. PICARD - J. DESHAYES	Rupture de modèles : loi asymptotique des statistiques de tests et des estimateurs du maximum de vraisemblance
M. POURCHALTCHI - D. REVUZ	Sur le schéma de remplissage pour les processus récurrents
R. SCHOTT	Marches aléatoires sur les espaces homogènes
M. YOR	Sur un processus associé aux temps locaux browniens

Ces exposés se trouvent dans le numéro 71 des Annales Scientifiques de l'Université de Clermont.

La frappe du manuscrit a été assurée par les Départements de Strasbourg, Berkeley et Boulder et nous remercions pour leur soin et leur efficacité les secrétaires qui se sont chargées de ce travail délicat.

Nous exprimons enfin notre gratitude à la Société Springer Verlag qui permet d'accroître l'audience internationale de notre Ecole en accueillant une nouvelle fois ces textes dans la collection Lecture Notes in Mathematics.

P.L. HENNEQUIN
Professeur à l'Université de Clermont II
B.P. n° 45
F-63170 AUBIERE

LISTE DES AUDITEURS

Mr. AHMED M.	Université de Paris VI
Mr. AZEMA J.	Université de Paris VI
Mr. BAKRY D.	Université de Strasbourg
Mr. BALDI P.	Institut de Mathématique à Pise (Italie)
Mr. BEN AROUS G.	Ecole Normale à Paris
Mr. BERTHUET R.	Université de Clermont-Ferrand II
Mr. BIRGE L.	Université de Paris VII
Mme CHALEYAT-MAUREL M.	Université de Paris VI
Mle CHEVET S.	Université de Clermont-Ferrand II
Mr. COMETS F.	Université de Paris XI
Mr. CURIEL J.	Université Nationale Autonome de Mexico
Mr. DESHAYES J.	Université de Paris VII
Mr. DESLAURIERS G.	Ecole Polytechnique de Montréal (Canada)
Mr. DOZZI M.	Université de Berne (Suisse)
Mr. EDER G.	Université de Linz (Autriche)
Mr. EHRHARD A.	Université de Strasbourg
Mme ELIE L.	Université de Paris VII
Mr. EMERY M.	Université de Strasbourg
Mr. FOURT G.	Université de Clermont-Ferrand II
Mr. GALLARDO L.	Université de Nancy I
Mr. GLOVER J.	Université de Rochester à New-York (U.S.A.)
Mr. GOLDBERG J.	I.N.S.A. de Lyon
Mr. GRAVERSEN S.E.	Université de Aarhus (Danemark)
Mme JULIA O.	Université de Barcelone (Espagne)
Mr. KERKYACHARIAN G.	Université de Nancy I
Mr. LEDOUX M.	Université de Strasbourg
Mr. MAISONNEUVE B.	Université des Sciences Sociales de Grenoble
Mr. MESSULAM P.	Ecole Normale à Paris
Mr. MOGHA G.V.	Université de Clermont-Ferrand II
Mr. MORALES P.	Université de Sherbrooke (Canada)
Mr. NDUMU NGU M.	Faculté des Sciences de Yaoundé (Cameroun)
Mr. NUALART D.	Université de Barcelone (Espagne)
Mle PICARD D.	Université de Paris XI
Mr. PICARD J.	Ecole Normale à Paris
Mr. PIERRE LOTI VIAUD D.	Université de Paris XI
Mr. POIRION F.	Université de Paris VI

Mr. REVUZ D.	Université de Paris VII
Mr. SANTIBANEZ J.	Université de Strasbourg
Mme SANZ SOLE M.	Université de Barcelone (Espagne)
Mr. SCHOTT R.	Université de Nancy I
Mr. SINTES BLANC A.	Université de Barcelone (Espagne)
Mr. WINTZ A.	Université de Strasbourg
Mr. YOR M.	Université de Paris VI

TABLE DES MATIERES

D.W. STROOCK : "SOME APPLICATIONS OF STOCHASTIC CALCULUS TO PARTIAL DIFFERENTIAL EQUATIONS"

M. WEBER : "ANALYSE INFINITESIMALE DE FONCTIONS ALEATOIRES

REGULARITE DE FONCTIONS ALEATOIRES NON GAUSSIENNES

PAR X. FERNIQUE

INTRODUCTION

On se propose de montrer dans ce cours que la plupart des propriétés des fonctions aléatoires gaussiennes découvertes ces quinze dernières années s'appliquent à des classes plus larges de fonctions aléatoires qui peuvent donc être utilisées simplement. On étudie dans le premier chapitre les structures de majoration ; les nombres liant entropie et fonctions aléatoires y jouent un rôle clef comme l'ont montré entre autres les travaux de R.M. Dudley dans le cas gaussien et de G. Pisier dans le cas non gaussien. Dans le second chapitre, on étudie certaines structures liées à l'indépendance et à la symétrie ; les travaux de De Acosta, Marcus et Pisier sont à l'origine de ce chapitre.

Le lecteur constatera l'importance des propriétés des familles symétriques de variables aléatoires à valeurs vectorielles. Nous l'invitons à se reporter aux pages 46 à 55 de l'article de M.B. Marcus et G. Pisier "Random Fourier series with applications to harmonic Analysis".

Dans tout ce cours et sauf mention expresse, les variables aléatoires seront construites sur un espace probabilisé complet noté $(\Omega, \mathcal{G}, P)$.

Ce cours ne contient guère de résultats originaux. Dans le domaine qu'il traite, on a vu déjà ou on verra prochainement paraître des articles ou d'autres cours présentant des résultats nouveaux ou des synthèses d'exposition et dus principalement à M. Marcus et G. Pisier qui ont été ici très largement utilisés. Alors que leurs publications visent plutôt à appliquer l'étude des fonctions aléatoires en Mathématiques Pures, je tente ici de montrer qu'elles sont aussi assez simples à manier pour les Mathématiques Appliquées. Par leur existence même, ces deux points de vue complémentaires prouvent me semble-t-il que la théorie des fonctions aléatoires atteint sa maturité.

CHAPITRE I

STRUCTURES DE MAJORATION DES FONCTIONS ALEATOIRES.

Continuité des trajectoires des écarts aléatoires, applications aux fonctions.

Sommaire : Nous étudions la régularité des fonctions aléatoires

$$D = \{D(\omega; s,t)\ ,\ \omega \in \Omega\ ,\ (s,t) \in T \times T\}$$

vérifiant pour tout triplet (s,t,u) d'éléments de T les relations

$$0 = D(\omega; t,t) \leq D(\omega; s,t) = D(\omega; t,s) \leq D(\omega; t,u) + D(\omega; u,s)\ ,\ \text{p.s.}$$

Nous donnons des conditions suffisantes, en termes d'entropie ou de mesures majorantes, pour que toute version séparable de D ait p.s. ses trajectoires continues. Nous appliquons ensuite les résultats aux fonctions aléatoires de la forme $\varphi(d(X(s),X(t)))$ où φ est une fonction sous-additive sur $\mathbb{R}^+$ et X une fonction aléatoire sur T à valeurs dans un espace métrique (P,d) ; ceci donne des conditions suffisantes pour la continuité des trajectoires de X qui regroupent et même améliorent les conditions précédemment connues dans le seul cas $P = \mathbb{R}$. Nous donnons enfin des conditions nécessaires ayant des formes voisines. Le cas des fonctions aléatoires stables à valeurs dans un espace de Banach séparable est spécialement étudié.

1. Introduction, Notations, Enoncé du résultat principal.

1.1. Nous notons (T,δ) un espace muni d'un écart continu ; sauf mention expresse, T sera compact. Pour tout $u>0$, $N(u)$ sera le plus petit nombre de δ-boules fermées de rayon u recouvrant T (l'usage voudrait que l'on utilise plutôt les δ-boules ouvertes, ça ne change pas grand chose ; je n'y vois que des inconvénients techniques) ; soit Φ une fonction de Young, c'est-à-dire une fonction positive d'une variable positive, continue, paire, convexe et vérifiant :

$$\lim_{x \to 0} \frac{\Phi(x)}{x} = 0\ ,\ \lim_{x \to \infty} \frac{\Phi(x)}{x} = \infty\ ;$$

nous notons $\mathcal{E}(\Phi)$ la classe des fonctions aléatoires sur T, séparables pour δ, $X = \{X(\omega,t), \omega\in\Omega, t\in T\}$ à valeurs dans $\mathbb{R}$ vérifiant :

(1) $$\forall (s,t)\in T\times T, \quad E\,\Phi\{\frac{|X(s)-X(t)|}{\delta(s,t)}\}\leq 1 .$$

Dans certains cas particuliers, R.M. Dudley ([1]), $\Phi(x) = e^{x^2}-1)$, C. Nanopoulos et P. Nobelis ([8], $\Phi(x) = e^{x^\alpha}-1)$, G. Pisier ([9], $\Phi(x) = x^p$, $p>1$) ont montré par des méthodes successives différentes et adaptées à ces cas que si l'intégrale $\int_0 \Phi^{-1}(N(u))du$ est convergente, alors tout élément X de $\mathcal{E}(\Phi)$ a p.s. ses trajectoires continues ; ces résultats amélioraient des résultats d'un type voisin valables dans $\mathbb{R}$ ou dans $\mathbb{R}^d$ seulement et associés aux fonctions x^p ([4], [5], [6]). Nous présentons ici le résultat général pour toute fonction de Young Φ.

Dans une note récente ([10]) et dans le cours parallèle, M. Weber montre que toute fonction aléatoire séparable X liée à une variable aléatoire X_o par la relation :

(2) $$\forall (s,t)\in T\times T, \ \forall u>0, \ P\{|X(s)-X(t)|\geq u\delta(s,t)\}\leq P\{|X_o|\geq u\}$$

a, sous certaines conditions d'intégrabilité liant X_o à N p.s. ses trajectoires bornées et est p.s. continue en chaque point de T ; il donne alors des évaluations de $\sup_T|X|$. Nous généralisons et précisons ce résultat. Soit en effet $X_o = \{X_o(\omega), \omega\in]0,1]\}$ une fonction positive et décroissante sur $\{]0,1],d\omega\}$; nous notons $\mathcal{F}(X_o)$ la classe des fonctions aléatoires sur T, séparables pour δ, $X = \{X(\omega,t), \omega\in\Omega, t\in T\}$ à valeurs dans $\mathbb{R}$ et vérifiant :

(3) $$\forall (s,t)\in T\times T, \ \forall u>0, \ E\{|X(s)-X(t)|I_{|X(s)-X(t)|\geq u}\}\leq \delta(s,t)\int_0^{P\{|X(s)-X(t)|\geq u\}} X_o(\omega)d\omega ;$$

nous montrons que sous certaines conditions d'intégrabilité liant X_o à N, tout élément X de $\mathcal{F}(X_o)$ a p.s. ses trajectoires continues ; le résultat concernant la classe $\mathcal{E}(\Phi)$ en est un corollaire.

1.2. Les hypothèses (1), (2) ou (3) associent à la fonction aléatoire X sur T la fonction aléatoire $D(s,t) = |X(s) - X(t)|$ sur $T \times T$; les manipulations techniques des preuves des résultats annoncés opèrent uniquement sur D, mais nécessitent pour tout couple (s,t) l'intégrabilité de $D(s,t)$; il s'agit là d'une condition très restrictive sur la loi de X. En fait, ces manipulations peuvent opérer sur d'autres fonctions aléatoires sur $T \times T$, par exemple $|X(s) - X(t)|^{\alpha}$, $\alpha \in]0,1]$, suffisamment sous-additive. Cette remarque nous amène à situer l'étude dans un cadre plus général.

Soit $D = \{D(\omega; s,t), \omega \in \Omega, (s,t) \in T \times T\}$ une fonction aléatoire sur $T \times T$; nous disons que D <u>est un écart ou une pseudo-distance aléatoire</u> si pour tout triplet (s,t,u) d'éléments de T, elle vérifie :

(4) $$0 = D(\omega; t,t) \leq D(\omega; s,t) = D(\omega; t,s) \leq D(\omega; t,u) + D(\omega; u,s) \text{ , p.s.}$$

Pour toute fonction X_o, positive et décroissante sur $]0,1]$, nous notons $F(X_o)$ la classe des écarts aléatoires séparables D vérifiant :

(5) $$\forall (s,t) \in T \times T \text{ , } \forall u > 0 \text{ , } E\{D(s,t) I_{D(s,t) \geq u}\} \leq \delta(s,t) \int_o^{P\{D(s,t) \geq u\}} X_o(\omega) d\omega \text{ ;}$$

avec ces notations, le résultat principal sera le suivant :

THEOREME 1.2. : <u>On suppose que l'intégrale</u> $\iint_{\{o < \omega \leq 1 < N(u)\}} X_o(\frac{\omega}{N(u)}) d\omega du$ <u>est finie, alors tout élément</u> D <u>de</u> $F(X_o)$ <u>a p.s. ses trajectoires continues sur</u> $T \times T$ <u>et vérifie, pour toute partie mesurable</u> A <u>de l'espace d'épreuves</u> :

(6) $$E\{[\sup_{T \times T} D] I_A\} \leq 8 \iint_{\{N(u) > 1 \text{ , } o < \omega \leq P(A)\}} X_o(\omega / N(u)) d\omega du \text{ .}$$

<u>Remarque</u> : Soient D un écart aléatoire séparable sur un espace (T, δ) et X_o une variable aléatoire positive qu'on peut supposer réalisée sur $]0,1]$ et décroissante ; supposons-les reliés par la propriété :

(2') $\forall\ (s,t)\in T\times T\ ,\ \forall\ u>0\ ,\ P\{D(s,t)>u\delta(s,t)\}\leq P\{X_o>u\}$;

on a alors par intégration et pour toute fonction f positive croissante sur R^+ :

$$E\{f(\frac{D(s,t)}{\delta(s,t)}\ I_{\delta(s,t)\neq o})\}\leq Ef(X_o)\ .$$

En particulier, D vérifie la propriété (5) et appartient donc à $F(X_o)$. Une partie des résultats qualitatifs de M. Weber apparaît alors comme corollaire du théorème 1.2 ; les résultats quantitatifs sont de nature différente.

1.3. Alors que dans les travaux précédents, les outils de majoration étaient l'inégalité de Borel-Cantelli, l'inégalité de Jensen ou celle de Young, nous utiliserons ici des techniques bien classiques sur les réarrangements qui se révèlent particulièrement adaptées ; elles ont fait précédemment des apparitions partielles dans le domaine des fonctions aléatoires ([3], [7]) et il me semble qu'elles méritent une utilisation systématique ; nous énonçons ci-dessous leur forme classique, leur adaptation utile et une propriété de variations :

LEMME 1.3 : (a) <u>Soient</u> D <u>une variable aléatoire positive intégrable et</u> $\bar{D}$ <u>la fonction équimesurable décroissante sur</u> $]0,1]$ <u>qui lui est associée ; pour toute variable aléatoire</u> f <u>à valeurs dans</u> $[0,1]$, <u>on a alors</u> :

$$E\{D.f\}\leq\int_{o\leq x\leq E(f)}\bar{D}(x)dx\ .$$

(b) <u>Soit</u> X_o <u>une fonction positive décroissante intégrable sur</u> $]0,1]$; <u>pour tout élément</u> D <u>de</u> $F(X_o)$ <u>et pour les mêmes v.a.</u> f , <u>on a aussi</u> :

(7) $$\forall\ (s,t)\in T\times T\ ,\ E\{D(s,t)f\}\leq\delta(s,t)\int_{o<\omega\leq E(f)}X_o(\omega)d\omega\ .$$

<u>Enfin pour tout entier positif</u> n <u>et tout nombre</u> $a\in[0,1]$, <u>on a</u> :

(8) $$\sup\{\sum_{k=1}^{n}\int_o^{p_k}X_o(\omega)d\omega,(p_k)\in[0,1]^n,\ \sum_1^n p_k=a\}=n\int_o^{\frac{a}{n}}X_o(\omega)d\omega\ .$$

2. Démonstration du lemme 1.3 et du théorème 1.2.

2.1. Le résultat (a) du lemme étant rappelé seulement pour sa forme classique et sa parenté avec (b), le résultat (8) par ailleurs se déduisant immédiatement de la concavité de $\{\int_0^x X_o(\omega)d\omega,\ x\in[0,1]\}$ et de la symétrie de la fonction

$$\{\sum_{k=1}^{n}\int_0^{p_k} X_o(\omega)d\omega,\ (p_k)\in[0,1]^n\},$$

nous ne démontrons que le résultat (7) ; nous omettrons les variables s et t.

Notons pour commencer sous les hypothèses (b) du lemme que pour tout nombre réel M et tout couple (λ,μ) de nombres positifs de somme 1, on a :

$$\forall\ \omega\in\Omega,\ (D(\omega)-M)f(\omega)\le(D(\omega)-M)^+\le(D(\omega)-M)\left[\lambda I_{D\ge M}(\omega)+\mu I_{D>M}(\omega)\right];$$

si on choisit M, λ et μ de sorte que :

$$E(f)=\lambda P\{D\ge M\}+\mu P\{D>M\},$$

on obtiendra alors par intégration en ω :

$$E\{D.f\}=ME(f)+E\{(D-M)f\}\le\lambda E\{DI_{D\ge M}\}+\mu\lim_{u\downarrow M}E\{DI_{D\ge u}\};$$

puisque D appartient à $F(X_o)$, ce dernier se majore par :

$$\delta\{\lambda\int_0^{P\{D\ge M\}}X_o(\omega)d\omega+\mu\int_0^{P\{D>M\}}X_o(\omega)d\omega\},$$

et la concavité de $\{\int_0^x X_o(\omega)d\omega,\ x\in[0,1]\}$ majore ensuite par :

$$\delta\int_0^{\lambda P\{D\ge M\}+\mu P\{D>M\}}X_o(\omega)d\omega,$$

c'est le résultat (7).

2.2. La démonstration du théorème aura deux étapes. Dans la première, nous établissons des majorations explicites des trajectoires permettant sous les hypothèses du théorème de montrer qu'elles sont p.s. majorées sur $T\times T$ et vérifient les inégalités (6). La seconde étape utilise des approximations de D par des espérances conditionnelles ; les majorations de la première étape montreront en effet la convergence uniforme presque sûre de certaines de ces espérances.

Dans la première étape, on pourra d'ailleurs se limiter au cas où $P(A)$ est positif, sinon les deux membres de (6) sont nuls, et même où $A = \Omega$; on peut en effet énoncer :

LEMME 2.2.1 : Soient X_o une fonction positive décroissante intégrable sur $]0,1]$ et D un écart aléatoire appartenant à $F(X_o)$ d'espace d'épreuves (Ω,P) ; soit de plus A un sous-ensemble mesurable de Ω non négligeable. On note Q la probabilité $\frac{I_A.P}{P(A)}$ et X'_o la fonction $\{X_o(\omega P(A)), \omega \in]0,1]\}$. Dans ces conditions, l'écart aléatoire D défini sur l'espace d'épreuves (Ω,Q) appartient à $F(X'_o)$ et on a :

$$E_Q\{[\sup_{T\times T} D]\} = \frac{1}{P(A)} E_P\{[\sup_{T\times T} D] I_A\},$$

$$\iint_{\{N(u)>1\}} X'_o\left(\frac{\omega}{N(u)}\right) d\omega\, du = \frac{1}{P(A)} \iint_{\{N(u)>1,\ o<\omega\le P(A)\}} X_o\left(\frac{\omega}{N(u)}\right) d\omega\, du .$$

Démonstration : les égalités indiquées résultant du changement de variables, il suffit de vérifier que D sur (Ω,Q) appartient à $F(X'_o)$. Soient donc un élément (s,t) de $T\times T$ et u positif, on a par définition de Q :

$$E_Q\{D(s,t) I_{D(s,t)>u}\} = \frac{1}{P(A)} E\{D(s,t) I_{A\cap\{D(s,t)>u\}}\} ;$$

ce dernier membre se majore à partir du lemme 1.3.(b), on obtient :

$$\frac{1}{P(A)}\, \delta(s,t) \int_{o\le\omega\le P\{A\cap\{D(s,t)>u\}\}} X_o(\omega) d\omega = \delta(s,t) \int_o^{Q\{D(s,t)>u\}} X'_o(\omega) d\omega$$

et le résultat du lemme est donc établi.

2.2.2. Première étape.

Nous notons u_o la borne supérieure sur R^+ de $\{u : N(u) > 1\}$; pour tout entier n, soit S_n une partie de T de cardinal $N(\frac{u_o}{2^n})$ telle que $\{B(s,\frac{u_o}{2^n}), s\in S_n\}$ recouvre T ; nous notons g_n une application de S_{n+1} dans S_n vérifiant :

$$\forall\, s \in S_{n+1}\,,\ s \in B(g_n(s)\,,\frac{u_o}{2^n})\ ;$$

nous choisissons un entier N et pour tout entier $k \in [0,N]$, nous notons f_k l'application de S_{N+1} dans S_k définie par la composition $g_k \circ g_{k+1} \circ \dots \circ g_N$; nous choisissons un entier $J \leq N$ et nous notons τ une application mesurable de Ω dans S_{N+1} vérifiant :

$$D(\tau, f_J(\tau)) = \sup_{t \in S_{N+1}} D(t, f_J(t))\ ;$$

dans ces conditions, on a immédiatement :

$$(9)\qquad E \sup_{S_{N+1} \times S_{N+1}} |D(s,t) - D(f_J(s), f_J(t))| \leq$$

$$2 \sum_{j=J,\ t \in S_{j+1}}^{N} \sum E\{D(t, g_j(t)) I_{\{f_{j+1}(\tau) = t\}}\}\ ;$$

C'est à ce point de la démonstration que les inégalités des réarrangements et plus précisément le lemme 1.3.(b) interviennent de manière cruciale ; chaque terme du second membre de (9) se majore en effet à partir de (7) puisque D appartient à $F(X_o)$; les distances $\delta(t, g_j(t))$ se majorent indépendamment de t et on obtient :

$$\forall\, j \in [J,N],\ \sum_{t \in S_{j+1}} E\{D(t, g_j(t)) I_{\{f_{j+1}(\tau) = t\}}\} \leq \frac{u_o}{2^j} \sum_{t \in S_{j+1}} \int_o^{P\{f_{j+1}(\tau) = t\}} X_o(\omega)d\omega\ ;$$

la somme des bornes supérieures d'intégration vaut alors 1 de sorte que le lemme 1.3.(b), inégalité (8) montre que le maximum du second membre sous cette condition est atteint quand ces bornes sont toutes égales. En regroupant, on obtient :

$$(10)\quad E \sup_{S_{N+1} \times S_{N+1}} |D(s,t) - D(f_J(s), f_J(t))| \leq 2 \sum_{j=J}^{N} \frac{u_o}{2^j} N(\frac{u_o}{2^{j+1}}) \int_o^{1/N(u_o/2^{j+1})} X_o(\omega)d\omega\,.$$

On prend alors $J = 0$ et on utilise l'évaluation intégrale des sommes, l'inégalité (10) fournit :

$$E\{\sup_{S_{N+1}\times S_{N+1}} D(s,t)\} \le 8 \iint_{o<u\le u_o,\, o<\omega\le 1} X_o\left(\frac{\omega}{N(u)}\right) d\omega\, du .$$

Soient alors $\varepsilon > 0$ et Δ une partie finie arbitraire de T de cardinal n_o ; nous choisissons N de sorte que $\frac{u_o}{2^N}$ soit inférieur à $\dfrac{\varepsilon}{2n_o \int_o^1 X_o(\omega)d\omega}$; on a :

$$E\{\sup_{\Delta\times\Delta} D(s,t)\} \le 2n_o \sup_{\delta(s,t) < \frac{u_o}{2^N}} E|D(s,t)| + E\{\sup_{S_{N+1}\times S_{N+1}} D(s,t)\} ;$$

l'inégalité (7) montre que le premier terme du second membre est majoré par ε ; on sait majorer le second terme ; faisant tendre ε vers zéro, on obtient :

$$E\{\sup_{\Delta\times\Delta} D(s,t)\} \le 8 \iint_{o<u\le u_o,\, o<\omega\le 1} X_o\left(\frac{\omega}{N(u)}\right) d\omega\, du ,$$

et l'inégalité (6) pour $\Omega = A$ s'en déduit par la séparabilité de D et pour les autres valeurs de A par le lemme 2.2.1. On en déduit la majoration presque sûre des trajectoires si l'intégrale majorante est finie ; on en déduit aussi dans ce cas la continuité presque sûre de D en tout point (t_o,t_o) de la diagonale de $T\times T$ et donc en tout autre point en appliquant (6) dans les suites $\{B(t_o,\delta_n)\times B(t_o,\delta_n), \delta_n \downarrow 0\}$. Ceci termine la première étape de la démonstration (et même la démonstration si D appartient à une classe pour laquelle la continuité presque sûre implique la p.s. continuité des trajectoires).

2.3. Pour la seconde étape, nous notons $\{\mathcal{G}_n, n\in \mathbb{N}\}$ une suite croissante de tribus finies engendrant sur Ω la même tribu que D ; il en existe puisque D est séparable. Pour tout couple (s,t) d'éléments de T et tout entier n, nous notons $D_n(s,t)$ une version de $E\{D(s,t)|\mathcal{G}_n\}$ que nous pouvons choisir puisque $\mathcal{G}_n$ est finie assez régulière pour que D_n soit un écart aléatoire et ait toutes ses trajectoires continues sur (T,δ) à partir des relations (5). Nous notons $D'_n(s,t)$ la différence $|D_n(s,t) - D(s,t)|$; puisque D appartient à $F(X_o)$, alors pour tout élément A de $\mathcal{G}$, on a :

$$E\{D_n(s,t)I_A\} = E\{D(s,t)E\{I_A|\mathcal{G}_n\}\} = E\{D(s,t)f\},$$

où f à valeurs dans $[0,1]$ a $P(A)$ pour espérance ; le lemme 1.3.(b) montre que ceci vaut $\delta(s,t)\int_0^{P(A)} X_o(\omega)d\omega$, c'est-à-dire que D_n est aussi un élément de $F(X_o)$ auquel on peut appliquer les résultats de la première étape de la preuve. Dans ces conditions, fixant les entiers n, N et $J \leq N$, on a :

$$E\{\sup_{S_{N+1}\times S_{N+1}} D'_n\} \leq E\{\sup_{S_J\times S_J} |D-D_n|\} + E\{\sup_{S_{N+1}\times S_{N+1}} |D-D(f_J,f_J)|\} +$$

$$+ E\{\sup_{S_{N+1}\times S_{N+1}} |D_n-D_n(f_J,f_J)|\}.$$

Les deux derniers termes se majorent par l'inégalité (10) ; laissant n et J fixes, utilisant la séparabilité de D'_n et sa continuité dans L^1 comme à la fin de la première étape, on en déduit :

$$E\{\sup_{T\times T} D'_n\} \leq N^2\left(\frac{u_o}{2^J}\right) \sup_{S_J\times S_J} E(D'_n) + 16 \iint_{o<\omega\leq 1,\ o<u\leq\frac{u_o}{2^J}} X_o\left(\frac{\omega}{N(u)}\right) d\omega\, du.$$

Pour tout $\varepsilon>0$, on peut donc sous l'hypothèse du théorème choisir J de sorte que le dernier terme soit majoré, indépendamment de n, par $\frac{\varepsilon}{2}$; le théorème de convergence des martingales dans L^1 permet alors, J étant ainsi fixé, l'ensemble S_J étant donc un ensemble fini fixe, de choisir n_o tel que :

$$\forall\, n\geq n_o, \forall\, t\in S_J,\ \forall\, s\in S_J,\ E\{D'_n(s,t)\} \leq \frac{\varepsilon}{2N^2\left(\frac{u_o}{2^J}\right)},$$

et par suite :

$$\forall\, n\geq n_o, E\{\sup_{T\times T} D'_n\} \leq \varepsilon.$$

Ceci signifie que :

$$\lim_{n\to\infty} E\{\sup_{T\times T} D'_n\} = 0,$$

et les techniques habituelles des sous-martingales montrent que la suite $\{(D'_n = |D_n-D|),\ n\in\mathbb{N}\}$ converge uniformément p.s. vers zéro. Ceci implique le résultat du théorème.

Remarque : Le point important dans la deuxième étape est qu'il suffit d'y utiliser le théorème de convergence des martingales dans L^1 et non pas dans un espace vectoriel lié à X_o.

3. Applications.

Dans ce paragraphe, nous énonçons les corollaires du théorème qui fournissent et étendent les résultats précédents cités dans l'Introduction.

COROLLAIRE 3.1 : Soit Φ une fonction de Young, on suppose que l'intégrale $\int_{N(u)>1} \Phi^{-1}(N(u))du$ est finie ; soit de plus D un écart aléatoire séparable (resp. X une fonction aléatoire séparable) tel que :

(11) $$\forall (s,t) \in T\times T \ , \ E\,\Phi\{\frac{D(s,t)}{\delta(s,t)}\} \quad (\text{resp. } E\,\Phi\{\frac{|X(s)-X(t)|}{\delta(s,t)}\}) \leq 1 \ .$$

Dans ces conditions, D (resp. X) a p.s. ses trajectoires continues et vérifie :

(12) $$\forall A \in \mathcal{G}, E\{\sup_{T\times T} D.I_A\} \ (\text{resp.} E \sup_{T\times T} |X(s)-X(t)|I_A) \leq 8P(A)\int_{N(u)>1} \Phi^{-1}\{\frac{N(u)}{P(A)}\}\,du \ .$$

Démonstration (dans le cas de l'écart aléatoire D) : L'hypothèse (11) permet à partir des inégalités d'Orlicz, de majorer $E\{D(s,t)I_A\}$ pour tout couple (s,t) d'éléments de T et tout élément A de $\mathcal{G}$:

(13) $$E\{D(s,t)I_A\} \leq \delta(s,t)P(A)\Phi^{-1}\{\frac{1}{P(A)}\} \ .$$

Puisque Φ est une fonction de Young, la fonction $\{p\Phi^{-1}(\frac{1}{p}) \ , \ p \in]0,1]\}$ est positive, croissante, concave ; le comportement de Φ à l'infini montre que $\lim_{p\downarrow o} p\Phi^{-1}(\frac{1}{p})$ est nul. Il existe donc une fonction positive et décroissante X_o sur $]0,1]$ telle que :

$$\forall \ p \in]0,1] \ , \int_o^p X_o(\omega)d\omega = p\Phi^{-1}(\frac{1}{p}) \ ,$$

et l'inégalité (13) signifie que D appartient à $F(X_o)$. Par ailleurs, la construction de X_o montre que l'intégrale $\iint_{\{N(u)>1 \ , \ o<\omega\leq P(A)\}} X_o(\frac{\omega}{N(u)})d\omega du$

s'écrit exactement $\int_{N(u)>1} P(A)\Phi^{-1}(\frac{N(u)}{P(A)})du$; dans ces conditions, l'application à D du théorème 1.2 est justifiée et fournit le corollaire.

Remarque 3.1.1 : Supposons que X soit une fonction aléatoire gaussienne séparable et normalisée sur (T,δ) c'est-à-dire que pour tout couple (s,t) d'éléments de T , on ait :

$$EX(t) = 0 \ , \ E(X(s)-X(t))^2 = \delta^2(s,t)$$

de sorte que :

$$E\{\exp(\frac{3}{8}\frac{(X(s)-X(t))^2}{\delta^2(s,t)})-1\} = 1 \ ;$$

le corollaire indique que si l'intégrale $\int_{N(u)>1}\sqrt{\log(1+N(u))}du$ est fini, alors p.s. les trajectoires de X sont continues et de plus :

$$(13) \ \forall A\in\mathcal{G} \ , \ E\{\sup_{T\times T}|X(s)-X(t)|I_A\} < 8\sqrt{\frac{8}{3}}\,P(A)\int_{N(u)>1}\sqrt{\log(1+\frac{N(u)}{P(A)})}\,du \, .$$

Si le résultat qualitatif est bien connu, le résultat quantitatif est singulièrement simple et maniable.

Remarque 3.1.2 : Supposons de même que X soit une fonction aléatoire séparable sur (T,δ) vérifiant pour un nombre $p>1$,

$$\forall\ (s,t)\in T\times T \quad EX(t) = 0 \ , \ E|X(s)-X(t)|^p \le \delta^p(s,t) \, .$$

Le corollaire indique que si l'intégrale $\int_{N(u)>1} N(u)^{\frac{1}{p}}du$ est convergente, alors p.s. les trajectoires de X sont continues et de plus :

$$(14) \qquad \forall A\in\mathcal{G} \ , \ E\{\sup_{T\times T}|X(s)-X(t)|I_A\} \le 8\ P(A)^{1-\frac{1}{p}}\int_{N(u)>1} N(u)^{\frac{1}{p}}du \, .$$

Là encore la formule (14) peut être utile ; en utilisant par exemple l'ensemble $A = \{|X(s)-X(t)|>u\}$ et l'inégalité de Čebičev, elle fournit :

$$(15) \qquad \forall\ u>0 \ , \ P\{\sup_{T\times T}|X(s)-X(t)|>u\} \le (\frac{8}{u})^p\int_{N(u)>1} N(u)^{\frac{1}{p}}du \, .$$

COROLLAIRE 3.2 : Soit X une fonction aléatoire sur T à valeurs dans un espace polonais (P,d) ; on suppose que $d(X(s),X(t))$ est séparable. Soit de plus X_o une fonction positive décroissante sur $]0,1]$; on suppose que $d(X(s),X(t))$ appartient à $F(X_o)$ et que l'intégrale $\iint_{o<\omega<1<N(u)} X_o(\omega/N(u))\,d\omega\,du$ est finie. Dans ces conditions, X a p.s. ses trajectoires continues sur P.

Ce corollaire ne nécessite pas de démonstration puisque c'est la réduction à un cas particulier du théorème.

COROLLAIRE 3.3. Soient $\alpha\in]0,1]$, Φ une fonction de Young et X une fonction aléatoire sur T à valeurs dans un espace de Banach séparable $(E,\|.\|)$; on suppose $\|X(s)-X(t)\|$ séparable et les deux conditions suivantes vérifiées :

$$\text{(a)}\quad \int_{N(u)>1} \Phi^{-1}(N(u))du<\infty\ ,\quad \text{(b)}\quad \forall (s,t)\in T\times T\ ,\ E\,\Phi\{\frac{\|X(s)-X(t)\|^{\alpha}}{\delta(s,t)}\}\leq 1\ ;$$

Dans ces conditions, X a p.s. ses trajectoires continues et vérifie :

$$(16)\qquad \forall A\in\mathcal{G}\ ,\ E\{\sup_{T\times T}\|X(s)-X(t)\|^{\alpha}\ I_A\}\leq 8P(A)\int_{N(u)>1}\Phi^{-1}(\frac{N(u)}{P(A)})\,du\ .$$

Ce corollaire résulte immédiatement des précédents puisque $(E,\|.\|^{\alpha})$ est un espace polonais. On constatera qu'il nécessite seulement des hypothèses faibles d'intégrabilité sur les variables aléatoires $\|X(s)-X(t)\|$.

3.3. Application aux fonctions aléatoires stables. Dans la suite de ce paragraphe, nous appliquons les résultats précédents aux fonctions aléatoires stables générales d'indice $\alpha\in]0,2[$. Nous nous limitons au cas des fonctions aléatoires symétriques puisque toute fonction aléatoire séparable continue en probabilité et dont la symétrisée a p.s. ses trajectoires continues, a p.s. elle-même ses trajectoires continues par les inégalités fortes de symétrisation, la continuité en probabilité contrôlant les médianes des accroissements. Soit X une fonction aléatoire séparable sur T, symétrique stable d'indice $\alpha\in]0,2[$; supposons pour commencer que X soit à valeurs dans $\mathbb{R}$; il existe alors une fonction d positive sur $T\times T$ telle que les accroissements $|X(s)-X(t)|$

aient chacun même loi que $d(s,t)|\theta_\alpha|$ où θ_α est une v.a. de fonction caractéristique $\exp\{-|u|^\alpha\}$. Si $\alpha \in]1,2[$, $d(s,t)$ est proportionnel à $E|X(s)-X(t)|$ si bien que d est un écart sur $T \times T$. Si $\alpha \in]0,1]$, pour tout $\beta \in]0,\alpha[$, $d^\beta(s,t)$ est proportionnel à $E|X(s)-X(t)|^\beta$ si bien que d^β et même d^α sont des écarts sur $T \times T$. Dans l'un et l'autre cas, nous notons $N_d(u)$ le cardinal minimal des parties S de T telles que la famille $\{t \in T, d(s,t) \le u\}$, $s \in S$ soit un recouvrement de T. Dans le cas $\alpha \in]1,2[$, $N_d(u)$ est égal au nombre $N(u)$ défini par l'écart d ; dans le cas $\alpha \in]0,1]$, le nombre $N(u)$ défini par l'écart d^α est égal à $N_d(u^{1/\alpha})$. Nous rappelons que dans tous les cas, on a :

(17) $$0 < \lim_{M\to\infty} M^\alpha P\{|\theta_\alpha| > M\} < \infty ;$$

dans le cas $\alpha \in]1,2[$, l'inégalité (14) fournit :

(18) $$E\{|\theta_\alpha| I_{|\theta_\alpha|>M}\} \le C_\alpha P\{|\theta_\alpha| > M\}^{1-\frac{1}{\alpha}} ;$$

dans le cas $\alpha \in]0,1]$, la même inégalité fournit pour tout $\beta \in]0,\alpha[$:

(18') $$E\{|\theta_\alpha|^\beta I_{|\theta_\alpha|>M}\} \le C_{\alpha,\beta} P\{|\theta_\alpha| > M\}^{1-\frac{\beta}{\alpha}} ,$$

les constantes ne dépendant que de leurs indices.

3.3.1. Si la fonction aléatoire X symétrique stable d'indice $\alpha \in]0,2[$ prend ses valeurs dans un espace de Banach $(B, \| \|)$, la situation est moins classique ; nous supposons la fonction aléatoire $\|X(s)-X(t)\|$ séparable. Les variables aléatoires $X(s)-X(t)$ à valeurs dans B sont alors aussi symétriques stables d'indice α et la forme de leurs lois est précisée par le lemme suivant :

LEMME 3.3.1 : <u>Soit</u> θ <u>une v.a. symétrique stable d'indice</u> $\alpha \in]0,2[$ <u>à valeurs dans un espace de Banach séparable</u> $(B, \| \|)$; <u>on note</u> m <u>le quartile supérieur de</u> $\|\theta\|$. <u>Alors pour tout</u> $\beta \in]0,\alpha[$, $\|\theta\|^\beta$ <u>est intégrable et vérifie</u> :

(19) $$\forall M \in \mathbb{R}^+ ,\ E\{\|\theta\|^\beta I_{\|\theta\|>M}\} \le C_{\alpha,\beta} P\{\|\theta\| > M\}^{1-\frac{\beta}{\alpha}} m^\beta .$$

<u>Pour tout couple</u> $\beta,\beta' \in]0,\alpha[$, <u>il existe donc des constantes</u> $C^{\alpha}_{\beta,\beta'}$, $C^{\alpha}_{\beta',\beta}$ <u>telles que</u> :

(20) $$0 < \frac{1}{C^{\alpha}_{\beta'\beta}} \leq \frac{(E\|\theta\|^{\beta})^{1/\beta}}{(E\|\theta\|^{\beta'})^{1/\beta'}} \leq C^{\alpha}_{\beta\beta'} < \infty$$

(les constantes ne dépendant que de leurs indices).

<u>Démonstration</u> : Soit $\{\theta_{n}, n \in \mathbb{N}\}$ une suite de copies de θ indépendantes ; la stabilité implique que $\sum_{1}^{n} \theta_k$ a même loi que $n^{1/\alpha}\theta$; l'inégalité de Levy montre :

$$\forall\, n>0 \;,\; \forall\, M \geq 0 \;,\; P\{\sup_{1}^{n}\|\theta_k\| > Mn^{1/\alpha}\} \leq 2P\{\|\theta\| > M\} \;;$$

l'inégalité de Borel-Cantelli implique alors :

$$\forall\, n>0 \;,\; \forall\, M \geq m \;,\; nP\{\|\theta\| > Mn^{1/\alpha}\} \leq 4(\ln 2)P\{\|\theta\| > M\} \;;$$

par interpolation, on en déduit :

(21) $$\forall\, M \geq m \;,\; \forall\, T \geq 1 \;,\; P\{\|\theta\| > MT\} \leq \frac{8\ln 2}{T^{\alpha}} P\{\|\theta\| > M\} \;,$$

l'intégrabilité annoncée en résulte.

Pour établir l'inégalité (19), supposons d'abord $M \geq m$; par intégration en T , l'inégalité (21) fournit une constante $C_{\alpha,\beta}$ telle que :

$$\forall\, M \geq m \,,\; E\{\|\theta\|^{\beta}\, I_{\|\theta\| > M}\} \leq C_{\alpha,\beta}\, M^{\beta}\, P\{\|\theta\| > M\} \;,$$

et l'inégalité (21) fournit encore :

$$\forall\, M \geq m \;,\; M \leq \frac{(2\ln 2)^{1/\alpha}\, m}{P\{\|\theta\| > M\}^{1/\alpha}} \;,$$

en reportant dans l'avant-dernière inégalité, on en déduit (19) dans ce premier cas. Pour $M \leq m$, on aura :

$$E\{\|\theta\|^{\beta}\, I_{\|\theta\| > M}\} \leq m^{\beta}\, P\{\|\theta\| > M\} + E\{\|\theta\|^{\beta}\, I_{\|\theta\| > m}\} \;,$$

le second membre se majore par :

$$m^\beta P\{\|\theta\|>M\}^{1-\frac{\beta}{\alpha}} + C_{\alpha,\beta}\, m^\beta P\{\|\theta\|>m\}^{1-\frac{\beta}{\alpha}},$$

d'où la formule générale (19) avec un facteur numérique un peu plus grand. La formule (19), la définition de m et l'inégalité de Cebicev fournissent ensuite :

$$E\|\theta\|^\beta \le C_{\alpha,\beta}\, m^\beta \le 4C_{\alpha,\beta}\, E\|\theta\|^\beta ,$$

qui suffit à établir (20).

Dans ces conditions, nous fixons un $\beta \in]0,\alpha[\cap]0,1]$ et nous munissons $T\times T$ de la fonction $d(s,t) = \{E\|X(s)-X(t)\|^\beta\}^{1/\beta}$. Si $\alpha \in]1,2[$, nous choisissons en fait $\beta = 1$ et d est un écart ; si $\alpha \in]0,1]$, d^α est un écart. Dans l'un et l'autre cas, $N_d(u)$ sera défini comme précédemment dans le cas réel.

3.3.2. Dans le cas $\alpha \in]1,2[$, les inégalités (18) ou (19) fournissent :

$$E\{\|X(s)-X(t)\| I_{\|X(s)-X(t)\|>M}\} \le C_\alpha\, d(s,t) P\{\|X(s)-X(t)\|>M\}^{1-\frac{1}{\alpha}} ;$$

l'application du théorème 1.2 à l'écart aléatoire $\|X(s)-X(t)\|$ et à la fonction $X_o(\omega) = \dfrac{C}{\omega^{1/\alpha}}$ implique alors :

COROLLAIRE 3.3.2 : __Soit__ X __une fonction aléatoire symétrique stable d'indice__ $\alpha \in]1,2[$ __sur un ensemble__ T __à valeurs dans un espace de Banach séparable ; on définit sur__ T __la topologie associée à__ $E\|X(s)-X(t)\|$ __et on suppose__ $\|X(s)-X(t)\|$ __séparable. On suppose aussi que l'intégrale__ $\int N(u)^{1/\alpha} du$ __est convergente. Dans ces conditions,__ X __a p.s. ses trajectoires continues et bornées ; pour toute partie mesurable__ A __de l'espace d'épreuves, on a__ :

$$(22)\qquad E\{\sup_{T\times T} \|X(s)-X(t)\| I_A\} \le K(\alpha) P(A)^{1-\frac{1}{\alpha}} \{\int_{\{N(u)>1\}} N(u)^{1/\alpha} du\}$$

__et donc pour tout nombre__ $M>0$:

$$(23)\qquad P\{\sup_{T\times T} \|X(s)-X(t)\|>M\} \le \Big[\frac{K(\alpha)}{M} \int_{\{N(u)>1\}} N(u)^{1/\alpha} du\Big]^\alpha .$$

3.3.3. Dans le cas $\alpha \in]0,1]$, fixant $\beta \in]0,\alpha[$, Les inégalités (18') ou (19) fournissent :

$$E\{\|X(s)-X(t)\|^{\beta} I_{\|X(s)-X(t)\|>M}\} \leq C\, d^{\beta}(s,t) P\{\|X(s)-X(t)\|>M\}^{1-\frac{\beta}{\alpha}} ;$$

l'application du théorème 1.1 à l'écart aléatoire $\|X(s)-X(t)\|^{\beta}$ et à la fonction $X_o(\omega) = \frac{C}{\omega^{\beta/\alpha}}$ implique alors :

COROLLAIRE 3.3.3 : <u>Soit</u> X <u>une fonction aléatoire symétrique stable d'indice</u> $\alpha \in]0,1]$ <u>sur un ensemble</u> T <u>à valeurs dans un espace de Banach séparable. On définit sur</u> T <u>la topologie associée à la convergence en probabilité pour</u> X <u>et on suppose que</u> $\|X(s)-X(t)\|$ <u>est séparable. On suppose aussi qu'il existe</u> $\beta \in]0,\alpha[$ <u>tel que</u> $\int N_d(u)^{\beta/\alpha} u^{\beta-1} du$ <u>soit convergente. Dans ces conditions, p.s. les trajectoires de</u> X <u>sont continues et bornées ; pour toute partie</u> A <u>de l'espace d'épreuves, on a</u> :

$$(24) \qquad E\{\sup_{T\times T} \|X(s)-X(t)\|^{\beta} I_A\} \leq K(\alpha,\beta) \int_{\{N(u)>1\}} N_d(u)^{\beta} u^{\beta-1} du\, P(A)^{1-\frac{\beta}{\alpha}}$$

<u>et donc pour tout nombre</u> M :

$$(25) \qquad P\{\sup_{T\times T} \|X(s)-X(t)\|>M\} \leq \frac{K(\alpha,\beta)^{\frac{\alpha}{\beta}}}{M^{\alpha}} \{\int_{N(u)>1} N_d(u)^{\beta/\alpha} u^{\beta-1} du\}^{\frac{\alpha}{\beta}} .$$

<u>Remarque</u> : L'inégalité de Hölder montre que l'hypothèse de convergence est d'autant moins forte que β est plus proche de α . Quelque soit le choix de β , l'inégalité (25) donne des évaluations du même ordre, le meilleur possible.

4. <u>Variante : mesures majorantes.</u>

Les techniques employées dans les paragraphes précédents se prêtent bien à l'utilisation des mesures majorantes suivant les schémas de majoration de [3]. Notant, pour toute variable aléatoire positive D , par $\bar{D}$ la fonction équimesurable décroissante sur $]0,1]$ associée, nous définissons pour toute probabilité π sur T et tout écart aléatoire séparable et intégrable D sur $T\times T$, la fonction F_{π} sur $\mathbb{R}^+$ par :

$$(26)\quad \forall\, u>0\,,\ F_\pi(u)=\int d_\pi(t)\Big[\sup_{\substack{s\in B(t,u)\\ s'\in B(t,u)}} \frac{1}{\pi B(s,\frac{u}{2^7})}\int_o^{\pi B(s,\frac{u}{2^7})} \frac{\bar{D}(s,s')(\omega)}{\delta(s,s')}\,d\omega\Big]\,.$$

Dans ces conditions, nous avons :

THEOREME 4.1 : <u>On suppose qu'on a</u>

$$(27)\qquad \sup_{\pi\in\mathcal{M}^{+1}(T)} \int_o^{u_o} F_\pi(u)\,du<\infty\ ;$$

<u>alors l'écart D a p.s. ses trajectoires bornées sur</u> $T\times T$ <u>et vérifie</u> :

$$(28)\qquad E\{\sup_{T\times T} D\}\le 2 \sup_{\pi\in\mathcal{M}^{+1}(T)} \int_o^{u_o} F_\pi(u)\,du\,.$$

<u>Si de plus on a</u> :

$$(29)\qquad \lim_{\varepsilon\to o}\ \sup_{\pi\in\mathcal{M}^{+1}(T)} \int_o^{\varepsilon} F_\pi(u)\,du=0\,,$$

<u>alors l'écart D a p.s. ses trajectoires continues.</u>

<u>Démonstration</u> : Elle suit la démarche des alinéas 22 et 23 à quelques détails techniques près. Le nombre u_o a la même signification. Pour tout entier $n\ge o$, nous notons S'_n une partie maximale de T vérifiant :

$$\forall\,(s,s')\in S'_n\times S'_n\,,\ s\ne s' \Rightarrow B(s,\frac{u_o}{4^n})\cap B(s',\frac{u_o}{4^n})=\emptyset\,,$$

de sorte que, par la maximalité, $\{B(s,\frac{2u_o}{4^n})\,,\ s\in S'_n\}$ est un recouvrement de T. Nous notons g'_n une application de S'_{n+1} dans S'_n telle que :

$$\forall\, s\in S'_{n+1}\,,\ s\in B(g'_n(s)\,,\frac{2u_o}{4^n})\,,$$

nous choisissons un entier $N>0$ et pour tout entier $k\in[0,N]$, nous posons :

$$f'_k=g'_k\circ g'_{k+1}\circ\cdots\circ g'_N\ ;$$

la suite des dénominateurs $\{4^n\,,\ n\in\mathbb{N}\}$ a été choisie ici pour assurer que :

$$\forall\, k \in [0,N]\,,\ \forall\, t \in T\,,\ \delta(t, f_k'(t)) \le \frac{2u_o}{3.4^{k-1}} < \frac{u_o}{4^{k-1}}\,,$$

de sorte qu'en fonction de la construction des S_k', on ait :

$$(30) \qquad \forall\, k \in [0,N]\,,\ \forall\, s \in S_k'\,,\ B\Big(s\,, \frac{u_o}{3.4^k}\Big) \cap S_{N+1}' \subset (f_k')^{-1}\{s\}\,.$$

Opérant maintenant comme en 2.2, nous choisissons un entier $J \le N$ et nous notons τ une application mesurable de Ω dans S_{N+1}' telle que :

$$D(\tau, f_j'(\tau)) = \sup_{t \in S'_{N+1}} D(t, f_j'(t))\ ;$$

on a, comme en (9) :

$$E \sup_{S_{N+1}' \times S_{N+1}'} |D(s,t) - D(f_J'(s), f_J'(t))| \le 2 \sum_{j=J}^{N} \sum_{t \in S_{j+1}'} \int_o^{P\{f_{j+1}'(\tau) = t\}} D(t, g_j'(t))(\omega)\, d\omega\,.$$

Nous introduisons ici la loi μ de τ, probabilité discrète sur T concentrée sur S_{N+1}' ; le second membre ci-dessus se majore par :

$$2 \int_{t \in S_{N+1}'} \Big\{ \sum_{j=J}^{N} \frac{2u_o}{4^j} \frac{1}{\mu\{f_{j+1}'(s) = f_{j+1}'(t)\}} \int_o^{\mu\{f_{j+1}'(s) = f_{j+1}'(t)\}} \frac{D(f_j'(t), f_{j+1}'(t)(\omega)}{\delta(f_j'(t), f_{j+1}'(t))}\, d\omega \Big\}\, d\mu(t)\,.$$

Utilisant le fait que $\frac{1}{x}\int_o^x \bar{D}(\omega)d\omega$ est décroissant en x et la propriété (26), on en déduit :

$$E \sup_{S_{N+1}' \times S_{N+1}'} |D - D(f_J', f_J')| \le 2 \sum_{j=J}^{N} \frac{2u_o}{4^j} F_\mu\Big(\frac{2u_o}{3.4^{j-1}}\Big)\,.$$

La démonstration complète se termine alors sans modification notable.

5. Propriétés réciproques.

Dans ce paragraphe, nous précisons dans plusieurs situations successives en quels sens les résultats précédents sont bons. L'espace (T, δ)

ne sera pas nécessairement compact ; nous le supposerons pourtant borné : en effet, si (T,δ) est un espace métrique séparable et si X_o est une fonction positive décroissante non identiquement nulle sur $]0,1]$, alors la fonction aléatoire $\{\delta(t,t_o)X_o(\omega)\ ,\ t\in T\ ,\ \omega\in\,]0,1]$ appartient à $\mathfrak{F}(X_o)$ et n'est p.s. bornée sur T que si (T,δ) est borné. Le cas où X_o est bornée est simple :

PROPOSITION 5.1 : On suppose X_o bornée sur $]0,1]$: sous les conditions ci-dessus, toute fonction aléatoire appartenant à $\mathfrak{F}(X_o)$ a p.s. ses trajectoires bornées et continues sur (T,δ).

Démonstration : Nous supposons X_o bornée par M et X vérifiant (3) ; on a alors :

$$\forall (s,t)\in T\times T\ ,\ \forall \rho>1\ ,\ E\{|X(s)-X(t)|\, I_{|X(s)-X(t)|>\rho\delta(s,t)}\}\leq \delta(s,t) M P\{|X(s)-X(t)| > \rho\delta(s,t)\}\ ;$$

L'inégalité de Čebičev implique donc :

$$P\{|X(s)-X(t)|>M\delta(s,t)\} = 0$$

et la séparabilité de X implique le résultat.

THEOREME 5.2. Soient (T,δ) un espace séparable et X_o une fonction positive décroissante intégrable sur $]0,1]$; on suppose que toute fonction aléatoire X appartenant à $\mathfrak{F}(X_o)$ a p.s. ses trajectoires bornées. Dans ces conditions, on a aussi

$$\sup_{u\in\,]0,1]} u\,X_o\left(\frac{1}{N(u)}\right) < \infty\ ; \tag{31}$$

en particulier, si X_o est non bornée, (T,δ) est précompact.

Remarque : On notera la parenté entre le résultat (3.1) et la minoration de Sudakov dans le cas gaussien ([2], 2.3.1) ; les hypothèses par contre sont différentes, la minoration de Sudakov utilisant les inégalités de Slépian conclut à partir des propriétés d'une seule fonction aléatoire.

La démonstration du théorème se fonde sur les deux lemmes 5.2.1 et 5.2.2 :

LEMME 5.2.1 : Si toute fonction aléatoire X appartenant à $\mathfrak{F}(X_o)$ a p.s. ses trajectoires bornées, il existe un nombre réel M tel que :

(32) $$\forall X \in \mathfrak{F}(X_o) , P\{\sup_{T\times T} |X(s)-X(t)| > M\} \leq \frac{1}{8} .$$

Démonstration : Notons D le diamètre de (T,δ). Pour tout couple (s,t) d'éléments de T et tout élément X de $\mathfrak{F}(X_o)$, la propriété (3) et l'inégalité de Čebičev impliquent que la médiane $\mu(|X(s)-X(t)|)$ est majorée par $2D\int_0^{\frac{1}{2}} X_o(\omega)d\omega$. Supposons alors la conclusion du lemme fausse. Pour tout entier n, il existerait alors un élément X_n de $\mathfrak{F}(X_o)$ tel que :

$$P\{\sup_{T\times T} |X_n(s)-X_n(t)| > 4^n + 2D\int_0^{\frac{1}{2}} X_o(\omega)d\omega\} > \frac{1}{8}$$

et donc :

$$P\{\sup_{T\times T} |X_n(s)-X_n(t)-\mu(X_n(s)-X_n(t))| > 4^n\} > \frac{1}{8} ;$$

Notons $\bar{X}_n$ une fonction aléatoire symétrisée de X_n et séparable ; les inégalités fortes de symétrisation fournissent alors :

$$2P\{\sup_{T\times T} |\bar{X}_n(s)-\bar{X}_n(t)| > 4^n\} > \frac{1}{8} .$$

Soient $(Y_n, n\in \mathbb{N})$ une suite de copies indépendantes des $\bar{X}_n$ et t_o un élément de T ; alors la série $\sum_{n=1}^{\infty} (Y_n - Y_n(t_o))2^{-(n+1)}$ dont les termes vérifient :

$$\forall (s,t)\in T\times T, \forall A\in \mathcal{A}, E\{\frac{|Y_n(s)-Y_n(t)|}{2^{n+1}} I_A\} \leq \frac{2}{2^{n+1}}\,\delta(s,t)\int_0^{P(A)} X_o(\omega)d\omega$$

converge, p.s. en chaque point de T, vers une fonction aléatoire U séparable appartenant à $\mathfrak{F}(X_o)$. Comme ses termes sont symétriques et indépendants, les inégalités de Lévy impliquent alors pour tout entier n :

$$4P\{\sup_{T\times T}|U(s)-U(t)|>2^{n-1}\}\geq 2P\{\sup_{T\times T}|Y_n(s)-Y_n(t)|>4^n\}>\frac{1}{8}\ ;$$

Ceci signifie que les trajectoires de U ne sont pas p.s. bornées d'où l'absurdité ; le lemme est établi.

LEMME 5.2.2. Soient $u>0$ et S une partie finie de T tels que $\{B(s,\frac{u}{2}),\ s\in S\}$ soient disjointes ; soient de plus s_o un élément de S et $\{x(s),\ s\in S\}$ une suite de v.a. indépendantes distribuées comme X_o et indexées par S. On leur associe la fonction aléatoire X définie sur T par :

$$X(t)=0 \quad \text{si} \quad t\notin B(S,\tfrac{u}{2}) \quad \text{ou} \quad t\in B(s_o,\tfrac{u}{2}),$$

$$X(t)=\frac{1}{2}\,\delta(t,T\setminus B(s,\tfrac{u}{2}))x(s) \quad \text{si} \quad t\in B(s,\tfrac{u}{2}) \quad \text{et} \quad s\in S,\ s\neq s_o\ ;$$

Dans ces conditions, X appartient à $\mathfrak{F}(X_o)$.

Démonstration : On doit prouver que pour tout couple (s,t) d'éléments de T et tout élément A de $\mathcal{G}$, on a :

$$E\{|X(s)-X(t)|I_A\}\leq \delta(s,t)\int_o^{P(A)} X_o(\omega)d\omega\,. \tag{33}$$

On remarquera pour cela que pour tout élément s de S et tout $A\in\mathcal{G}$, on a

$$E\{x(s)I_A\}\leq \int_o^{P(A)} X_o(\omega)d\omega$$

et on distinguera suivant les positions respectives de s et t dans T. Si s et t appartiennent tous deux à $B(s_o,\frac{u}{2})\cup \complement B(S,\frac{u}{2})$, le premier membre de (33) est nul et l'inégalité vérifiée. Si s appartient à $B(s_o,\frac{u}{2})\cup\complement B(S,\frac{u}{2})$ et si t appartient à $B(t',\frac{u}{2})$ où t' est un élément de S différent de s_o le premier membre de (33) est inférieur à $\frac{1}{2}\,\delta(t,s)\,E\{x(t')I_A\}$ de sorte que (33) est aussi vérifiée. Si s appartient à $B(s',\frac{u}{2})$ et si t appartient à $B(t',\frac{u}{2})$ où s' et t' sont des éléments de S différents entre eux et de s_o, alors on a :

$$\delta(s,T\setminus B(s',\tfrac{u}{2})) \leq \delta(s,t) ,$$

$$\delta(t,T\setminus B(t',\tfrac{u}{2})) \leq \delta(s,t) ,$$

de sorte que le premier membre de (33) est inférieur à

$$\frac{1}{2}\,\delta(s,t)[E\{[x(t')+x(s')]I_A\}]$$

et l'inégalité (33) est encore vérifiée. Si enfin s et t appartiennent à $B(s',\frac{u}{2})$ où s' est un élément de S différent de s_o, on a :

$$|\delta(t,T\setminus B(s',\tfrac{u}{2}))-\delta(s,T\setminus B(s',\tfrac{u}{2}))|\leq \delta(s,t)$$

de sorte que le premier membre de (33) est inférieur à $\frac{1}{2}\,\delta(s,t)E\{x(s')I_A\}$; l'inégalité (33) est toujours vérifiée ; ceci résume les différentes éventualités à l'ordre de s et t près de sorte que le lemme est établi.

5.2.3. <u>Démonstration du théorème 5.2.</u> Supposons ses hypothèses vérifiées et construisons le nombre M et la fonction aléatoire X suivant les schémas du lemme 5.2.1 et du lemme 5.2.2. On aura donc :

$$\frac{1}{8}\geq P\{\sup_{s\in S}|X(s)-X(s_o)|>M\}\geq P\{\sup_{\substack{s\in S\\ s\neq s_o}}\frac{u}{4}\,x(s)>M\} ;$$

on aura aussi, en utilisant l'indépendance des $x(s)$ et leurs lois et en notant N le cardinal de S :

$$P\{\sup_{\substack{s\in S\\ s\neq s_o}} x(s)\geq X_o(\tfrac{1}{N})\}\geq 1-(1-\tfrac{1}{N})^{N-1} .$$

Si nous choisissons maintenant S maximal de sorte que son cardinal N soit supérieur ou égal à $N(u)$, les deux dernières inégalités impliquent pour peu que $N(u)$ soit supérieur ou égal à 2 :

$$\frac{u}{4}\,X_o(\frac{1}{N(u)})\leq M ,$$

et donc le résultat du théorème.

5.3. Le théorème suivant généralise les résultats de Hahn Klass ([4]), Kono ([6]) et Pisier ([9]) établis dans le cas où X_o est une fonction puissance ; nous y utiliserons, comme Pisier, les propriétés particulières des séries trigonométriques à coefficients monotones.

THEOREME 5.3 : Soit X_o une fonction positive, décroissante, intégrable sur $]0,1]$; on pose $Y_o(u) = \int_0^u X_o(\omega)d\omega$ et on suppose qu'il existe $\alpha > 0$ tel que $\frac{1}{u^\alpha}\int_0^u Y_o(\sigma)\frac{d\sigma}{\sigma}$ soit croissante. Soit de plus φ une fonction positive, croissante, sous-additive et continue sur $[0,1]$; on suppose que l'intégrale $\iint_{\{0<\omega\leq 1<N(u)\}} X_o(\frac{\omega}{N(u)})d\omega du$ est divergente où $N(u)$ est associée à la distance δ définie par φ sur $[0,1]$. Il existe alors une fonction aléatoire séparable X sur $([0,1],\delta)$ appartenant à $\mathfrak{Z}(X_o)$ et ayant p.s. ses trajectoires non bornées.

La démonstration utilisera le lemme élémentaire suivant :

LEMME 5.3.1 : Soit n un entier >0, sur $[0,1]$ muni de la mesure de Lebesgue, on définit les variables aléatoires X et X_h, $h\in[0,1]$, par :

$$\forall\ \omega\in[0,1],\ X(\omega) = \Big|\sum_{2^n\leq k<2^{n+1}} e^{i2k\pi\omega}\Big| ,$$

$$\forall\ h\in[0,1],\ X_h(\omega) = |X(\omega)-X(\omega+h)| ;$$

alors les variables aléatoires équimesurables décroissantes associées vérifient :

$$\bar{X}(\omega)\leq C\ \inf(2^n,\frac{1}{2\omega}) ,$$

$$\bar{X}_h(\omega)\leq C\ \inf(1,2^n h)\inf(2^n,\frac{1}{2\omega})$$

où C est une constante absolue.

Démonstration : Le nombre des termes de X montre immédiatement qu'elle est majorée par 2^n ; en additionnant les termes, on majore aussi X par $\frac{1}{\sin(\pi\omega)}$ et donc, sur $]0,\frac{1}{2}]$, par $\frac{1}{2\omega}$. La première formule en résulte. Deux calculs successifs du même type montrent que le module de la dérivée de X se majore sur le même intervalle par 2^{2n} aussi bien que par $\frac{C2^n}{\omega}$ et ceci majore $\bar{X}_h(\omega)$ par $C2^n h \inf(2^n, \frac{1}{2\omega})$; comme $|\bar{X}_h(\omega)|$ se majore aussi par $2|\bar{X}(\omega)|$, on en déduit la deuxième formule.

5.3.2. Démonstration du théoreme 5.3. On considère sur $\Omega = [0,1]$ et $T = [0,1]$, la série aléatoire :

$$\sum_{k=1}^{\infty} \varphi(\frac{1}{2^k})[Y_o(\frac{1}{2^k}) - \frac{1}{2^k} X_o(\frac{1}{2^k})] \sum_{n=2^k}^{2^{k+1}-1} e^{i2\pi n(\omega+t)} ;$$

ses coefficients sont positifs et même décroissants puisque X_o est décroissant ; ils tendent vers zéro de sorte que cette série converge en tout couple (ω,t) de somme non nulle. Nous notons $X(\omega,t)$ sa somme.

Soit A une partie mesurable de l'espace d'épreuves, supposons $P(A) \in]\frac{1}{2^{N+1}}, \frac{1}{2^N}]$; le lemme 5.3.1 montre qu'on a en posant $b_n = Y_o(\frac{1}{2^n}) - \frac{1}{2^n} X_o(\frac{1}{2^n})$:

$$E\{|X(t) - X(s)|I_A\} \leq C \sum_{k=1}^{N} \varphi(\frac{1}{2^k}) b_k \inf(1, 2^k|t-s|) 2^{k-N}$$

$$+ C \sum_{k=N+1}^{\infty} \varphi(\frac{1}{2^k}) b_k \inf(1, 2^k|t-s|)(k+2-N) ;$$

la croissance et la sous-additivité de φ montrent qu'on a :

$$\forall\, k \geq 1, \forall\, (t,s) \in T \times T,\ \varphi(\frac{1}{2^k}) \inf(1, 2^k|t-s|) \leq 2\varphi(|t-s|) ,$$

on en déduit :

$$E\{|X(t) - X(s)|I_A\} \leq C\varphi(|t-s|)\{\sum_{1}^{N} 2^n b_n P(A) + \sum_{N+1}^{\infty} (n+2-N) b_n\} ,$$

si bien qu'à une constante multiplicative près, X appartiendra à $\mathfrak{F}(X_o)$ si on a :

(34) $$\sum_{1}^{N} 2^n b_n \leq C_1 2^N Y_o(\frac{1}{2^N})$$

(35) $$\sum_{N+1}^{\infty} \sum_{n}^{\infty} b_k \leq C_2\, Y_o\left(\frac{1}{2^N}\right)$$

et la forme particulière de X démontrera le théorème si on a de plus :

(36) $$\sum_{k=1}^{\infty} 2^k \varphi\left(\frac{1}{2^k}\right) b_k = +\infty .$$

Or le premier membre de (34) se majore par $2\int_{1/2^N}^{1} \frac{Y(u)-uX(u)}{u^2}\,du$ qui s'intègre pour donner (34). Au premier membre de (35) les sommes $\sum_n^{\infty} b_k$ se majorent en fonction de l'hypothèse de l'énoncé par $\frac{1}{\log 2}\int_o^{1/2^{n-1}} Y_o(u)\,\frac{du}{u}$ et donc par $\frac{2}{\alpha}\, Y_o\left(\frac{1}{2^{n-1}}\right)$; en réitérant cette majoration, on obtient (35). Enfin le premier membre de (36) se minore par $\int_o^{1/2} \frac{Y_o(\sigma)}{\sigma}\, d\varphi(\sigma) - 2Y_o\left(\frac{1}{2}\right)\varphi\left(\frac{1}{2}\right)$; ce terme intégral se minore, puisque u est inférieur ou égal à $\varphi\left(\frac{1}{N(u)}\right)$, par $\int_o^{\varphi(\frac{1}{2})} N(u) Y_o\left(\frac{1}{N(u)}\right) du$; l'hypothèse de divergence de l'énoncé implique donc (36) et finalement le résultat du théorème.

Remarque 5.3.3 : Le théorème 5.3 ne s'applique pas pour $X_o(t) = \frac{1}{t(\log \frac{1}{t})^n}$. Il s'applique pour $X_o(t) = \frac{1}{t^{\alpha}}$, $0<\alpha<1$; il couvre donc les classes de fonctions aléatoires vérifiant les inégalités du genre :

$$E\left\{\left|\frac{X(s)-X(t)}{\varphi(|s-t|)}\right|^p\right\} \leq 1 \ , \ p>1 ;$$

on obtient alors en corollaire les théorèmes de Hahn-Klass $(p=2)$, Kono $(p>2)$, Ibragimov $(1<p<\infty)$; il s'applique aussi pour $X_o(t) = (\log\frac{1}{t})^{\alpha}$, $\alpha>0$; on obtient alors en corollaire les théorèmes sous-gaussiens ou exponentiels. En un certain sens, il résume donc et simplifie les différents théorèmes connus ; on doit pourtant remarquer qu'il s'agit d'un théorème relativement faible : les fonctions aléatoires construites ne sont pas du type $\sum f_n(t)\theta_n(\omega)$ où les θ_n seraient indépendantes. Elles peuvent donc simultanément avoir p.s. leurs trajectoires non bornées (c'est le cas) et non continues et être p.s. continues en chaque point, c'est aussi le cas. Cette remarque s'applique d'ailleurs aux exemples partiels construits par les auteurs cités ci-dessus dans les cas non exponentiels.

THEOREME 5.6 : Soient $\alpha \in]0,2[$ et f une fonction positive décroissante sur $]0,1]$; on suppose que l'intégrale $\int f(u)u^{\alpha-1}du$ est divergente. Il existe alors un ensemble T, une fonction aléatoire X symétrique stable d'indice α sur T p.s. à trajectoires non bornées et telles que :

(37) $$\forall\, u>0\,,\; N(u) \le f(u)\,.$$

Démonstration : On suppose pour simplifier que pour tout $n \ge 0$, $f(\frac{1}{2^{n/\alpha}})$ est un nombre entier ; on note S_n un ensemble de cardinal $f(\frac{1}{2^{n/\alpha}})$ et g_n une application de S_{n+1} sur S_n. On note $(T, \varphi_n, n \in \mathbb{N})$ la limite projective du système $\{S_n, g_n, n \in \mathbb{N}\}$ et $\{\theta_s, s \in S_n, n \in \mathbb{N}\}$ une suite symétrique stable d'indice α à composantes indépendantes. On définit la fonction aléatoire X sur T en posant :

(38) $$X(t) = \sum_{n=0}^{\infty} \frac{1}{2^{\frac{n+1}{\alpha}}}\, \theta \circ \varphi_n(t)\,.$$

On vérifie alors que $d(t,t')$ est inférieur ou égal à $\frac{1}{2^{n/\alpha}}$ si et seulement si $\varphi_n(t) = \varphi_n(t')$ de sorte que pour tout $u>0$, on a $N(u) \le f(u)$. Par ailleurs, si p.s. les trajectoires de X sont bornées, l'inégalité de Lévy montre que la somme $\sum_{n=0}^{\infty} f(\frac{1}{2^{n/\alpha}})\, P\{|\theta_\alpha| > M\, 2^{\frac{n+1}{\alpha}}\}$ est finie pour tout M assez grand ; ceci implique la convergence de $\sum_{n=0}^{\infty} f(\frac{1}{2^{n/\alpha}})\, \frac{1}{2^n}$; c'est contraire à l'hypothèse du théorème ; les trajectoires de X ne sont donc pas p.s. bornées. On notera d'ailleurs que la forme particulière de X implique alors qu'il existe au moins un élément t_o de T tel que X ne soit pas p.s. continu en t_o.

Remarque : Le théorème 5.6 et les corollaires 3.3.2 et 3.3.3 n'apportent pas de solution définitive pour la régularité des trajectoires des fonctions aléatoires stables. Si $f(u)$ est de la forme u^{-p}, les énoncés sont bons puisque l'hypothèse (37) implique la p.s. continuité des trajectoires si α est supérieur à p (corollaire 3.3.2 et 3.3.3) et est effectivement compatible avec leur irrégularité (théorème 5.6) si α est inférieur ou égal à p. Par contre si $f(u)$ est de la forme $u^{-\alpha}(\log \frac{1}{u})^{-\beta}$, l'hypothèse (37) implique la p.s. continuité des trajectoires si β est supérieur à α et à 1 et n'est certainement compatible avec leur irrégularité (théorème 5.6) que si β est inférieur ou égal à 1.

6. Fonctions aléatoires stables et propriétés de Slépian ([11]).

On pourrait croire à la lecture des pages précédentes que toutes les propriétés simples des fonctions aléatoires gaussiennes peuvent être étendues à des classes très larges d'autres fonctions aléatoires au prix éventuel de manipulations techniques difficiles. Il est pourtant une classe de propriétés qui résiste jusqu'ici à toute extension ; il s'agit des propriétés de Slépian qui expriment que certaines fonctionnelles gaussiennes sont fonction monotone de la distance associée ([2], lemme 2.1.1, théorème 2.1.2). Nous allons montrer dans ce paragraphe que ce type de propriété ne peut sous aucune forme être étendue aux fonctions aléatoires symétriques stables d'indice $\alpha \in]0,2[$. L'élément essentiel de l'étude sera un mode de construction de certaines fonctions aléatoires stables.

6.1. Exemple de constructions de fonctions aléatoires stables.

PROPOSITION 6.1 : Soient $\alpha \in]0,2[$ et G une fonction aléatoire gaussienne sur un ensemble T ; soit de plus U l'ensemble des applications de T dans $\mathbb{R}$ à support fini ; alors la fonction Φ_α sur U définie par :

$$\Phi_\alpha(u) = \exp\{-(E|\sum_{t\in T} u(t)G(t)|^2)^{\alpha/2}\}$$

est la fonction caractéristique d'une fonction aléatoire G_α symétrique stable d'ordre α ; pour tout couple (s,t) d'éléments de T, on a :

$$d_{G_\alpha}(s,t) = d_G(s,t) .$$

On suppose de plus G et G_α séparables sur (T,d_G) ; dans ces conditions, pour que G_α ait p.s. ses trajectoires continues bornées sur (T,d_G), il faut et il suffit que G ait la même propriété.

Démonstration : Les propriétés des transformées de Laplace montrent qu'il existe une probabilité μ_α sur $\mathbb{R}^+$ telle que :

$$\forall\, t \in \mathbb{R} \quad \exp\{-|t|^\alpha\} = \int \exp\{-\frac{t^2x^2}{2}\}\, d\mu_\alpha(x) .$$

On note alors x_α une variable aléatoire positive de loi μ_α et indépendante de la fonction aléatoire G. Notons Φ la fonction caractéristique de $x_\alpha G$; par intégrations successives, on obtient :

$$\forall\, u \in U,\ \Phi(u) = E\exp\{i \sum_{t\in T} u(t)x_\alpha G(t)\} = E\exp\{-\frac{(x_\alpha)^2}{2} E|\sum_{t\in T} u(t)G(t)|^2\} =$$

$$= \int \exp\{-\frac{x^2}{2} E|\sum_{t\in T} u(t)G(t)|^2\} d\mu_\alpha(x) = \Phi_\alpha(u)$$

ceci établit la première affirmation ; les autres résultent immédiatement de la construction de G_α.

6.2. Irrégularité de fonctions aléatoires stables et entropie ou distance associée.

6.2.1. On sait que si $\alpha = 2$, la fonction N_X permet d'obtenir des conditions suffisantes ([1]) et nécessaires si X est stationnaire ([2]) pour la régularité des trajectoires de X sur T. Si α est inférieur à 2, la même fonction permet d'obtenir des conditions suffisantes (corollaires 3.3.2 et 3.3.3) pour la même régularité ; elle permet aussi (théorème 5.6) d'obtenir des conditions nécessaires voisines en un sens faible précisé dans leur énoncé. L'exemple suivant montre qu'il n'est pas possible, contrairement au cas gaussien, de renforcer le sens de cette nécessité :

Exemple 6.2.1 : Soit G une fonction aléatoire gaussienne stationnaire sur $\mathbb{R}$ continue en probabilité ; nous posons $T = [0,1]$; nous supposons que la fonction N_G associée vérifie simultanément :

$$\int_0 \sqrt{\log N_G(u)}\,du < \infty \;,\; \int_0 N_G(u)u^{\alpha-1}du = \infty \;;$$

la première de ces conditions montre que G et G_α ont p.s. leurs trajectoires continues et bornées sur $[0,1]$. Par contre la deuxième condition permet de construire un autre ensemble T' et une autre fonction aléatoire X' symétrique stable de même indice α sur T', p.s. à trajectoires non bornées et vérifiant pourtant :

$$\forall u > 0 \,,\, N_{X'}(u) \leq N_{G_\alpha}(u) \,.$$

Cet exemple montre donc que la régularité des trajectoires des fonctions aléatoires X symétriques stables d'ordre $\alpha < 2$ n'est pas une fonction monotone de N_X.

6.2.2. L'exemple précédent pourrait laisser espérer l'existence de paramètres plus précis que N_X caractérisant la régularité des trajectoires à partir des seules lois marginales des accroissements. C'est bien entendu le cas pour $\alpha = 2$, puisque la loi d'une fonction aléatoire gaussienne est déterminée, à une translation près par les seules lois marginales des accroissements et donc par le système $\{d(s,t)\,,\, s \in T\,,\, t \in T\}$. Au contraire, ce n'est pas le cas pour $\alpha < 2$, comme le montre l'exemple suivant.

<u>Exemple 6.2.2</u>. Nous notons g une application de $\mathbb{N}^*$ dans $\mathbb{N}^*$ et f la fonction définie par $f(n) = \prod_{k=1}^{n} g(k)$. Pour tout $n \in \mathbb{N}^*$, nous notons S_n un ensemble de cardinal $f(n)$ et φ_n une application de S_{n+1} dans S_n telle que pour tout $s \in S_n$, le cardinal de $\varphi_n^{-1}(s)$ soit égal à $g(n+1)$. Enfin, nous notons $(T; \psi_n\,,\, n \in \mathbb{N})$ la limite projective de $(S_n; \varphi_n\,,\, n \in \mathbb{N})$. Soient $\{\lambda(s)\,,\, s \in S_n\,,\, n \in \mathbb{N}\}$ et $\{\theta(s)\,,\, s \in S_n\,,\, n \in \mathbb{N}\}$ des suites de v.a. indépendantes symétriques réduites, identiquement distribuées, respectivement gaussiennes et stables d'indice α. Nous leur associons trois fonctions aléatoires sur T en posant :

$$G(t) = \sqrt{\frac{3}{8}} \sum_{n=1}^{\infty} \frac{1}{2^n} \lambda \circ \psi_n(t) \text{ , } X_1(t) = G_\alpha(t) \text{ ,}$$

$$X_2(t) = [\frac{1}{2}(1 - \frac{1}{2^\alpha})]^{1/\alpha} \sum_{n=1}^{\infty} \frac{1}{2^n} \theta \circ \psi_n(t) \text{ ;}$$

on vérifie immédiatement que X_1 et X_2 sont symétriques stables d'indice α ; de plus, pour tout couple (s,t) d'éléments de T, soit n le plus grand entier tel que $\psi_n(s) = \psi_n(t)$, on a $d_{X_1}(s,t) = d_{X_2}(s,t) = \frac{1}{2^{n+1}}$; ceci signifie que X_1 et X_2 définissent un même écart sur T et donc que pour chaque couple (s,t), les accroissements $X_1(s) - X_1(t)$ et $X_2(s) - X_2(t)$ ont la même loi. L'espace (T,d) est compact par construction et G est stationnaire de sorte que X_1 aura p.s. ses trajectoires continues ou bornées si et seulement si $\Sigma \frac{1}{2^n} \sqrt{\log N(\frac{1}{2^n})}$ est convergente, c'est-à-dire si et seulement si $\Sigma \frac{1}{2^n} \sqrt{\log g(n)}$ l'est. Par ailleurs si les trajectoires de X_2 sont p.s. bornées, les inégalités de Lévy impliquent comme dans la démonstration du théorème 5.6. la convergence de la série $\sum_{n=1}^{\infty} f(n)2^{-n\alpha}$. Il résulte donc de ces deux évaluations que si $g(n)$ est égal à 4 pour tout n, les trajectoires de X_2 seront p.s. non bornées et celles de X_1 p.s. continues. Nous pouvons donc résumer : <u>Pour tout</u> $\alpha \in]0,2[$, <u>on peut construire un espace compact métrisable</u> (T,d) <u>et deux fonctions aléatoires</u> X_1 <u>et</u> X_2 <u>symétriques stables d'ordre</u> α <u>sur</u> T <u>telles que</u> :

(i) <u>pour tout couple</u> (s,t) <u>d'éléments de</u> T, $X_1(s) - X_1(t)$ <u>et</u> $X_2(s) - X_2(t)$ <u>ont même loi et on a</u> $d_{X_1}(s,t) = d_{X_2}(s,t) = d(s,t)$.

(ii) X_1 <u>est p.s. à trajectoires continues sur</u> (T,d).

(iii) X_2 <u>est p.s. à trajectoires non bornées sur</u> (T,d).

7. <u>Conclusion</u>.

L'étude présente montre que les propriétés d'entropie introduites par R.M. Dudley pour l'analyse des fonctions aléatoires gaussiennes ont un champ d'application beaucoup plus vaste. Ces méthodes d'entropie ont vu longtemps leur utilisation liée à des propriétés de convexité dans des espaces vectoriels. C'est inutile pour la majoration des trajectoires comme le montre M. Weber ([10]) aussi bien que pour leur continuité ; nous le montrons ici : elles s'appliquent dans des espaces métriques grâce à des propriétés métriques. Notre étude montre aussi que si pendant ces 15 dernières années, le domaine des fonctions aléatoires gaussiennes a pu sembler particulièrement simple, il apparaît maintenant sous beaucoup d'aspects comme le prototype du domaine des fonctions aléatoires plus générales qui ont les mêmes propriétés simples. En particulier, l'étude de la régularité des fonctions aléatoires stables nous paraît prometteuse.

On notera que l'étude n'aborde pas l'évaluation des modules de continuité uniforme des trajectoires. Dans ce domaine, on ne dispose pas de résultats généraux satisfaisants et les raisons n'en sont pas claires pour l'instant.

REFERENCES DU PREMIER CHAPITRE

[1] DUDLEY R.M. Sample functions of the gaussian process, Ann. of Prob., 1, 1973, 66-103.

[2] FERNIQUE X. Régularité des trajectoires des fonctions aléatoires gaussiennes, Lecture Notes, Springer, 480, 1-91.

[3] FERNIQUE X. Caractérisation de processus à trajectoires majorées ou continues, Lecture Notes, Springer, 649, 691-706.

[4] HAHN M.G., KLASS M.J. Sample continuity of square-integrable processes, Ann. of Prob., 5, 1977, 361-370.

[5] IBRAGIMOV I.A. Properties of sample functions for stochastic processes and embedding theorems, Theory Prob. Appl., 18, 1973, 442-453.

[6] KONO N. Sample path properties of stochastic processes, J. Math. Kyoto Univ., 20-2, 1980, 295-313.

[7] MARCUS M.B. Continuity and the central limit theorem for random trigonometrical series, Z. Wahrscheinlichkeitsth., 42, 1978, 35-56.

[8] NANOPOULOS C., NOBELIS P. Régularité et propriétés limites de fonctions aléatoires, Lecture Notes, Springer, 649, 567-690.

[9] PISIER G. Conditions d'entropie assurant la continuité de certains processus, Séminaire Analyse fonctionnelle, 1979-1980, XIII-XIV, Ecole Polytechnique.

[10] WEBER M. Une méthode élémentaire pour l'étude de la régularité d'une large classe de fonctions aléatoires, CR Acad. Sc. Paris, 292, I, 599-602.

[11] A. EHRHARD et X. FERNIQUE Fonctions aléatoires stables irrégulières, C.R. Acad. Sc. Paris, 292, I, 999-1001.

CHAPITRE II

FONCTIONS ALEATOIRES ET STRUCTURES D'INDEPENDANCE. LES FONCTIONS ALEATOIRES DE TYPE INTEGRAL.

0. Introduction, Mesures aléatoires à valeurs indépendantes ou symétriques.

0.1. Les études réalisées entre 1960 et 1975 sur les fonctions aléatoires gaussiennes ont paru montrer qu'elles étaient exceptionnelles par leur simplicité. En fait, certaines de leurs propriétés (lois zéro-un, intégrabilité par exemple) ont été étendues depuis 10 ans à d'autres fonctions aléatoires comme les séries $\Sigma f_n \lambda_n(\omega)$ à termes indépendants. Nous nous proposons de mettre ici en évidence des classes plus larges de fonctions aléatoires simples. Prenons par exemple un phénomène modélisé par l'équation de la corde vibrante homogène sur $[0,1]$ fixe à ses extrémités et partant de la position d'équilibre avec une répartition aléatoire $V = \{V(\omega,x)\ ,\ \omega \in \Omega\ ,\ x \in]0,1[\}$ de vitesses initiales ; il peut être souhaitable de pouvoir analyser le phénomène pour une classe de ces répartitions assez large pour pouvoir contenir des répartitions poissonniennes de fonction caractéristique $E\{\exp(i<V,\varphi>)\} = \exp\{-\int_0^1 (1-\cos\varphi(x))d\mu(x)\}$ aussi bien que des répartitions gaussiennes de fonction caractéristique $\exp\{-\frac{1}{2}\int_0^1 \varphi^2(x)d\mu(x)\}$. La classe étudiée ici sera définie par intégration de fonctions certaines par rapport à des mesures aléatoires. Elle contiendra donc des familles très larges de solutions d'équations différentielles ou aux dérivées partielles linéaires à coefficients constants non aléatoires et à données initiales aléatoires. Nous rappelons pour commencer quelques résultats relatifs à ces mesures aléatoires.

0.2.1. Soit $(\mathcal{U},\mathcal{B})$ un espace lusinien muni de sa tribu canonique. Une mesure aléatoire sur $\mathcal{U}$ à valeurs indépendantes (resp. symétriques) est une fonction aléatoire sur $\mathcal{B}$ vérifiant les propriétés suivantes :

(1) $\forall\ b \in \mathcal{B}\ ,\ P\{|m(b)| < \infty\} = 1$,

(2) Pour toute suite $\{b_n, n \in \mathbb{N}\}$ d'éléments de $\mathcal{B}$ disjoints, la

suite $\{m(b_n), n \in \mathbb{N}\}$ est indépendante (resp. symétrique) et sa série $\sum_{\mathbb{N}} m(b_n)$ converge en probabilité vers $m(\bigcup_{\mathbb{N}} b_n)$.

On remarquera que dans l'un et l'autre cas, les séries $\sum_{\mathbb{N}} m(b_n)$ associées à des éléments de $\mathcal{B}$ disjoints convergent presque sûrement. C'est évident dans le cas de l'indépendance ; dans le cas de la symétrie, si $\{\varepsilon_n, n \in \mathbb{N}\}$ désigne une suite de Rademacher indépendante de m, la série $\sum \varepsilon_n m(b_n)$ convergeant en probabilité, pour presque tout ω de l'espace d'épreuves de M la série $\sum \varepsilon_n m(\omega, b_n)$ converge en probabilité ; à termes indépendants, elle converge presque sûrement sur l'espace d'épreuves de $\{\varepsilon_n, n \in \mathbb{N}\}$ et le théorème de Fubini permet de conclure à la convergence presque sûre sur l'espace d'épreuves produit ; la série $\sum m(b_n)$ associée aux mêmes lois par la symétrie converge aussi presque sûrement.

Remarquons aussi ([3], th. I.1.2 et th. III 2.1) que puisque $\mathcal{U}$ est lusinien, il existe un isomorphisme de $\mathcal{U}$ sur un borélien $\mathcal{U}_o$ de $[0,1]$ qui associe aux mesures aléatoires à valeurs indépendantes (resp. symétriques) sur $\mathcal{U}$, les mesures aléatoires à valeurs indépendantes (resp. symétriques) sur $[0,1]$ nulles sur les parties de $C\mathcal{U}_o$; on pourra donc toujours supposer que $\mathcal{U} = [0,1]$.

Nous rappelons brièvement comment on définit la variable aléatoire $\int f dm = \{\int f(x) m(\omega, dx), \omega \in \Omega\}$ associée à une fonction mesurable et bornée f et à une mesure aléatoire à valeurs symétriques ou symétrique à valeurs indépendantes. Nous nous appuyerons sur les deux lemmes suivants :

LEMME 0.2.2 : <u>Soient</u> $(\Omega, \mathcal{C})$ <u>un espace mesurable et</u> φ <u>une fonction réelle positive sur</u> $\mathcal{C}$; <u>on suppose qu'il existe un nombre</u> $k > 0$ <u>tel que pour toute partition mesurable</u> $\{A_n, n \in \mathbb{N}\}$ de Ω, on ait :

(3) $$\forall\, b \in \mathcal{C}, \quad k \sum \varphi(b \cap A_n) \leq \varphi(b) \leq \sum \varphi(b \cap A_n) \;;$$

<u>il existe alors une mesure positive bornée</u> μ <u>sur</u> $(\Omega, \mathcal{C})$ <u>telle que</u> :

(4) $$\forall\, b \in \mathcal{C}, \quad k\mu(b) \leq \varphi(b) \leq \mu(b) .$$

LEMME 0.2.3 : <u>Pour toute suite symétrique</u> $\{X_j, 1 \le j \le n\}$ <u>de variables aléatoires et tout</u> $u > 0$, <u>on a</u> :

(5) $$E \inf(u^2, (\sum_{j=1}^{n} X_j)^2) \le \sum_{j=1}^{n} E \inf(u^2, X_j^2) .$$

<u>De plus, il existe un nombre</u> $k > 0$ <u>tel que pour toute suite</u> $\{X_j, 1 \le j \le n\}$ <u>de v.a. indépendantes et symétriques et tout</u> u <u>assez grand pour que</u> $E \inf(1, \frac{1}{u^2}(\sum_{j=1}^{n} X_j)^2) \le \frac{1}{8}$, <u>on ait</u> :

(6) $$k \sum_{j=1}^{n} E \inf(u^2, X_j^2) \le E \inf(u^2, (\sum_{j=1}^{n} X_j)^2) .$$

<u>Démonstration du lemme 0.2.2</u> : Notons $\mathfrak{A}$ l'ensemble des partitions mesurables $A = \{A_n, n \in \mathbb{N}\}$ de Ω ; définissons sur $\mathcal{C}$ la fonction μ en posant :

(7) $$\forall b \in \mathcal{C}, \quad \mu(b) = \sup_{A \in \mathfrak{A}} \sum_{\mathbb{N}} \varphi(b \cap A_n) ;$$

les relations (3) et (7) impliquent $\varphi \le \mu \le \frac{1}{k} \varphi$ de sorte que μ est positive bornée et vérifie la relation (4) ; il reste à montrer que c'est une mesure. Pour cela, soient A un élément de $\mathfrak{A}$ et b un élément de $\mathcal{C}$; pour tout $\varepsilon > 0$, la définition (7) implique l'existence d'un élément A' de $\mathfrak{A}$ tel que :

$$\mu(b) \le \varepsilon + \sum_{j \in \mathbb{N}} \varphi(b \cap A'_j) ;$$

la relation (3) implique alors :

$$\mu(b) \le \varepsilon + \sum_{j \in \mathbb{N}} \sum_{n \in \mathbb{N}} \varphi(b \cap A'_j \cap A_n) ,$$

c'est-à-dire par la définition des $\mu(b \cap A_n)$:

(8.1) $$\mu(b) \le \varepsilon + \sum_{n \in \mathbb{N}} \sum_{j \in \mathbb{N}} \varphi(b \cap A_n \cap A'_j) \le \varepsilon + \sum_{n \in \mathbb{N}} \mu(b \cap A_n) .$$

Inversement, pour tout $n \in \mathbb{N}$, la définition de $\mu(b \cap A_n)$ implique l'existence d'un élément A'_n de $\mathfrak{A}$ tel que :

$$\mu(b \cap A_n) \le \frac{\varepsilon}{2^{n+1}} + \sum_{j \in \mathbb{N}} \varphi(b \cap A_n \cap A'^{n}_j) ,$$

et la définition de $\mu(b)$ implique alors :

(8.2) $$\sum_{n\in\mathbb{N}} \mu(b\cap A_n) \leq \varepsilon + \sum_{(n,j)\in\mathbb{N}^2} \varphi(b\cap A_n\cap A'^n_j) \leq \varepsilon+\mu(b) ,$$

de sorte que (8.1) et (8.2) fournissent la σ - additivité de μ ; c'est une mesure.

<u>Démonstration du lemme 0.2.3.</u> (a) Pour établir la propriété (5), il suffit utilisant la symétrie des lois et une récurrence évidente, de montrer que pour tout couple (x,y) de réels positifs et tout couple $(\varepsilon,\varepsilon')$ de variables de Rademacher indépendantes, on a :

$$E\inf(1,(\varepsilon x+\varepsilon' y)^2) \leq \inf(1,x^2)+\inf(1,y^2) ;$$

or le premier membre est inférieur à $\inf(1,x^2+y^2)$ de sorte que cette inégalité est évidente. (b) Pour démontrer la propriété (6) sous l'hypothèse indiquée, on peut se réduire au cas $u = 1$ et $E\inf(1,(\sum_{j=1}^{n} X_j)^2) \leq \frac{1}{8}$; nous notons alors $\{f_j, 1\leq j\leq n\}$ et f les fonctions caractéristiques des termes et de leur somme. Les inégalités élémentaires liant les moments tronqués et les fonctions caractéristiques des variables symétriques permettent d'écrire :

(9) $$\sum_{j=1}^{n} E\inf(1,X_j^2) \leq 7 \sum_{j=1}^{n} \int_0^1 (1-f_j(t))dt ,$$

aussi bien qu'inversement :

$$\forall\, t\in[0,1] ,\ 1-f(t) \leq 2(1+t^2)E\inf(1,(\sum_{j=1}^{n} X_j)^2) \leq \frac{1}{2} ,$$

et donc aussi :

$$\forall\, t\in[0,1] ,\ \sum_1^n (1-f_j(t)) \leq \ln\frac{1}{\prod_1^n f_j(t)} \leq \ln\frac{1}{f(t)} \leq 2\ln 2(1-f(t)) ,$$

ou encore :

$$\forall\, t\in[0,1] ,\ \sum_1^n (1-f_j(t)) \leq 4\ln 2(1+t^2)E\inf(1,(\sum_{j=1}^{n} X_j)^2) .$$

En reportant dans (9), on en déduit (6) avec $k = \frac{3}{112 \ln 2}$ ou plus simplement $k = \frac{1}{30}$.

0.3. Intégration par rapport à une mesure aléatoire à valeurs symétriques.

0.3.1. Soient μ une mesure positive et m une mesure aléatoire à valeurs symétriques sur $(\mathcal{U}, \mathcal{B})$, nous dirons que μ est une mesure de contrôle pour m si :

$$(10) \qquad \forall\, b \in \mathcal{B}, \quad E \inf(1, m^2(b)) \leq \mu(b)\ ;$$

nous dirons que μ est une mesure de contrôle strict pour m si :

$$(11) \qquad \exists\, c>0\ ,\ \forall\, b \in \mathcal{B}\ ,\ c\mu(b) \leq E \inf(1, m^2(b)) \leq \mu(b)\ .$$

Nous ignorons si toute mesure aléatoire à valeurs symétriques possède une mesure de contrôle bornée ou une mesure de contrôle strict ; nous pouvons seulement énoncer.

PROPOSITION 0.3.2 : (a) Toute mesure de contrôle strict est bornée ; deux mesures de contrôle strict d'une même mesure aléatoire sont équivalentes.
(b) Toute mesure aléatoire du second ordre (i.e $E\,|\,m^2(\mathcal{U})\,| < \infty$) à valeurs symétriques possède une mesure de contrôle bornée.
(c) Toute mesure aléatoire à valeurs indépendantes et symétriques possède une mesure de contrôle strict.

Démonstration : Les résultats (a) sont immédiats ; (b) Si m est du second ordre, l'application $\{E\, m^2(b)\ ,\ b \in \mathcal{B}\}$ est une mesure de contrôle bornée pour m ; (c) Soient m une mesure aléatoire à valeurs indépendantes et symétriques, il existe un nombre u_o assez grand pour que $E\{\inf(1\ ,\ \frac{m^2(\mathcal{U})}{u_o^2})\}$ soit inférieur à $\frac{1}{16}$; les inégalités de Lévy impliquent alors :

$$\forall\, b \in \mathcal{B}\ ,\ E \inf(1\ ,\ \frac{m^2(b)}{u_o^2}) \leq \frac{1}{8}\ .$$

La propriété (2) des mesures aléatoires à valeurs indépendantes et symétriques et le lemme 0.2.3 montrent donc que pour toute partition mesurable $\{A_n, n \in \mathbb{N}\}$ de Ω, on a :

$$k \sum_{n=1}^{\infty} E \inf(u_o^2, m^2(b \cap A_n)) \leq E \inf(u_o^2, m^2(b)) \leq \sum_{n=1}^{\infty} E \inf(u_o^2, m^2(b \cap A_n)) .$$

Il existe alors (lemme 0.2.2) une mesure positive bornée μ_o telle que :

$$\forall b \in \mathcal{B} , k\mu_o(b) \leq E \inf(1 , \frac{m^2(b)}{u_o^2}) \leq \mu_o(b)$$

on en déduit le résultat annoncé en posant :

$$\mu = \sup(1, u_o^2)\mu_o , c = k \inf(u_o^2 , \frac{1}{u_o^2}) > 0 .$$

0.3.3. Soient m une mesure aléatoire à valeurs symétriques et μ <u>une mesure de contrôle pour</u> m ; soit de plus $f = \sum_1^n f_i I_{b_i}$ une fonction étagée mesurable sur $\mathcal{U}$; en appliquant la propriété (5) du lemme 0.2.3, on a :

$$\forall \varepsilon > 0 , E \inf(1, (\int f dm)^2) \leq E \inf(1, (\int_{|f| \leq \varepsilon} f dm)^2) + \sum_{|f_i| > \varepsilon} \sup(1, f_i^2) E \inf(1, m^2(b_i)) .$$

En appliquant le lemme de contraction des v.a. symétriques pour le premier terme du second membre et la relation (10) pour le second terme, on en déduit :

$$(12) \qquad E \inf(1, (\int f\, dm)^2) \leq 2 E \inf(1, \varepsilon^2 m^2(\mathcal{U})) + \sup(1 , \frac{1}{\varepsilon^2}) \int f^2 d\mu .$$

La relation (12) montre que si la suite des fonctions étagées mesurables $\{f_n, n \in \mathbb{N}\}$ converge dans $L^2(\mu)$ vers une fonction f, alors la suite associée $\{\int f_n d_m, n \in \mathbb{N}\}$ converge en probabilité dans (Ω, P) ; on pose alors $\int f\, dm = \lim \int f_n dm$. Dans ces conditions, pour tout $f \in L^2(\mu)$, $\int f\, dm$ est une variable aléatoire vérifiant la relation (12).

Inversement d'ailleurs pour toute fonction étagée mesurable f, on a aussi avec les mêmes notations si μ est <u>une mesure de contrôle strict pour</u> m

$$c\int \inf(1,f^2)d\mu \leq \sum_1^n E\inf(1, f_i^2 m^2(b_i))$$

et donc en appliquant la formule (6) du lemme 0.2.3 dès que $E\inf(1,(\int fdm)^2) \leq \frac{1}{8}$:

(13) $$ck\int \inf(1,f^2)d\mu \leq E\inf(1,(\int fdm)^2)$$

et ceci aussi par prolongement si f appartient à $L^2(\mu)$. La formule (13) montre donc que pour toute suite $\{f_n, n\in\mathbb{N}\}$ d'éléments de $L^2(\mu)$, si les variables aléatoires associées $\{\int f_n dm, n\in\mathbb{N}\}$ convergent en probabilité dans (Ω,P), alors la suite $\{f_n, n\in\mathbb{N}\}$ converge en mesure dans $(\mathcal{U},\mu)$.

0.3.4. Les propriétés classiques des processus à accroissements indépendants montrent que toute mesure aléatoire symétrique m à valeurs indépendantes est la somme de trois mesures aléatoires symétriques à valeurs indépendantes m_1, m_2, m_3 mutuellement indépendantes et possédant les propriétés suivantes :

(1) m_1 est de la forme $\Sigma\mu_n(\omega)\delta_{x_n}$ où les x_n décrivent une suite non aléatoire dans $\mathcal{U}$, les δ_{x_n} sont les mesures de Dirac associées et $\Sigma\mu_n$ est une série p.s. convergente de v.a. symétriques indépendantes.

(2) m_2 est gaussienne sans partie discrète.

(3) m_3 est poissonnienne sans partie discrète : il existe une mesure positive π sur $\mathbb{R}^*\times\mathcal{U}$ vérifiant :

(3a) $$\iint_{\lambda\in\mathbb{R}^*, x\in\mathcal{U}} \inf(1,\lambda^2)d\pi(\lambda,x) < \infty \; ; \; \forall x_o\in\mathcal{U}, \int_{\lambda\in\mathbb{R}^*} d\pi(\lambda,x_o) = 0 .$$

(3b) Pour toute fonction f étagée mesurable sur $\mathcal{U}$, on a :

$$E\{\exp(i\int fdm_3)\} = \exp\{\iint[\cos(\lambda f(x)) - 1]d\pi(\lambda,x)\} .$$

Dans une telle situation, nous noterons $\{m_1^n, n\in\mathbb{N}\}$ la suite des termes de m_1, $\{m_2^n, n\in\mathbb{N}\}$ une décomposition de m_2 en termes indépendants de rang 1 et $\{m_3^n, n\in\mathbb{N}\}$ la décomposition de m_3 en mesures poissonniennes mutuellement indépendantes associées respectivement aux mesures $\pi_n = I_{\{\frac{1}{n+1}\leq\lambda<\frac{1}{n}\}}\pi$.

On notera que pour presque tout $\omega, m_3^n(\omega)$ est <u>combinaison linéaire d'un nombre fini</u> aléatoire de mesures de Dirac.

1. <u>Les fonctions aléatoires de type intégral</u>.

1.0. <u>Définition</u> : Soient $m = \{m(\omega), \omega \in \Omega\}$ une mesure aléatoire à valeurs symétriques sur un espace lusinien $\mathcal{U}$ et $X = \{X(\omega,t), \omega \in \Omega, t \in T\}$ une fonction aléatoire séparable sur un espace métrique séparable (T, δ) ; nous dirons que X est <u>du type intégral associé à</u> m s'il existe une mesure de contrôle μ pour m et un ensemble $\{f_t, t \in T\}$ de fonctions sur $\mathcal{U}$ de carrés intégrables pour la mesure μ tels que X ait même loi temporelle que $\{\int f_t dm, t \in T\}$. Si l'ensemble $\{f_t, t \in T\}$ est de carrés équiintégrables, on dira que X est <u>du type équiintégral</u>.

1.0.1. <u>Remarque</u> : Les classes de fonctions aléatoires introduites ci-dessus sont symétriques. Nous nous limitons à ces classes pour des raisons techniques liées au maniement des mesures aléatoires. Rappelons comme au premier chapitre que les inégalités fortes de symétrisation montrent que pour qu'une fonction aléatoires séparable sur un espace métrique séparable ait p.s. des trajectoires régulières il faut et il suffit que les médianes de ses accroissements et p.s. les trajectoires de ses symétrisées séparables soient régulières.

1.1. <u>Exemples</u> (a) Soit $\{\lambda_n, n \in \mathbb{N}\}$ une suite symétrique de variables aléatoires ; notons $\{\delta_n, n \in \mathbb{N}\}$ la suite des mesures de Dirac sur $\mathbb{N}$. Alors $\Sigma \lambda_n \delta_n$ est une mesure aléatoire à valeurs symétriques sur $\mathbb{N}$ si et seulement si pour presque tout ω, $\Sigma \lambda_n^2(\omega)$ est convergente et donc en particulier si la série $\Sigma E \inf(1, \lambda_n^2)$ est convergente. Sous cette hypothèse, la mesure $\Sigma E \inf(1, \lambda_n^2) \delta_n$ est une mesure de contrôle bornée ; le lemme 0.2.3 montre que c'est une mesure de contrôle strict si la suite $\{\lambda_n, n \in \mathbb{N}\}$ est indépendante. Toute série $\{\Sigma f_n(t) \lambda_n(\omega), t \in T, \omega \in \Omega\}$ sera donc de type intégral associé à $\Sigma \lambda_n \delta_n$ si $\{\Sigma(1 + f_n^2(t)) E \inf(1, \lambda_n^2), t \in T\}$ est convergent et de type équiintégral si le même ensemble est uniformément convergent. En particulier, c'est le cas si les (λ_n) sont indépendants et symétriques et si $\sup_n \|f_n\|_T$ est fini. Pour $T = \mathbb{R}^d$ et $f_n(t) = \exp\{i < a_n, t >\}$, on obtient les séries trigonométriques à coefficients aléatoires classiques.

(b) Toute fonction aléatoire gaussienne centrée séparable sur $\mathbb{R}$ ou $\mathbb{R}^n$ stationnaire est de type équiintégral associé à sa mesure aléatoire spectrale, gaussienne à valeurs indépendantes et symétriques, ayant sa mesure spectrale pour moment du second ordre et donc pour mesure de contrôle strict.

(c) Soit $(\Omega, \mathcal{G}, P)$ un espace probabilisé que nous supposons lusinien de sorte que $L^2(P)$ est séparable. Notons $(\Omega', \mathcal{G}', P')$ une copie indépendante de $(\Omega, \mathcal{G}, P)$, $\{\lambda_n, n \in \mathbb{N}\}$ une suite gaussienne normale et $\{f'_n, n \in \mathbb{N}\}$ une base orthonormale de $L^2(P')$. On peut alors définir une fonction aléatoire sur $\mathcal{G}'$ en posant :

$$\forall A' \in \mathcal{G}' , \ p'(A') = \sum_n \left(\int_{\omega' \in A'} f_n(\omega') dP'(\omega') \right) \lambda_n \ ;$$

on vérifie facilement que p' est une mesure aléatoire gaussienne à valeurs indépendantes et symétriques sur Ω' ayant P' pour moment du second ordre et donc pour mesure de contrôle strict. Soit alors X une fonction aléatoire gaussienne centrée séparable, bornée en probabilité, sur un espace métrique (T, δ) ; nous pouvons supposer puisque X est séparable que son espace d'épreuves est lusinien ; nous le notons $(\Omega, \mathcal{G}, P)$ et nous utilisons la construction précédente de p' ayant P' pour mesure de contrôle strict. Les propriétés d'intégrabilité marginale de X borné en probabilité assurent que $\sup_{t \in T} \int X^4(\omega', t) dP'(\omega')$ est fini et donc que toute version séparable de $\{\int_{\Omega'} X(\omega', t) dp(\omega, d\omega') , \omega \in \Omega, t \in T\}$ est de type équiintégral associé à p . Comme le calcul de la loi temporelle de cette fonction aléatoire montre que pour toute famille finie $(t_1, \ldots, t_n)$ d'éléments de T , on a :

$$E\{\exp i \left[\sum_{j=1}^n a_j \int X(\omega', t_j) dp(\omega, d\omega') \right]\} = \exp\{ -\tfrac{1}{2} \int \left[\sum_{j=1}^n a_j X(\omega', t_j) \right]^2 dP(\omega') \} ,$$

ceci signifie que la fonction aléatoire gaussienne X est de type équiintégral associé à p.

(d) Pour toute mesure aléatoire m à valeurs indépendantes et symétriques sur R ou R^n, les intégrales $\{\int \exp i < t,x > dm(\omega,dx)\}$ définissent sur le même espace des fonctions aléatoires de type équiintégral puisque m possède une mesure de contrôle bornée. Ce n'est peut-être pas le cas si on suppose seulement m à valeurs symétriques. De la même manière, nous ignorons si toute fonction aléatoire stable d'indice inférieur à 2 centrée séparable bornée en probabilité est de type équiintégral.

1.2. Continuité en probabilité des fonctions aléatoires de type intégral.

Les propriétés des mesures aléatoires et en particulier les formules (12) et (13) de l'alinéa 0.3.3 permettent d'énoncer (sans démonstration) :

THEOREME 1.2 : Soit X une fonction aléatoire de type intégral associée aux intégrales $\{\int f(x,t)dm(\omega,dx)\ ,\ t \in T\}$. (a) Soit μ une mesure de contrôle pour m ; si l'application $t \to f_t$ de (T,δ) dans $L^2(\mu)$ est continue, alors X est continue en probabilité sur (T,δ). (b) Réciproquement soit μ une mesure de contrôle strict pour m ; si X est continue en probabilité sur (T,δ), alors l'application $t \to f_t$ de (T,δ) dans $L^o(\mu)$ est continue. (c) Enfin si X est de type équiintégral et si m possède une mesure μ de contrôle strict, X est continu en probabilité sur (T,δ) si et seulement si l'application $t \to f_t$ de (T,δ) dans $L^o(\mu)$ est continue.

1.3. Propriétés d'oscillation.

Dans ce paragraphe, nous étendons à certaines fonctions aléatoires de type intégral les propriétés simples des oscillations des fonctions aléatoires gaussiennes ([9], [11]) ; le théorème fondamental aura un cadre plus général :

THEOREME 1.3.1 : Soient (T,δ) un espace métrique séparable et X une fonction aléatoire séparable sur T. On suppose qu'il existe une série ΣX_n de fonctions aléatoires réelles indépendantes sur T telles que :

(a) pour tout $t \in T$, la série $\Sigma X_n(t)$ converge presque sûrement,

(b) pour tout $t \in T$, il existe un voisinage $\mathcal{V}$ de T sur lequel chaque X_n a p.s. ses trajectoires uniformément continues,

(c) la somme ΣX_n a même loi temporelle que X.

Dans ces conditions, il existe trois applications α, β_1, β_2 de T dans $\overline{R}^+$, une partie négligeable N de Ω et pour tout élément t de T une partie négligeable N_t de Ω telles que :

(14) $$\forall t \in T, \forall \omega \notin N_t, \liminf_{s \to t} X(\omega,s) = X(\omega,t) - \beta_1(t),$$

$$\limsup_{s \to t} X(\omega,s) = X(\omega,t) + \beta_2(t),$$

(15) $$\forall \omega \notin N, \forall t \in T, \limsup_{\substack{s \to t \\ s' \to t}} \{X(\omega,s) - X(\omega,s')\} = \alpha(t) = \beta_1(t) + \beta_2(t).$$

Démonstration : Elle utilise la notion d'oscillation ; soit f une fonction sur T à valeurs dans R ou $\overline{R}$, on appelle oscillation de f et on note $W(f)$ la fonction sur T à valeurs dans R ou $\overline{R}$ définie par :

(16) $$\forall t \in T, W(f,t) = \lim_{\varepsilon \downarrow 0} \sup_{\substack{\delta(s,t) < \varepsilon \\ \delta(s',t) < \varepsilon}} \{f(s) - f(s')\};$$

de la même manière, à tout $t \in T$ et tout $u > 0$, nous associons l'oscillation de f sur $\{\delta(s,t) \leq u\}$ définie par :

(17) $$V(f,t,u) = \lim_{\varepsilon \downarrow 0} \sup_{\begin{cases} \delta(s,t) < u \\ \delta(s',t) < u \\ \delta(s,s') < \varepsilon \end{cases}} \{f(s) - f(s')\};$$

on a immédiatement :

$$W(f,t) = \lim_{u \downarrow 0} V(f,t,u).$$

Dans la preuve qui suit, nous supposerons que Ω est P-complet et que X est précisément une version séparable de ΣX_n. Dans ces conditions, la séparabilité de X et l'hypothèse (b) montrent que $\limsup_{s\to t}\{X(t)-X(s)\}$, $\limsup_{s\to t}\{X(s)-X(t)\}$ et $V(X,t,u)$ sont des v.a. p.s. positives et $(\bigcap_n \mathcal{B}_n)$-mesurables où $\mathcal{B}_n$ est la tribu complète engendrée par $\{X_k, k\geq n\}$; elles sont donc dégénérées ; nous notons $\beta_1(t),\beta_2(t),\alpha(t,u)$ leurs valeurs presque sûres ; β_1 et β_2 vérifient par construction les relations (14). Soit de plus S une suite dense dans T ; il existe une partie négligeable N de Ω telle que :

$$(18) \qquad \forall\, \omega\notin N\,,\ \forall\, s\in S\,,\ \forall\, u\in \mathbb{Q}^*\,,\ V(X(\omega),s,u) = \alpha(s,u)\,.$$

Fixons un élément $\omega\notin N$ et un élément t de T ; les deux fonctions $\alpha(t,.)$ et $V(X(\omega),t,.)$ sont deux fonctions croissantes sur R^+ vérifiant :

$$(19) \qquad \forall\, u>0\,,\ \forall\, s\in T\,,\ V(X(\omega),t,u) \leq V(X(\omega),s,u+\delta(s,t))\,,$$
$$\alpha(s,u) \leq \alpha(t,u+\delta(s,t))\,.$$

Nous allons comparer leurs limites respectives $\alpha(t)$ et $W(X(\omega),t)$ à l'origine ; supposons par exemple que la première limite soit finie ; fixant $\varepsilon>0$, nous déterminons $\eta>0$ tel que :

$$(20) \qquad 0<u\leq\eta \Rightarrow \alpha(t,u)\leq\alpha(t)+\varepsilon\,,$$

nous choisissons un élément s de S et un nombre $u\in\,]0,\frac{\eta}{3}]$ tels que $\delta(s,t)$ soit inférieur à $\frac{\eta}{3}$ et $u+\delta(s,t)$ soit rationnel ; on aura alors en utilisant (18), (19) et (20) :

$$W(X(\omega),t) \leq V(X(\omega),t,u) \leq V(X(\omega),s,u+\delta(s,t)) \leq \alpha(s,u+\delta(s,t)) \leq \alpha(t,u+2\delta(s,t))$$
$$\leq \alpha(t)+\varepsilon\,,$$

ce qui signifie :

$$W(X(\omega),t) \leq \alpha(t)\ ;$$

on prouve de la même manière l'inégalité inverse si $W(X(\omega),t)$ est fini et donc dans tous les cas l'égalité. En tenant compte des relations (14), ceci fournit (15) et le théorème est établi.

COROLLAIRE 1.3.2 : Soit X une fonction aléatoire continue en probabilité vérifiant les hypothèses du théorème 1.3.1 ; on suppose de plus qu'il existe un sous-ensemble ouvert G de T, une partie dense S de G et un nombre $a>0$ tels que :

(a) $\forall\, t\in S$, $\beta_1(t)\geq a$,

ou (b) $\forall\, t\in S$, $\beta_2(t)\geq a$.

Dans ces conditions, on a plus précisément :

$$\forall\, t\in G,\ \alpha(t)=+\infty .$$

Ce corollaire signifie que sous les hypothèses indiquées, il existe une partie négligeable N de Ω telle que :

$$\forall\, \omega\notin N,\ \forall\, t\in G,\ \limsup_{\substack{s\to t\\ s'\to t}} X(\omega,s)-X(\omega,s')=+\infty ;$$

les trajectoires de X sont donc p.s. non bornées au voisinage de tout point de G.

Démonstration : Puisque S est dense dans G et X est continu en probabilité, on peut extraire de S une suite S' séparante pour la restriction de X à G ; on note N' la partie négligeable associée à S' dans cette séparation ; on note de plus N et $\{N_t, t\in T\}$ les parties négligeables définies dans l'énoncé du théorème 1.3.1. Supposons que X vérifie les hypothèses du corollaire et particulièrement (a) ; supposons de plus qu'il existe un élément t de G tel que $\alpha(t)$ soit fini, nous allons prouver la contradiction. Nous choisissons pour cela $\omega\notin N\cup N'\cup \bigcup_{t\in S'} N_t$; par définition de la séparabilité et le résultat (15) du théorème, pour tout $\varepsilon>0$, il existe deux éléments s et s' de S' tels que :

$$\delta(s,t)<\frac{\varepsilon}{2},\ \delta(s',t)<\frac{\varepsilon}{2},\ X(\omega,s)-X(\omega,s')>\alpha(t)-\frac{a}{4} ;$$

la relation (14) permet alors de construire un couple (u,u') d'éléments de l'ouvert G tels que :

$$\delta(s,u) < \frac{\varepsilon}{2}\ ,\ \delta(s',u') < \frac{\varepsilon}{2}\ ,\ X(\omega,u) - X(\omega,s) > \beta_2(s) - \frac{a}{4}\ ,$$

$$X(\omega,s') - X(\omega,u') > \beta_1(s') - \frac{a}{4}\ ;$$

le couple (u,u') vérifie alors aussi :

$$\delta(t,u) < \varepsilon\ ,\ \delta(t,u') < \varepsilon\ ,\ X(\omega,u) - X(\omega,u') > \alpha(t) + \beta_1(s') - \frac{3a}{4}\ ,$$

on en déduit :

$$\limsup_{\substack{u \to t \\ u' \to t}} \{X(\omega,u) - X(\omega,u')\} \geq \alpha(t) + \frac{a}{4}\ ,$$

ceci contredit la définition de α ; le corollaire est donc établi sous son hypothèse (a) et par symétrie sous son hypothèse (b).

COROLLAIRE 1.3.3 : Soit X une fonction aléatoire vérifiant les hypothèses du théorème 1.3.1 ; on suppose de plus que pour tout $t \in T$, $P\{\lim_{s \to t} X(s) = X(t)\}$ est non nulle. Dans ces conditions, X a p.s. ses trajectoires continues.

Démonstration : Nous y utilisons les notations du théorème 1.3.1. Pour tout élément t de T, l'hypothèse implique l'existence d'une partie non négligeable Ω_t de Ω telle que :

$$\forall\ \omega \in \Omega_t\ ,\ \liminf_{s \to t} \{X(\omega,s) - X(\omega,t)\} = \limsup_{s \to t} \{X(\omega,s) - X(\omega,t)\} = 0\ ;$$

choisissant un élément ω de Ω_t n'appartenant pas à N_t, on obtient en reportant dans (12) :

$$\forall\ t \in T,\ \beta_1(t) = \beta_2(t) = 0\ ;$$

la relation (15) signifie alors que pour tout $\omega \notin N$, la trajectoire $X(\omega)$ est continue.

Nous indiquons maintenant le champ d'application du théorème 1.3.1 :

THEOREME 1.3.4 : (a) Soient X et Y deux fonctions aléatoires indépendantes sur un espace métrique (T,δ) vérifiant les hypothèses du théorème 1.3.1, alors toute version séparable de leur somme les vérifie aussi. (b) Soient

$\{f_n, n \in \mathbb{N}\}$ une suite de fonctions continues sur un espace localement compact métrisable T et $\{\lambda_n, n \in \mathbb{N}\}$ une suite de v.a. indépendantes. On suppose que pour tout $t \in T$, la série $\Sigma f_n(t)\lambda_n$ converge p.s. et on note X une version séparable de sa somme, alors X vérifie les hypothèses du théorème 1.3.1.

(c) Soient m une mesure aléatoire symétrique à valeurs indépendantes sur un espace lusinien $\mathcal{U}$ et μ une mesure de contrôle pour m ; soient de plus (T, δ) un espace localement compact métrisable et $\{t \to f_t\}$ une application de T dans $L^2(\mu)$. On suppose que la mesure extérieure $\mu^*\{x \in \mathcal{U} : t \to f_t(x) \notin C(T)\}$ est nulle. Dans ces conditions, toute fonction aléatoire séparable sur T ayant mêmes lois temporelles que $\{\int f_t dm, t \in T\}$ vérifie les hypothèses du théorème 1.3.1.

Démonstration : Dans ces différents cas, il suffit de mettre en évidence des décompositions indépendantes ΣX_n vérifiant les hypothèses (a) et (b) du théorème 1.3.1. Dans le cas (b), on pose $X_n = \lambda_n f_n$; dans le cas (c), on pose $X_n = \int f(dm_1^n + dm_2^n + dm_3^n)$ où m_1^n, m_2^n, m_3^n ont la signification indiquée en 0.2.4.

1.3.5. Dans le théorème suivant, nous précisons une situation où les résultats du théorème 1.3.1 sont particulièrement puissants :

THEOREME 1.3.5. Soit X une fonction aléatoire séparable de type intégral de la forme $\{\int e^{i<x,t>} dm(\omega,dx), \omega \in \Omega, t \in R^n\}$ où m est une mesure aléatoire symétrique à valeurs indépendantes. On suppose qu'il existe un élément t_o de R^n tel que $P\{\limsup_{h \to o}|X(t_o+h)| < \infty\}$ soit positive. Dans ces conditions, X a p.s. ses trajectoires continues sur R^n.

Démonstration : Nous notons U et V les parties réelles et imaginaires de X ; le théorème 1.3.4 montre que U et V vérifient les hypothèses du théorème 1.3.1 et la symétrie de M montre qu'on a $\beta_1^U = \beta_2^U = \frac{\alpha^U}{2}$ et $\beta_1^V = \beta_2^V = \frac{\alpha^V}{2}$. Soit alors r un nombre positif, l'égalité :

$$U(t+r) = \int \{\cos<x,t>\cos<x,r>\}dm(x) - \int \{\sin<x,t>\sin<x,r>\}dm(x)$$

permet puisque $|\cos <x,r>|$ et $|\sin <x,r>|$ sont inférieurs à 1 pour les deux termes du second membre d'utiliser les lemmes de contraction et on en déduit, pour tout couple (σ,τ) d'éléments de $\mathbb{R}^n$:

$$P\{\lim_{h\to o} U(\sigma+h) - U(\sigma) > u\} \le 2P\{\limsup_{h\to o} U(\tau+h) - U(\tau) > \frac{u}{2}\}$$

$$+ 2P\{\limsup_{h\to o} V(\tau+h) - V(\tau) > \frac{u}{2}\}$$

et ceci implique, avec la relation semblable pour V :

$$\sup(\alpha^U(\sigma),\alpha^V(\sigma)) \le 2\sup(\alpha^U(\tau),\alpha^V(\tau)) ,$$

qui montre que si $\alpha^U(\sigma)$ ou $\alpha^V(\sigma)$ était strictement positif, alors $\inf_{\mathbb{R}^n} \sup(\alpha^U,\alpha^V)$ serait aussi strictement positif ; soient dans ces conditions U' et V' des copies, indépendantes entre elles, de U et V et Y une version séparable de leur somme ; le théorème 1.3.4 (a) implique que Y vérifie les hypothèses du théorème 1.3.1 ; les inégalités de Lévy montreraient donc qu'en tout point τ de $\mathbb{R}^n$, $\alpha^{U'+V'}(\tau)$ serait supérieur ou égal à $\sup(\alpha^{U'}(\tau),\alpha^{V'}(\tau))$. Le corollaire 1.3.2 impliquerait alors qu'en tout élément t de $\mathbb{R}^n$ $\alpha^{U'+V'}(t)$ et donc $\sup(\alpha^U(t),\alpha^V(t)$ seraient infinis. Ceci est contradictoire avec l'hypothèse sur t_o : α^U et α^V sont donc nuls en tout point, c'est le résultat du théorème.

Remarque : on notera qu'en général X n'est pas stationnaire.

2. Approximation des fonctions aléatoires de type intégral.

THEOREME 2.1 : Soient m une mesure aléatoire symétrique à valeurs indépendantes sur un espace lusinien $\mathcal{U}$ et μ une mesure de contrôle pour m ; soient de plus (T,δ) un espace compact métrisable et $\{t \to f_t\}$ une application de T dans $L^2(\mu)$. Soient enfin X une version séparable de $\{\int f_t dm, t \in T\}$ et pour tout entier n, X_n une version séparable de $\{\int f_t d(m_1^n + m_2^n + m_3^n), t \in T\}$ au sens 0.2.4. On suppose que X a p.s. ses trajectoires continues ; dans ces conditions, la série ΣX_n converge p.s. vers X au sens de la convergence uniforme sur T.

Démonstration : C'est une propriété générale des séries de vecteurs aléatoires indépendants et symétriques. Pour détailler, nous prenons quelques précautions liées au fait qu'il n'est pas évident que les sommes partielles $\sum_{k=o}^{n} X_k$ soient séparables : l'hypothèse sur les trajectoires de X implique pour tout $\varepsilon > 0$, l'existence de $\eta > 0$ et $M < \infty$ tels que pour toute partie finie S de T, on ait :

$$P\{\sup_{\substack{\delta(s,t)<\eta \\ s,t \in S \times S}} \frac{|X(s)-X(t)|}{\varepsilon} \vee \sup_{s \in S} \frac{|X(s)|}{M} > 1\} < \varepsilon ;$$

les inégalités de Lévy en dimension finie impliquent alors :

$$\forall n \in \mathbb{N},\ P\{\sup_{\substack{\delta(s,t)<\eta \\ s,t \in S \times S}} \frac{|X_n(s)-X_n(t)|}{\varepsilon} \vee \sup_{s \in S} \frac{|X_n(s)|}{M} > 1\} < 2\varepsilon ,$$

de sorte que la séparabilité de X_n montre que p.s. ses trajectoires sont continues et il en est de même des sommes partielles $\sum_{k=o}^{n} X_k$. Les inégalités de Lévy applicables alors directement sur T montrent :

$$\forall n \in \mathbb{N},\ P\{\sup_{\delta(s,t)<\eta} \frac{|\sum_1^n (X_k(s)-X_k(t))|}{\varepsilon} \vee \sup_T \frac{|\sum_1^n X_k|}{M} > 1\} < 2\varepsilon$$

de sorte que les sommes partielles $\{\sum_1^n X_k,\ n \in \mathbb{N}\}$ forment un ensemble relativement compact pour la convergence en loi dans $C(T)$. Le théorème de Ito et Nisio montre alors que la série $\sum_1^\infty X_k$ converge p.s. dans $C(T)$ vers X ; c'est le résultat.

On démontre suivant un schéma identique :

THEOREME 2.2 : Soient m une mesure aléatoire symétrique à valeurs indépendantes sur un espace lusinien $\mathcal{U}$ et $\mathcal{P} = \{p_n,\ n \in \mathbb{N}\}$ une partition mesurable et dénombrable de $\mathcal{U}$; on note μ une mesure de contrôle pour m. Soient de plus (T, δ) un espace compact métrisable et $\{t \to f_t\}$ une application de T dans

$L^1(\mu)$. Soient enfin X une version de $\{\int f_t dm, t \in T\}$ ayant p.s. ses trajectoires continues et pour tout entier n , X_n une version séparable de $\{\int_{P_n} f_t dm, t \in T\}$. Dans ces conditions, la série ΣX_n converge p.s. vers X au sens de la convergence uniforme sur T .

3. Majoration des lois, propriétés d'intégrabilité.

Les propriétés d'intégrabilité des vecteurs aléatoires gaussiens ([13], [4], [2]) ont déjà été adaptées aux séries de vecteurs aléatoires indépendants et intégrables ([10]). Elles s'étendent aussi aux fonctions aléatoires de type intégral et peuvent dans ce cas donner lieu à des évaluations très maniables. Le schéma d'étude, différent des schémas gaussiens, dérive des techniques de Yurinskii ([17]), Kuelbs ([12]), de Acosta ([1]) dans le domaine des propriétés limites des vecteurs aléatoires. Le lemme fondamental sera le suivant :

LEMME 3.1 : Soient $\mathcal{B}_1, \mathcal{B}_2, \mathcal{B}_3$ trois tribus indépendantes et X une v.a. intégrable et mesurable par rapport à la tribu engendrée par $\mathcal{B}_1$ et $\mathcal{B}_3$; alors X a même espérance conditionnelle relativement à $\mathcal{B}_1$ ou relativement à la tribu engendrée par $\mathcal{B}_1$ et $\mathcal{B}_2$.

Démonstration : Les arguments habituels de réduction montrent qu'il suffit de prouver le résultat si X est positive étagée ou même si X est l'indicatrice I_A d'un élément $A = A_1 \cap A_3$, $A_1 \in \mathcal{B}_1$, $A_3 \in \mathcal{B}_3$ de la semi-algèbre engendrée par $\mathcal{B}_1$ et $\mathcal{B}_3$; on a alors :

$$E\{X|\mathcal{B}_1\} = I_{A_1}.P(A_3) \text{ p.s.}$$

et on doit montrer que pour tout élément B de la tribu engendrée par $\mathcal{B}_1$ et $\mathcal{B}_2$, on a :

$$E\{XI_B\} = E\{I_{A_1}.P(A_3)I_B\} ;$$

il suffit en fait de vérifier cette égalité si $B = B_1 \cap B_2$, $B_1 \in \mathcal{B}_1$, $B_2 \in \mathcal{B}_2$ appartient à la semi-algèbre engendrée par $\mathcal{B}_1$ et $\mathcal{B}_2$. Le calcul montre alors que les deux nombres à comparer valent $P(A_1 \cap B_1)P(B_2)P(A_3)$ et sont donc égaux, ce qui établit le lemme.

THEOREME 3.2 : Soit X une fonction aléatoire de type intégral associée aux intégrales $\{\int_{\mathcal{U}} f(x,t)dm(\omega,dx)\ ,\ t \in T\}$; on suppose que la fonction f, à valeurs réelles ou complexes, vérifie :

(21) $$\forall\ x \in \mathcal{U}, \forall\ t \in T,\ |f(x,t)| \leq 1\ ;$$

on suppose aussi que X a p.s. ses trajectoires bornées sur T. Dans ces conditions, pour tout $p > 0$, $\sup_T |X|^p$ est intégrable si et seulement si $|m(\mathcal{U})|^p$ l'est. De plus, on a :

(22) $$\sigma^2\{\inf(u, \sup_T(X))\} \leq C_o\, E \inf(u^2, |m(\mathcal{U})|^2) \quad \text{pour tout } u > 0 \text{ tel que}$$
$$E \inf\left(1, \frac{1}{u^2}|m(\mathcal{U})|^2\right) \leq \frac{1}{8},$$

(23) $$E\{|\sup_T|X| - E\sup_T|X||^p\} \leq C_p E|m(\mathcal{U})|^p \quad \text{pour tout } p \geq 2 \text{ tel que}$$
$$E|m(\mathcal{U})|^p < \infty,$$

où les constantes C_p ne dépendent que de leurs indices et $C_2 = 1$.

Remarque 3.2.1 : La formule (2.2) s'utilise quand $|m(\mathcal{U})|$ est peu intégrable, la formule (23) dans le cas contraire ; l'une et l'autre sont des formules bilatères qui donnent l'ordre de grandeur de la concentration de $\sup_T|X|$ autour de certain paramètre de sa loi, donc aussi de tout autre paramètre de cette loi, sa médiane par exemple. Elles évaluent cette concentration à partie de la seule loi de $|m(\mathcal{U})|$; elles ne permettent pas d'évaluer les paramètres de centrage, ce sera dans certains cas l'objet des paragraphes 4 et 5. L'aspect qualitatif du théorème est une simple extension des théorèmes précédents sur les séries de vecteurs aléatoires indépendants. L'aspect quantitatif est plus important par sa simplicité ; on pourra constater au cours de la preuve du théorème comment la formule (22) apparemment plus compliquée que (23) puisque le centrage y dépend de u se manie sans peine.

3.3. Démonstration du théorème 3.2.

3.3.1. Supposons pour commencer que X vérifie la propriété (22) ; alors, en notant u_o un nombre assez grand pour que $E\inf(1, \frac{1}{u_o^2}|m(\mathcal{U})|^2) \leq \frac{1}{8}$, on a pour tout $u \geq u_o$ et tout $x > 0$:

$$P\{|\inf(u, \sup_T |X|) - E\inf(u, \sup_T |X|)| \geq x\} \leq \frac{C_o}{x^2} E\inf(u^2, |m(\mathcal{U})|^2) ;$$

Soit alors μ une médiane de $\sup_T |X|$; si μ est inférieur à u, c'est aussi une médiane de $\inf(u, \sup_T |X|)$ de sorte que l'inégalité ci-dessus implique :

$$\forall u \geq u_1 = \sup(\mu, u_o), |\mu - E\inf(u, \sup_T |X|)|^2 \leq 2C_o E\inf(u^2, |m(\mathcal{U})|^2) ;$$

en reportant dans (22), on en déduit :

$$(22)' \qquad \forall u \geq u_1, E\inf(u^2, \sup_T |X|^2) \leq 4C_o E\inf(u^2, |m(\mathcal{U})|^2) + 3u_1^2 .$$

Dans ces conditions, la propriété élémentaire des moments de toute v.a. λ :

$$\forall p \in]0,2[, E|\lambda|^p = \frac{p(2-p)}{2} \int_o^\infty E\inf(u^2, \lambda^2) u^{p-3} du ,$$

montre immédiatement par intégration de (22') l'intégrabilité annoncée si $p \in]0,2[$ et donc en particulier si $p = 1$; plus précisément, on a :

$$\forall p \in]0,2] \quad E\sup_T |X|^p \leq 4C_o E|m(\mathcal{U})|^p + 3u_1^p .$$

3.3.2. Supposons maintenant l'espace T fini. Dans ces conditions, toutes les intégrabilités de l'énoncé sont évidentes puisque les propriétés de contraction des v.a. symétriques montrent, sous l'hypothèse (21) :

$$\forall u \geq 0, P\{\sup_T |X| > u\} \leq 2\,\mathrm{Card}(T) P\{|m(\mathcal{U})| > u\} ;$$

il suffit donc dans ce cas pour établir le théorème de prouver les propriétés

(22) et (23). Nous supposerons pour commencer la famille de fonctions f étagée de sorte qu'il existe une partition finie $\{b_i \,,\, 1 \le i \le n\}$ de $\mathcal{U}$ et une suite $\{\varphi_i \,,\, 1 \le i \le n\}$ de fonctions sur T telles que :

$$\forall\, t \in T \,,\, \int f_t \, dm = \sum_{i=1}^{n} \varphi_i(t) m(b_i) \,,\, |\varphi_i(t)| \le 1 \,.$$

Notons $\mathcal{B}_k$ la tribu engendrée par les $\{m(b_j) \,,\, 1 \le j \le k\}$; l'indépendance de ces v.a. et le lemme 3.1 montrent qu'on a :

$$\forall\, u \ge 0 \,,\, (E^{\mathcal{B}_k} - E^{\mathcal{B}_{k-1}})[\inf(u, \sup_T |X - \varphi_k m(b_k)|)] = 0 \,,$$

et aussi, si $|m(\mathcal{U})|$ est intégrable :

$$(E^{\mathcal{B}_k} - E^{\mathcal{B}_{k-1}})[\sup_T |X - \varphi_k m(b_k)|] = 0 \;;$$

nous posons alors :

$$\eta_k = \inf(u, \sup_T |X|) - \inf(u, \sup_T |X - \varphi_k m(b_k)|) \,,$$

$$\zeta_k = \sup_T |X| - \sup_T |X - \varphi_k m(b_k)| \,,$$

on en déduit dans le premier cas :

$$\forall\, u \ge 0 \,,\, \inf(u, \sup_T |X|) - E \inf(u, \sup_T |X|) = \sum_{k=1}^{n} (E^{\mathcal{B}_k} - E^{\mathcal{B}_{k-1}}) \eta_k \,,$$

et dans le second cas :

$$\sup_T |X| - E \sup_T |X| = \sum_{k=1}^{n} (E^{\mathcal{B}_k} - E^{\mathcal{B}_{k-1}}) \zeta_k \,. \tag{24}$$

On intègre alors les carrés des deux membres en utilisant le fait que les $(E^{\mathcal{B}_k} - E^{\mathcal{B}_{k-1}})$ sont des opérateurs de projections mutuellement orthogonales ; on obtient suivant les cas

$$\forall\, u \ge 0 \,,\, \sigma^2 \inf(u, \sup_T |X|) \le \sum_{k=1}^{n} E|\eta_k|^2 \le \sum_{k=1}^{n} E \inf(u^2, |m(b_k)|^2) \,,$$

$$\sigma^2 \sup_T |X| \le \sum_{k=1}^{n} E|\zeta_k|^2 \le \sum_{k=1}^{n} E|m(b_k)|^2 \;;$$

le lemme 0.2.3 donne donc dans ce cas (T fini, f étagée) les relations (22) et nous avons aussi établi (23) pour $p = 2$ avec $C_2 = 1$. Pour $p > 2$, les inégalités de Burkholder permettent à partir de (24) d'écrire :

$$E|\sup_T|X| - E\sup_T|X||^p \leq A_p E[\sum_{k=1}^{n} |(E^{\mathcal{B}_k} - E^{\mathcal{B}_{k-1}})\zeta_k|^2]^{p/2} ;$$

en majorant $|\zeta_k|$ par $|m(b_k)|$, on en déduit :

$$\{E|\sup_T|X| - E\sup_T|X||^p\}^{\frac{1}{p}} \leq B_p\{E[\sum_1^n|m(b_k)|^2]^{\frac{p}{2}}\}^{\frac{1}{p}} + B_p\{\sum_1^n E|m(b_k)|^2\}^{\frac{1}{2}} ;$$

puisque la famille $\{(m(b_k)), 1 \leq k \leq n\}$ est symétrique, les relations de Khintchine donnent alors la formule (23) pour tout $p > 2$ (T fini, f étagée) .

3.3.3. Supposons toujours l'espace T fini, mais la fonction f , non nécessairement étagée, vérifiant les inégalités (21). Il existe alors une suite $\{f^n, n \in \mathbb{N}\}$ de fonctions étagées vérifiant (21) telles que :

$$\forall\, t \in T\, , \int f_t^n dm \xrightarrow[n\to\infty]{p.s} \int f_t\, dm \text{ et donc } \sup_T \int f^n\, dm \xrightarrow[n\to\infty]{p.s} \sup_T \int f\, dm ;$$

pour tout $u \geq 0$, la suite $(\inf\{u, \sup_T \int f^n dm\}, n \in \mathbb{N})$ majorée par u converge alors dans $L^2(P)$ vers $\inf\{u, \sup_T \int f dm\}$ et la formule (22) dans ce cas en résulte ; supposons de plus qu'il existe $p \geq 2$ tel que $E|m(\mathcal{U})|^p$ soit fini, alors la formule (22') montre quand u y tend vers l'infini que $E\{\sup_T|\int f^n dm|^2\}$ est majorée si bien que $\sup_T \int f^n\, dm$ converge vers $\sup_T \int f\, dm$ dans $L^1(P)$; le lemme de Fatou donne alors :

$$E|\sup_T|X| - E\sup_T|X||^p \leq \liminf_{n\to\infty} E\{|\sup_T|\int f^n dm| - E\sup_T|\int f^n dm||^p\}$$

et le résultat 3.3.1 implique alors la propriété (23) dans ce cas aussi.

3.3.4. Le résultat général d'intégrabilité pour $p < 2$ et l'inégalité (22) pour T arbitraire se déduisent du résultat 3.3.3 appliqué aux parties finies de T et de la séparabilité de X . Ceci montre que s'il existe $p \geq 2$ tel que $E|m(\mathcal{U})|^p$ soit fini, alors $E\sup_T|X|$ est aussi fini de sorte que l'inégalité

(23) se déduit ensuite des mêmes arguments.

3.4. Le résultat du théorème 3.3. est utile même dans le cas où X est une fonction aléatoire gaussienne stationnaire. En le combinant avec les résultats de C. Borell, on obtient par exemple ([7]).

COROLLAIRE 3.4 : Soient X une fonction aléatoire gaussienne centrée séparable et stationnaire sur R^d et T une partie de R^d sur laquelle X soit p.s. bornée ; on a alors pour tout $x \geq 2$:

$$(25) \qquad P\{|\sup_T |X| - E\sup_T |X|| \geq x\sqrt{E|X(0)|^2}\} \leq \frac{1}{\sqrt{2\pi}} \int_{x-2}^{\infty} e^{-\frac{u^2}{2}}\, du .$$

Démonstration : Puisque X est stationnaire, on peut écrire :

$$X(t) = \int \cos tx\, dm(x) + \int \sin tx\, d\bar{m}(x) ,\ E|m(R^d)|^2 = E|\bar{m}(R^d)|^2 = 1 ,$$

où m et $\bar{m}$ sont deux mesures aléatoires gaussiennes symétriques à valeurs indépendantes sur R^d, mutuellement indépendantes et de même loi. Appliquant le théorème 3.3, on peut donc majorer la variance de $\sup_T |X|$ par $2E|X(0)|^2$. Soit alors S une suite séparante pour X sur T ; nous notons A la partie de l'espace de Banach séparable $\ell^\infty(S)$ définie par :

$$A = \{x \in \ell^\infty(S) : |\sup_{\mathbb{N}} |x_n| - E\sup_T |X|| \leq 2\sqrt{E|X(0)|^2}\} ;$$

en confondant X et sa restriction à S, l'inégalité de Čebičev montre :

$$P\{X \notin A\} \leq \frac{1}{2} ,$$

et l'inégalité de Borell implique alors :

$$\forall\, t \geq 0 ,\ P\{X \in A + t\theta_X\} \geq \frac{1}{\sqrt{2\pi}} \int_{-\infty}^{t} e^{-\frac{u^2}{2}}\, du$$

où θ_X est une partie de $\ell^\infty(S)$ vérifiant :

$$\forall\, x \in \theta_X ,\ \sup|x| \leq \sqrt{E|X(0)|^2} .$$

Les propriétés de A et θ_X et la séparabilité de X fournissent donc l'inégalité annoncée.

4. Etude locale des trajectoires.

4.0. Dans ce paragraphe, nous présentons certains résultats de Marcus ([14]), Fernique ([5]), Marcus et Pisier ([15], [16]) sur la régularité des trajectoires des fonctions aléatoires X de type intégral de la forme

$$\{\int_{x\in\mathbb{R}^n} e^{i<x,t>}dm(\omega,dx)\,,\ t\in\mathbb{R}^n\}$$

où m est une mesure aléatoire sur $\mathbb{R}^n$. Bien que les résultats des auteurs cités ne soient pas liés à la symétrie, nous supposerons que m est symétrique, cette situation suffit en effet pour présenter la technique générale et l'essentiel du résultat (cf. remarque 1.0.1). Par contre nous ne supposons pas que m ait des moments. Les théorèmes 1.3.1, 1.3.4(c) et 1.3.5. montrent que si m est symétrique à valeurs indépendantes et X est borné au voisinage d'un point t_o avec probabilité non nulle, alors X a p.s. ses trajectoires continues sur $\mathbb{R}^n$; nous visons à préciser des conditions suffisantes pour cette propriété. Nous utiliserons essentiellement les résultats gaussiens sous la forme suivante :

THEOREME 4.0.1 : (a) Soit X une fonction aléatoire séparable sur $\mathbb{R}^n$ à valeurs réelles ou complexes ; notons δ un écart mesurable sur $\mathbb{R}^n$ stable par translation ; on suppose que :

$$\forall\, z\in\mathbb{C}\,,\ \forall\,(s,t)\in\mathbb{R}^d\times\mathbb{R}^d\,,\ E|\exp[z(X(s)-X(t))]|\le\exp[\frac{\delta^2(s,t)}{2}|z|^2]\,.$$

Dans ces conditions, on a pour tout $T>0$:

$$(26)\quad E\sup_{\substack{|s|\le T\\|t|\le T}}|X(s)-X(t)|\le A\sqrt{n}\{D(T)+\int_0^\infty\sqrt{\log\frac{(2T)^n}{\lambda\{s:|s|\le T,\delta(0,s)\le u\}}}\,du\}\,,$$

où λ est la mesure de Lebesgue normalisée, A une constante absolue et $D(T)=\sup_{|s|\le T}\delta(0,s)$.

(b) Pour tout entier $n\ge1$, il existe un nombre $B>0$ tel que pour toute fonction aléatoire gaussienne X stationnaire sur $\mathbb{R}^n$ centrée à trajectoires continues, on ait :

$$(27)\quad \forall\, T>0\,,\ E \sup_{|t|\le T} X(t) \ge B\{D_X(T) + \int_0^\infty \sqrt{\log \frac{(2T)^n}{\lambda\{s : |s|\le T, \delta_X(0,s)\le u\}}}\, du\}\,,$$

où δ_X et D_X sont définis par :

$$\delta_X^2(s,t) = E|X(s)-X(t)|^2\,,\ D_X(T) = \sup_{|s|\le T} \delta_X(0,s)\,.$$

Démonstration : (a) Le théorème 6.1.1 de [6] fournit la majoration :

$$E \sup_{\substack{|s|\le T\\ |t|\le T}} |X(s)-X(t)| \le 300 \sup_{|t|\le T} \int_0^{2D(T)} \sqrt{\log(1+\frac{(4T)^n}{\lambda\{s : |s|\le 2T, \delta(s,t)\le u\}})}\, du\,;$$

dans ce dernier terme, puisque $|t|\le T$, le dénominateur se minore par $\lambda\{s : |s|\le T, \delta(0,s)\le u\}$; ceci donne la majoration (26) avec $A\le 640$.

(b) Supposons que la propriété annoncée soit fausse ; alors pour tout entier $k\ge 1$, il existe un nombre $T_k>0$ et une fonction aléatoire gaussienne X_k stationnaire sur $\mathbb{R}^n$ centrée et à trajectoires continues tels que :

$$4^k E \sup_{|t|\le T_k} X_k(t) \le D_{X_k}(T_k) + \int_0^\infty \sqrt{\log \frac{(2T_k)^n}{\lambda\{s : |s|\le T_k, \delta_k(0,s)\le u\}}}\, du\,;$$

on peut supposer la suite $\{X_k, k\ge 1\}$ indépendante et par homogénéité :

$$T_k = 1\,,\ E \sup_{|t|\le 1} X_k(t) = 1\,,\ D_k(1)\le \sqrt{2\pi}\,;$$

on définit alors une fonction aléatoire gaussienne stationnaire Y sur $\mathbb{R}^n$ à trajectoires continues en posant :

$$Y(t) = \sum_{k=1}^\infty \frac{1}{2^k} X_k(t)\,,$$

et on a pour tout $k\ge 1$:

$$\int_0^{\sqrt{\pi}} \sqrt{\log(1+\frac{2^n}{\lambda\{s : |s|\le 1, \delta_Y(0,s)\le u\}})}\, du \ge \int_0^\infty \sqrt{\log \frac{2^n}{\lambda\{s : |s|\le 1, \delta_{X_k}(0,s)\le 2^k u\}}}\, du\,;$$

les hypothèses faites sur la suite $\{X_k, k \geq 1\}$ minorent ce dernier terme par $\frac{4^k - \sqrt{2\pi}}{2^k}$ qui tend vers l'infini avec k ; le premier membre est donc une intégrale divergente ; ceci est contradictoire ([5], th. 8.1.1) avec la continuité des trajectoires de Y, d'où l'absurdité et la minoration de l'énoncé.

4.1. Inégalités intégrales [5].

Dans cet alinéa, nous énonçons des inégalités, mises en lumière sous une autre forme ([14]) par M. Marcus et que nous utiliserons pour étudier les variations du second membre de (23) en fonction de δ :

LEMME 4.1.1. Soient T une v.a. positive sur $(\Omega, \mathcal{G}, P)$ et λ un nombre compris entre zéro et un ; on pose :

$$\varphi_\lambda(x) = (\lambda - x) I_{0 \leq x \leq \lambda}, \; A(\lambda) = \{f \in \mathcal{L}_0(\Omega, P) : 0 \leq f \leq 1, \int f dP = \lambda\},$$

$$F_T(u) = P\{T \leq u\}.$$

On a alors :

$$(28) \qquad \int_0^\infty \varphi_\lambda \circ F_T(u) du = \inf\{\int fTdP, \, f \in A(\lambda)\}.$$

Démonstration : (a) Posons $\theta = \sup\{u : F_T(u) > \lambda\}$ et notons I le premier membre de (28) ; on a alors :

$$I = \theta\{\lambda - P\{T < \theta\}\} + \int_{T(\omega) < \theta} T(\omega) dP(\omega).$$

On définit presque sûrement une variable aléatoire f_T en posant :

$$f_T(\omega) = 1 \text{ si } T(\omega) < \theta, \; f_T(\omega) = 0 \text{ si } T(\omega) > \theta,$$

$$f_T(\omega) = \frac{\lambda - P\{T < \theta\}}{P\{T = \theta\}} \text{ si } T(\omega) = \theta ;$$

on constate que f_T appartient à $A(\lambda)$ et que de plus I est égal à $\int f_T T dP$.

(b) Soit maintenant f un élément arbitraire de $A(\lambda)$, on a :

$$\int (f-f_T)TdP = \int_{\substack{F_T(u)\le\lambda \\ u\ge 0}} [F_T(u) - \int_{T(\omega)\le u} fdP]du + \int_{\substack{F_T(u)>\lambda \\ u\ge 0}} [\lambda - \int_{T(\omega)\le u} fdP]du .$$

Dans le second membre, les deux intégrandes sont positifs puisque f appartient à $A(\lambda)$; le résultat s'ensuit.

PROPOSITION 4.1.2 : Soient $\delta = \{\delta_\omega(t), \omega\in\Omega, t\in T\}$ une v.a. positive sur un produit $(\Omega, \mathfrak{A}, P)\times(T, \mathfrak{T}, \lambda)$ d'espaces probabilisés et $D = \{D(\omega), \omega\in\Omega\}$ une v.a. positive sur le premier facteur ; soit de plus φ une fonction $]0,1]\to R$ positive décroissante convexe, on a alors :

$$(29)\qquad \int_0^{E(D)} \varphi\circ\lambda\{t : E(\delta(t))\le u\}du \ge E[\int_0^D \varphi\circ\lambda\{t : \delta(t)\le u\}du] .$$

Démonstration : (a) Nous démontrons d'abord (29) si $D=+\infty$ et si φ est l'un des φ_m, $m\in[0,1]$. Utilisant les notations du lemme 4.1.1, nous introduisons la fonction f_T associée à $T = E(\delta)$, v.a. positive sur $(T, \mathfrak{T}, \lambda)$; on a alors :

$$\int \varphi_m\circ\lambda\{t : E\delta(t)\le u\}du = \inf_{f\in A(m)} \int f(t)E\delta(t)d\lambda(t) \ge$$

$$\ge E[\inf_{f\in A(m)} \int f(t)\,\delta(t)\,d\lambda(t)] \ge E[\int \varphi_m\circ\lambda\{t : \delta(t)\le u\}du] ;$$

c'est le résultat dans ce premier cas. (b) Si $D=+\infty$ et $\varphi(1) = 0$, alors φ peut s'écrire $\int\varphi_m d\pi(m)$ où π est une mesure positive sur $]0,1]$ et la formule (29) dans ce cas se déduit du résultat précédent par intégration par rapport à $\pi(m)$. (c) Si $\varphi(1) = 0$ et D arbitraire, la formule (29) se déduit du résultat (b) appliqué aux v.a. $\inf(\delta,D)$; le résultat général s'ensuit immédiatement.

4.2. Le théorème de continuité.

THEOREME 4.2.1 : Soit X une f.a. séparable sur R^n de type intégral de la forme $\{\frac{1}{2}\int \exp(2i<x,t>)\, dm(\omega,dx)\, ,\ \omega\in\Omega\, ,\ t\in\mathbb{R}^n\}$ où m est une mesure aléatoire sur $\mathbb{R}^n$ à valeurs symétriques ; soit μ une mesure de contrôle bornée pour m. On suppose que l'intégrale :

$$I(\mu) = \int_0 \sqrt{\log \frac{2^n}{\lambda\{s\in[-1,+1]^n : \int \sin^2 <x,s>d\mu(x) \le u^2\}}}\, du$$

est convergente. Alors X est p.s. à trajectoires continues et on a :

$$(30)\quad E\inf(1, \sup_{\substack{|s|\le 1\\ |t|\le 1}} |X(s)-X(t)|) \le K\sqrt{n}\{\sqrt{\mu(\mathbb{R}^n)} + \int_0^\infty \sqrt{\log \frac{2^n}{\lambda\{|s|\le 1, \int \sin^2 <x,s>d\mu(x)\le u^2\}}}\, du\}$$

où K est une constante absolue.

Démonstration : (a) Nous démontrerons d'abord que X vérifie l'inégalité (30) ; pour cela nous introduisons les notations suivantes : $\{\varepsilon_n, n\in\mathbb{N}\}$ est une suite de v.a. de Rademacher indépendantes entre elles et de m. Pour tout entier $k\ge 1$, A_k est une partition finie $\{a_{j,k}, 0\le j\le J(k)\}$ de $\mathbb{R}^n$ par des parties mesurables vérifiant :

$$\mu\{a^o_{o,k}\} \le \frac{2}{k}\|\mu\|\, ,\quad \sup_{j=1}^{J(k)} |a_{j,k}| \le \frac{1}{\sqrt{k}}\, ;$$

$\{x_{j,k}, 1\le j\le J(k)\}$ est une suite d'éléments respectifs des $a_{j,k}$; les fonctions aléatoires X_k, Y_k, Z_k sont définies par :

$$2X_k(t) = \sum_{1\le j\le J(k)} e^{2i<x_{j,k},t>}\, m(a_{j,k})\, ,$$

$$2Y_k(t) = \sum_{1\le j\le J(k)} e^{2i<x_{j,k},t>}\, \varepsilon_j \inf(1, |m(a_{j,k})|)\, ,$$

$$2Z_k(t) = \sum_{1\le j\le J(k)} e^{2i<x_{j,k},t>}\, \varepsilon_j [\,|m(a_{j,k})|-1]^+ .$$

Pour tout élément t de $\mathbb{R}^n$, on a :

$$\frac{1}{4}\int |e^{2i<x,t>} - \sum_{j=1}^{J(k)} e^{2i<x_{jk},t>} I_{x\in a_{jk}}|^2 d\mu \leq \frac{(|t|^2+1)\|\mu\|}{k}$$

et ceci montre (alinéa 0.3.3 relation (12)) que $\{X_k, k\geq 1\}$ converge en probabilité vers X sur $[-1,+1]^n$. La séparabilité de X montre donc qu'il suffit pour établir (30) d'établir :

$$(30') \quad \lim_{k\to\infty} E \inf(1, \sup_{\substack{|s|\leq 1\\ |t|\leq 1}} |X_k(s) - X_k(t)|) \leq K\sqrt{n}\,\{\sqrt{\mu(\mathbb{R}^n)} + $$
$$+ \int_0^\infty \sqrt{\log \frac{2^n}{\lambda\{|s|\leq 1, \int \sin^2 <x,s> d\mu(x) \leq u^2\}}}\, du\} .$$

On remarquera aussi que X_k a même loi que $Y_k + Z_k$. On utilisera enfin que l'hypothèse du théorème implique la convergence de l'intégrale

$$\int_0 \sqrt{\log \frac{2^n}{\lambda\{s\in[-1,+1]^n, \int \sin^2 <x,s> d\mu(x) + |s|^2 \leq u^2\}}}\, du .$$

(b) Nous majorons maintenant $E \inf(1, \sup_{\substack{|s|\leq 1\\ |t|\leq 1}} |X_k(s) - X_k(t)|)$; elle est inférieure à $P\{Z_k \neq 0\} + E \sup_{\substack{|s|\leq 1\\ |t|\leq 1}} |Y_k(s) - Y_k(t)|$; le premier terme s'évalue à partir de l'inégalité de Lévy et le deuxième terme par deux intégrations successives ; on obtient :

$$P\{Z_k \neq 0\} \leq 2P\{|m(\mathbb{R}^n)| \geq 1\} ,$$

$$2E \sup_{\substack{|s|\leq 1\\ |t|\leq 1}} |Y_k(s) - Y_k(t)| \leq \int dP(\omega) \int dP(\varepsilon) \sup_{\substack{|s|\leq 1\\ |t|\leq 1}} \sum_{j=1}^{J(k)} (e^{2i<x_{j,k},t>} - e^{2i<x_{j,k},s>}) \times$$
$$\times \inf(1, |m(\omega, a_{j,k})|)\, \varepsilon_k .$$

Pour effectuer la première intégration du second membre, on utilise le théorème 4.01 (a) ; en posant :

$$\delta_k^2(\omega;s,t) = \sum_{j=1}^{J(k)} \sin^2 <x_{j,k},t> \inf(1,m^2(\omega,a_{j,k})) ,$$

$$D_k^2(\omega) = \sum_{j=1}^{J(k)} \inf(1,m^2(\omega,a,j,k)) ,$$

la formule (26) fournit :

$$E \sup_{\substack{|s|\le 1\\ |t|\le 1}} |Y_k(s)-Y_k(t)| \le A\sqrt{n}\int dP(\omega)\{D_k(\omega) + \int_0^{D_k(\omega)} \sqrt{\log \frac{2^n}{\lambda\{s:|s|\le 1, \delta_k(\omega;o;s)\le u\}}}\, du\}.$$

Pour majorer le second membre, nous appliquons la proposition 4.1.2 à la fonction $\varphi(x) = \sqrt{\log(1+x)}$ qui est inférieure à $1+\sqrt{\log x}$ et on obtient :

$$(31)\quad E \sup_{\substack{|s|\le 1\\ |t|\le 1}} |Y_k(s)-Y_k(t)| \le 2A\sqrt{n}\{\sqrt{\mu(R^n)} + \int_0^{\sqrt{\mu(R^n)}} \sqrt{\log \frac{2^n}{\lambda\{s:|s|\le 1, \delta_k(o,s)\le u\}}}\, du\},$$

$$\delta_k^2(0,s) = \sum_{j=1}^{J(k)} \sin^2 <x_{j,k},(t)> \mu(a_{j,k}) ;$$

l'inégalité de Hölder permet d'écrire :

$$\delta_k(0,s) \le \sqrt{\int \sin^2 <x,s> d\mu(x)} + \sqrt{|s|^2 \frac{1}{k} \|\mu\|} ,$$

comme par ailleurs :

$$\lim_{k\to\infty} \delta_k(0,s) = \sqrt{\int \sin^2 <x,s> d\mu(x)} ,$$

on peut utiliser le théorème de convergence dominée en faisant tendre k vers l'infini dans la relation (31) ; ceci établit (30') et donc (30).

(c) Supposant m symétrique à valeurs indépendantes, nous démontrons maintenant que X a p.s. ses trajectoires continues. Nous choisissons pour cela un compact C de R^n tel que

$$K\sqrt{n}\{\sqrt{\mu(R^n \setminus C)} + \int_0^\infty \sqrt{\log \frac{2^n}{\lambda\{|s|\le 1, \int_{x\notin C} \sin^2 <x,s> d\mu(u) \le u^2\}}}\, du\} < 1$$

et nous décomposons X en X_C et X'_C suivant les deux intégrales aléatoires associées à C et son complémentaire ; la première intégrale est la transformée de Fourier d'une distribution aléatoire à support C compact et ses trajectoires sont p.s. indéfiniment différentiables et donc continues. On peut appliquer à la seconde intégrale le résultat (b) en substituant à μ sa trace sur le complémentaire de C ; l'inégalité (30) fournissant pour X'_C un second membre strictement inférieur à 1, X'_C a ses trajectoires bornées sur $\{|t|\leq 1\}$ avec probabilité positive et le théorème 1.3.5 montre qu'il a aussi p.s. ses trajectoires continues, d'où le résultat du théorème dans ce cas.

d) Nous établissons enfin le même résultat en supposant seulement m à valeurs symétriques. Pour tout entier $k>0$, nous choisissons un compact C_k de $\mathbb{R}^n$ tel que :

$$K\sqrt{n}\,\{\sqrt{\mu(\mathbb{R}^n\setminus C_k)}+\int_0^\infty\sqrt{\log\frac{2^n}{\lambda\{|s|\leq 1,\int_{x\notin C_k}\sin^2\langle x,s\rangle d\mu(u)<u^2\}}}\,du\}<\frac{1}{2^k}$$

et nous décomposons comme ci-dessus X en X_k et X'_k. Les X_k ont p.s. leurs trajectoires continues et l'application du résultat (b) aux X'_k montre que la série $\sum_k \inf(1, \sup_{\substack{|s|\leq 1\\|t|\leq 1}} |X'_k(s)-X'_k(t)|)$ converge p.s.

Ceci montre donc que $\{X_k, k\in\mathbb{N}\}$ converge p.s. uniformément vers X sur $\{|s|\leq 1\}$, d'où le résultat du théorème dans tous les cas.

<u>Remarque 4.2.2</u> : La formule (30) peut prendre des formes différentes, mais voisines ; soient X une fonction aléatoire et m une mesure aléatoire sur $\mathbb{R}^n$ à valeurs symétriques ayant une mesure de contrôle μ bornée ; on suppose X et m liés par les hypothèses du théorème et on note G une fonction aléatoire gaussienne centrée stationnaire sur $\mathbb{R}^n$ ayant μ pour mesure spectrale ; alors la formule (30) et le théorème 4.0.1 montrent qu'il existe une constante C_n ne dépendant que de la dimension telle que :

$$(31)\quad E\inf(1, \sup_{\substack{|s|\leq 1\\|t|\leq 1}}|X(s)-X(t)|)\leq 2P\{|m(\mathbb{R}^n)|\geq 1\}+C_n E\sup_{\substack{|s|\leq 1\\|t|\leq 1}}|G(s)-G(t)|\,.$$

Par homogénéité et passage à la limite, on en déduit plus précisément si μ est le moment du second ordre de m :

$$(32)\qquad E \sup_{\substack{|s|\leq 1\\|t|\leq 1}} |X(s)-X(t)| \leq C_n E \sup_{\substack{|s|\leq 1\\|t|\leq 1}} |G(s)-G(t)| .$$

4.2.3. Les majorations du théorème 4.2.1 peuvent être améliorées si on se restreint à utiliser des classes particulières de mesures aléatoires ; E. Giné, M. Marcus et G. Pisier ont obtenu des meilleures conditions suffisantes de régularité des trajectoires des fonctions aléatoires de type $\{\int e^{i<x,t>} dm(\omega,dx)\}$ où m est une mesure aléatoire à valeurs indépendantes, symétriques et stables d'indice $\alpha \in]0,2[$ ([8], [16]). M. Marcus et G. Pisier viennent d'ailleurs [18] de démontrer la nécessité de ces conditions de régularité ; nous renvoyons le lecteur à leur travail. Dans la généralité, le théorème 4.2.1 est le meilleur possible comme le montrent les énoncés suivants :

THEOREME 4.2.4 : Soit μ une mesure positive bornée sur R^n ; (a) soit M l'ensemble des mesures aléatoires symétriques à valeurs indépendantes contrôlées par μ ; pour tout $m \in M$, on note X_m une version séparable de $\{\int e^{i<x,t>} dm(x) , t \in R^n\}$; on suppose que pour tout $m \in M$, X_m a p.s. ses trajectoires continues ; dans ces conditions, l'intégrale $I(\mu)$ est convergente. (b) Pour tout $\alpha \in]0,2]$, on note M_α l'ensemble des mesures aléatoires à valeurs symétriques stables d'indice α contrôlées par μ ; on suppose qu'il existe $\alpha \in]0,2]$ tel que pour tout $m \in M_\alpha$, X_m ait p.s. ses trajectoires continues. Dans ces conditions, l'intégrale $I(\mu)$ est aussi convergente.

Démonstration : dans le cas (a), on choisit l'élément m_2 de M gaussien ayant μ pour moment du second ordre et on applique le théorème 4.0.1. Dans le cas (b), on choisit l'élément m_α de M_α obtenu en multipliant m_2 par une v.a. indépendante x_α ayant la loi particulière μ_α utilisée à la proposition 6.1 du premier chapitre.

THEOREME 4.2.5 : Soient $\{a_k, k \in \mathbb{N}\}$ une suite d'éléments de $\mathbb{R}^n$, $\{x_k, k \in \mathbb{N}\}$ une suite de v.a. réelles symétriques indépendantes et $\{\lambda_k, k \in \mathbb{N}\}$ une suite de v.a. de lois $\mathcal{N}(0,1)^{\mathbb{N}}$; on suppose que

$$\sup_{k \in \mathbb{N}} \frac{E\{\inf(1, x_k^2)\}}{[E\{\inf(1, |x_k|)\}]^2} = \rho^2$$

est fini ; on suppose aussi que la série $\frac{1}{2} \sum_{k \in \mathbb{N}} \exp(2i < a_k, t >) x_k$ p.s. convergente pour chaque t possède une version X de sa somme ayant p.s. ses trajectoires continues. Dans ces conditions, la série

$$\frac{1}{2} \sum_{k \in \mathbb{N}} \exp(2i < a_k, t >) \sqrt{E \inf(1, x_k^2)}\, \lambda_k$$

a aussi une version ayant p.s. ses trajectoires continues et l'intégrale

$$\int_0 \sqrt{\log \frac{2^n}{\lambda\{s \in [-1,+1]^n : \sum_{k \in \mathbb{N}} \sin^2 < a_k, s > E \inf(1, x_k^2) \le u^2\}}}\, du$$

est convergente.

Démonstration : Soient M un nombre positif et (ε_k) une suite de v.a. de Rademacher indépendantes mutuellement et des données précédentes ; nous définissons diverses fonctions aléatoires séparables en posant :

$$\forall k \in \mathbb{N} \quad G_{k,1}(t) = \frac{1}{2} \sum_{j=1}^{k} \exp(2i < a_j, t >) \sqrt{E \inf(1, x_j^2)}\, |\lambda_j| I_{|\lambda_j| \ge M}\, \varepsilon_j ,$$

$$G_{k,2}(t) = \frac{1}{2} \sum_{j=1}^{k} \exp(2i < a_j, t >) \sqrt{E \inf(1, x_j^2)}\, |\lambda_j| I_{|\lambda_j| \le M}\, \varepsilon_j ,$$

$$G_{k,t} = G_{k,1}(t) + G_{k,2}(t) ,$$

$$G(t) = \lim_{k \to \infty} G_k(t) , \text{ p.s.},$$

$$\overline{X}(t) = \frac{1}{2} \sum_{j=1}^{\infty} \exp(2i < aj, t >) \inf(1, |x_j|)\, \varepsilon_j , \text{ p.s. .}$$

Pour toute fonction aléatoire séparable U sur $[-1,+1]^n$, nous posons :

$$\|U\| = E \sup_{\substack{|s|\le 1\\ |t|\le 1}} |U(s)-U(t)| .$$

On notera que l'hypothèse sur X implique, par le théorème des 2 séries, la convergence de la série $\Sigma E \inf(1,x_k^2)$ de sorte que G et $\bar{X}$ sont de type intégral et associées à $\{\frac{1}{2}\int \exp(2i<x,t>)dm(x), t\in \mathbb{R}^n\}$ où les mesures aléatoires correspondantes à valeurs symétriques et indépendantes :

$$m_G = \tfrac{1}{2}\sum_1^\infty (\sqrt{E\inf(1,x_k^2)}\,|\lambda_k|\,\varepsilon_k)\,\delta_{a_k} ,$$

$$m_{\bar{X}} = \tfrac{1}{2}\sum_1^\infty (\inf(1,|x_k|)\,\varepsilon_k\,\delta_{a_k} ,$$

sont toutes deux de carré intégrable. La continuité des trajectoires de X , le lemme de contraction et le théorème 3.2 montrent alors que $\|\bar{X}\|$ est fini.

Pour tout $k\ge 0$, les lemmes de contraction et la définition de ρ impliquent :

$$\|G_{k,2}\| \le 2M\rho\|\tfrac{1}{2}\sum_{j=1}^{k} \exp(2i<a_j,t>)\,E\inf(1,|x_j|)\varepsilon_j\| \le 2M\rho\|\bar{X}\| .$$

Par ailleurs, le théorème 4.2.1 sous la forme (32) implique aussi :

$$\|G_{k,1}\| \le C_n\sqrt{E\{|\lambda|^2 I_{|\lambda|\ge M}\}}\,\|G_k\| ;$$

choisissant alors M de sorte que $C_n\sqrt{E\{|X|^2 I_{|\lambda|\ge M}\}}$ soit inférieur à $\frac{1}{2}$, on obtient :

$$\forall\, k\in\mathbb{N} ,\ \|G_k\| \le 4C_n\rho\|\bar{X}\| ;$$

ceci fournit, en utilisant la séparabilité de G :

$$\|G\| \le 4C_n\rho\|\bar{X}\| ,$$

G a donc p.s. ses trajectoires bornées sur $[-1,+1]^n$; le théorème 4.0.1 montre qu'il a aussi p.s. ses trajectoires continues ; le théorème 4.0.2 (b) conclut alors le théorème.

5. Etude asymptotique des trajectoires ([7]).

5.0. Dans ce paragraphe, nous ne présentons pas l'étude de l'ordre de grandeur à l'infini des fonctions aléatoires très régulières, renvoyant pour cela, dans les cas gaussiens ou non gaussiens, aux travaux classiques basés sur des hypothèses fortes de régularité locale et d'indépendance asymptotique. Nous étudions le comportement asymptotique des trajectoires de larges classes de fonctions aléatoires de type intégral. Nous précisons d'abord dans quelles conditions ces trajectoires restent bornées. Les mesures aléatoires m intervenant dans ce paragraphe seront toutes à valeurs symétriques et indépendantes.

THEOREME 5.1 : Soit X une fonction aléatoire de type intégral de la forme $\{\int_{\mathcal{U}} f(x,t)dm(\omega,dx)\ ,\ t\in R^n\}$; on suppose que la fonction f à valeurs réelles ou complexes vérifie :

$$\forall\ x\in\mathcal{U},\ \forall\ t\in R^n,\ |f(x,t)|\leq f(x,0)=1\ ;$$

on suppose aussi que X a p.s. ses trajectoires localement bornées. On note pour tout $t\in R^n$, $\mu(t)$ une médiane de $\sup\limits_{|s|\leq|t|}|X(s)|$. Dans ces conditions X a p.s. ses trajectoires bornées sur R^n si et seulement si μ est bornée sur R^n.

Démonstration : (a) Supposons que la médiane choisie $\mu(t)$ soit bornée sur R^n par M et fixons un nombre u_o assez grand pour que $E\inf(1,\frac{1}{u_o^2}|m(\mathcal{U})|^2)$ soit inférieur à $\frac{1}{8}$; utilisant alors la démonstration du théorème 3.2 et plus particulièrement l'inégalité (22'), nous obtenons :

$$\forall\ t\in R^n,\ \forall\ u\geq\sup(M,u_o), E\{\inf(u,\sup_{|s|\leq|t|}|X(s)|\}\leq[M+3\sqrt{C_o E\inf(u^2,|m(\mathcal{U})|^2}]\ ;$$

on en déduit :

$$\forall\ u\geq\sup(M,u_o), E\inf(1,\frac{1}{u}\sup_{R^n}|X|)\leq[\frac{M}{u}+3\sqrt{C_o E\inf(1,\frac{1}{u^2}|m(\mathcal{U})|^2)}]\ ;$$

Le premier membre tend donc vers zéro quand u tend vers l'infini, ceci signifie que $\sup_{\mathbb{R}^n}|X|$ est p.s. fini. (b) La réciproque est immédiate.

Dans le cas où les trajectoires de X ne sont pas p.s. bornées, elles gardent pourtant un comportement asymptotique simple :

THEOREME 5.2 : Soit X une fonction aléatoire de type intégral de la forme $\{\int_{\mathcal{U}} f(x,t)dm(\omega,dx)\,,\, t\in\mathbb{R}^n\}$; on suppose que la fonction f à valeurs réelles ou complexes vérifie :

$$\forall\, x\in\mathcal{U},\ \forall\, t\in\mathbb{R}^n,\ |f(x,t)|\leq 1\ ;$$

on suppose aussi que X a p.s. ses trajectoires localement bornées. On note pour tout $t\in\mathbb{R}^n$, $\mu(t)$ une médiane de $\sup_{|s|\leq|t|}|X(s)|$. On suppose enfin que μ n'est pas bornée sur $\mathbb{R}^n$. Dans ces conditions, on a :

$$\limsup_{|t|\to\infty}\frac{|X(t)|}{\mu(t)} = 1 \quad \text{p.s.} \tag{33}$$

Démonstration : Nous utilisons la décomposition de m présentée en 0.2.4 et nous posons :

$$m' = m - \Sigma\,\mu_n(\omega)I_{\{|\mu_n|\geq 1\}}\delta_{x_n} - m_3^o\,,$$

de sorte que m' est une mesure aléatoire symétrique à valeurs indépendantes de carré intégrable et que, par homogénéité, nous pouvons supposer $P\{m\neq m'\}$ inférieure à $\frac{1}{4}$. Nous posons $E\{|m'(\mathcal{U})|^2\} = M^2$ et nous notons X_1 et X_2 des fonctions aléatoires séparables ayant mêmes lois temporelles respectives que $\int f(x,t)dm'$ et $\int f(x,t)d(m-m')$ de sorte que X_1+X_2 ait même loi temporelle que X. On remarquera que $(m-m')$ est combinaison linéaire d'un nombre fini (aléatoire) de mesures de Dirac si bien que l'hypothèse sur f implique que X_2 a p.s. ses trajectoires bornées et X_1 a p.s. ses trajectoires localement bornées comme X. Sur toute partie bornée T de $\mathbb{R}^n$, on peut donc appliquer le théorème 3.2 ; on obtient :

$$P\{|\sup_T|X_1| - E\sup_T|X_1|| > 2M\}\leq\frac{1}{4}\,,$$

si bien que :

$$P\{|\sup_T|X| - E\sup_T|X_1|| > 2M\} \le \frac{1}{4} + P\{m \ne m'\} \le \frac{1}{2} ;$$

on en déduit pour tout élément t de $\mathbb{R}^n$

$$|\mu(t) - E\{\sup_{|s|\le|t|}|X_1(s)|\}| \le 2M$$

et ceci implique :

$$\limsup_{|t|\to\infty} \frac{|X(t)|}{\mu(t)} = \limsup_{|t|\to\infty} \frac{|X_1(t)|}{E\{\sup_{|s|\le|t|}|X_1(s)|\}} , \text{ p.s. .} \tag{34}$$

Nous calculons maintenant le second membre de la dernière relation. Pour tout entier positif k et tout nombre $\rho > 1$, nous notons pour cela $J_k(\rho)$ l'ensemble $\{t \in \mathbb{R}^n : E\{\sup_{|s|\le|t|}|X_1(s)|\} \in [\rho^k, \rho^{k+1}[\}$; on a bien entendu :

$$\rho^{k+1} - \rho^k \le E\{\sup_{J_k}|X_1|\} \le \rho^{k+1} ; \tag{35}$$

Le théorème 3.2 montre que pour tout $\varepsilon > 0$, on a :

$$\sum_{k=1}^{\infty} P\{|\sup_{J_k}|X_1| - E\sup_{J_k}|X_1|| > \varepsilon\rho^k\} \le \frac{M^2}{\varepsilon^2} \sum_{k=1}^{\infty} \frac{1}{\rho^{2k}} < \infty ,$$

on en déduit, en utilisant (32) et la définition de J_k :

$\exists\, k_o < \infty$ p.s. , $\forall\, k \ge k_o$, $\forall\, t \in J_k$,

$$|X_1(t)| \le \rho^k(\rho+\varepsilon) \le (\rho+\varepsilon)E \sup_{|s|\le|t|}|X_1(s)| ;$$

on en déduit aussi, à partir des mêmes propriétés :

$\exists\, k_1 < \infty$ p.s. , $\forall\, k \ge k_o$, $\exists\, t \in J_k$,

$$|X_1(t)| \ge \rho^{k+1}\left(1 - \frac{1+\varepsilon}{\rho}\right) \ge \left(1 - \frac{1+\varepsilon}{\rho}\right)E \sup_{|s|\le|t|}|X_1(s)| ,$$

et les deux dernières relations impliquent :

$$\forall\ \rho>1\ ,\ \forall\ \varepsilon>0\ ,\ 1-\frac{1+\varepsilon}{\rho}\leq \limsup_{|t|\to\infty}\frac{|X_1(t)|}{E\{\sup_{|s|\leq|t|}|X_1(s)|\}}\leq \rho+\varepsilon\ .$$

On fait tendre à gauche ρ vers l'infini, puis à droite ρ vers 1 et ε vers zéro ; on en déduit le résultat du théorème à partir de (34).

COROLLAIRE 5.2.1. Soit X une f.a. séparable sur $\mathbb{R}^n$ de type intégral de la forme $\{\frac{1}{2}\int \exp(2i<x,t>)dm(\omega,dx)\ ,\ \omega\in\Omega\ ,\ t\in\mathbb{R}^n\}$ où m est une mesure aléatoire symétrique à valeurs indépendantes sur $\mathbb{R}^n$; soit μ une mesure de contrôle strict pour m. On suppose que l'intégrale

$$\int_0 \sqrt{\log\frac{2^n}{\lambda\{s\in[-1,+1]^n : \int \sin^2<x,s>d\mu(x)\leq u^2\}}}\,du$$

est convergente, on suppose de plus que l'intégrale

$$J(t)=\int_0 \sqrt{\log\frac{(2t)^n}{\lambda\{s\in[-t,+t]^n : \int \sin^2<x,s>d\mu(x)\leq u^2\}}}\,du$$

est bornée sur $\mathbb{R}^+$; alors X a p.s. ses trajectoires bornées sur $\mathbb{R}^n$. Si au contraire $J(t)$ n'est pas bornée, alors $\limsup_{|t|\to\infty}\frac{|X(t)|}{J(|t|)}$ est p.s. majoré par une constante K qui ne dépend que de la dimension.

COROLLAIRE 5.2.2. Soient $\{a_k, k\in\mathbb{N}\}$ une suite d'éléments de $\mathbb{R}^n$, $\{x_k, k\in\mathbb{N}\}$ une suite de v.a. réelles symétriques indépendantes ; on suppose que $\sup_{k\in\mathbb{N}}\frac{E\inf(1,x_k^2)}{[E\inf(1,|x_k|)]^2}$ est fini ; soit de plus f une fonction strictement positive croissante sur $\mathbb{R}^+$. On suppose que la série Σx_k est p.s. convergente et on note X une version séparable de $\frac{1}{2}\Sigma\exp(2i<a_k,t>)x_k$. Dans ces conditions, pour que $\frac{|X(t)|}{f(|t|)}$ soit p.s. borné sur $\mathbb{R}^n$, il faut et il suffit que

$$\frac{1}{f(t)}\int_0 \sqrt{\log\frac{(2t)^n}{\lambda\{s\in[-t,+t]^n : \sum_{k\in\mathbb{N}}\sin^2<a_k,s>E\inf(1,x_k^2)\leq u^2\}}}\,du$$

le soit aussi.

REFERENCES DU CHAPITRE 2

[1] A. de ACOSTA — Inequalities for B - valued random vectors with applications to the strong law of large numbers.

[2] C. BORELL — The Brunn-Minkowski inequality in Gauss space, Inv. Math., 30, 1975, 207-216.

[3] C. DELLACHERIE et P.A. MEYER — Probabilités et Potentiel, Hermann Paris 1975, Actualités Sci. et Ind. 1372, et 1980 A.S.I 1385.

[4] X. FERNIQUE — Intégrabilité des vecteurs gaussiens, C.R. Acad. Sci., Paris, A, 270, 1970, pp. 1698-1699.

[5] X. FERNIQUE — Continuité et théorème central limite pour les transformées de Fourier des mesures aléatoires du second ordre, Z.W.v.G., 42, 57-66, 1978.

[6] X. FERNIQUE — Régularité des trajectoires des fonctions aléatoires gaussiennes, Lecture Notes in Math., 480, Springer 1975.

[7] X. FERNIQUE — L'ordre de grandeur à l'infini de certaines fonctions aléatoires, Colloque International C.N.R.S., St Flour 1980, à paraître.

[8] E. GINE et M.B. MARCUS — Some results on the domain of attraction of stable measures in C(K), manuscrit, 1980.

[9] K. ITO et M. NISIO — On the oscillation of Gaussian processes, Math. Scand., 22, 1968, p. 209-223.

[10] N.C. JAIN et M.B. MARCUS — Integrability of infinite sums of independent vector valued random variables, Trans. Amer. Math. Soc., 212, 1975, 1-36.

[11] N.C. JAIN et M.B. KALLIANPUR Norm convergent expansions for gaussian processes in Banach spaces. Proc. Amer. Math. Soc., 25, 1970, 890-895.

[12] J. KUELBS et J. ZINN Some stability results for vector valued random variables, Ann. Prob., 7, 1979, 75-84.

[13] H.J. LANDAU et L.A. SHEPP On the supremum of a gaussian process, Sankhya, A, 32, 1971, 369-378.

[14] M.B. MARCUS Continuity and the central limit theorem for random trigonometric series, Z.W., 42, 1978, 35-56.

[15] M.B. MARCUS et G. PISIER Necessary and sufficient conditions for the uniform convergence of random trigonometric series, Lecture Notes Series, Aarhus University, 50, 1978.

[16] M.B. MARCUS et G. PISIER Random Fourier series with applications to harmonic Analysis, preprint.

[17] V.V. YURINSKII Exponential bounds for large deviations, Th. Prob. Appl., 19, 1974, 154-155.

UNIVERSITE LOUIS PASTEUR
Institut de Recherche Mathématique Avancée
Laboratoire Associé au C.N.R.S.
rue du Général Zimmer
F-67084 STRASBOURG CEDEX

THE MINIMAX PRINCIPLE IN ASYMPTOTIC STATISTICAL THEORY *

PAR P. Warwick MILLAR

* Research partially supported by National Science Foundation Grant MCS 80-02698

I. Introduction

A number of concepts in asymptotic statistical theory can be regarded as specializations of an abstract minimax principle. Among such concepts, for example, are efficiency, robustness and δ_n-consistency. Recent studies of these fundamental concepts have employed, on the one hand, the powerful Hajek-Le Cam (1972) asymptotic minimax theorem, and, on the other hand, a systematic consideration of Gaussian experiments with parameter in Hilbert space (with the concommitant Wiener spaces). Chapters II-VI of these lecture notes develop the abstract decision theoretic tools and asymptotic expansions necessary for understanding these recent developments; the remaining chapters illustrate their use.

Here is a brief outline:

Chapter II. *Decision theoretic prerequisites*

The goal of this chapter is to provide the basic concept of convergence of statistical experiments; this notion, due to Le Cam (1964), underlies all the asymptotics of these notes. Topics: statistical experiments, randomizations and procedures as bilinear functionals; compactness of the collection of procedures; lower semi-continuity of the risk function; deficiency; various characterizations of convergence of statistical experiments.

Chapter III. *Two asymptotic optimality theorems*

This chapter gives the basic tools for the asymptotic decision theory to be employed throughout. Topics: Hajek-Le Cam (1972) asymptotic minimax theorem; Hajek-Le Cam convolution theorem

Chapter IV. *Some asymptotic expansions*

This chapter provides the key example of a sequence of experiments converging in the sense of Chapter II. Topics: quadratic mean differentiability; examples; convergence of the log likelihood stochastic process

Chapter V. *Gaussian shift experiments*

In many applications, statistical experiments converging in the sense defined in Chapter II have limits that are Gaussian shift experiments with parameter in a Hilbert space. These are analogues of the Gaussian measures on $\mathbb{R}^k$, with mean vector θ, covariance the identity, except that θ now belongs to a separable Hilbert space, and the measure itself is on a certain Banach space. This chapter provides the basic facts about such experiments. Topics: cylinder measure, characteristic functional, images of cylinder measures; canonical normal distribution on a Hilbert space; abstract Wiener space; absolute continuity and form of likelihood ratios; examples.

Chapter VI. *Optimality theory for Gaussian shift experiments*

This chapter provides key examples of experiments which converge to the Gaussian shift experiments defined in Chapter V. The minimax risk is then derived for the limit experiments. Then an infinite dimensional convolution theorem is proved. The results of this chapter are applied in all of the applications of subsequent chapters.

Chapter VII. *Classical parametric estimation*

The main goal is to estimate efficiently the parameter θ in a statistical experiment $\{P_\theta, \theta \in \Theta\}$ $\Theta \subset \mathbb{R}^k$. Two notions of efficiency are introduced, one based on an asymptotic minimax property, the other based on the convolution theorem. Finally, the classical 1-step MLE is introduced, and is shown to be efficient, in both senses.

Chapter VIII. *Optimality properties of the empirical distribution function*

This chapter provides the first substantial nonparametric applications of the theory. The main goal is to prove, via purely abstract methods, the famous Dvoretsky-Kiefer-Wolfowitz theorem (1955) and its variants; previous proofs of the DKW theorem have involved painful calculation. A convolution theorem is also proved, showing the efficiency of the empirical cdf in yet another respect.

Chapter IX. *Recent developments in the theory of asymptotically optimal nonparametric inference*

The basic method of proof detailed in Chapter VIII vis a vis the empirical cdf can be extended and modified so as to apply to a large number of other important nonparametric problems. This chapter gives several such examples. Topics: estimation of the spectral function of a stationary process, estimation of the quantile function, estimation of the cdf in censored data,

Chapter X. *Minimum distance procedures*

Minimum distance methods in a Hilbertian framework are introduced and a comprehensive rather abstract asymptotic representation of such procedures is derived. This is then applied in a variety of important problems to derive asymptotic normality. Special applications include minimum chi-square methods, Cramer-von Mises methods, minimum distance procedures based on quantile processes, on spectral function estimators, and so forth.

Chapter XI. *Robustness and the minimum distance concept*

This chapter proves asymptotic minimaxity of certain minimum distance estimators introduced in Chapter X. This optimality property is closely connected with robustness--and indeed can be used as a definition of robustness, a fact discussed in this chapter.

Chapter XII. *Optimal estimation of real nonparametric functionals.*

This chapter applies the general theory to construct a general framework that establishes the local asymptotic minimaxity of a broad class of functionals, including L, M, R and bootstrap functionals.

Chapter XIII. *Further applications of the asymptotic minimax theory.*

As two further illustrations of the utility of the abstract development of Chapters II-VI, this chapter discusses applications to (a) regression, (b) δ_n-consistency.

Chapter XIV. *Bibliographical Notes*

Chapter XV. *References*

II. Decision theoretic preliminaries

This chapter provides the decision theoretic tools necessary for proving the asymptotic minimax theorem of Chapter III. Section 1 reviews the basic notions of loss, risk and procedures; it is the decision theoretic framework of Wald (1950), as modified by Le Cam (1955,1964). Section 2 provides the basic notion of convergence of statistical experiments that underlies all of the asymptotic theory of these notes. Section 3 explains the decision theoretic significance of the convergence discussed in section 2; this development of section 3 is used in these notes only in discussion of the convolution theorem, and could be omitted at first reading.

1. Experiments, loss, risk

Let Θ be an index set. For each $\theta \in \Theta$ let P_θ be a probability on a measure space $(S,\mathcal{S})$. The collection $E = \{P_\theta, \theta\in\Theta\} = \{P_\theta, (S,\mathcal{S}), \theta\in\Theta\}$ is called an *experiment*. A *decision space* is defined to be a topological space D. Let $\mathcal{D}$ be a sigma field on D containing the Baire sigma field (in all cases treated in these notes, D is a separable metric space and $\mathcal{D}$ its Borel sets). A *procedure* b is a *Markov kernel* of $(S,\mathcal{S})/(D,\mathcal{D})$:

$$(1.1)\quad \begin{array}{l} \text{for each } x \in S,\ b(x,\cdot) \text{ is a probability on } (D,\mathcal{D}) \\ \text{for each } A \in \mathcal{D},\ b(\cdot,A) \text{ is } \mathcal{S}\text{-measurable.} \end{array}$$

Such Markov kernels are the familiar "randomized decision rules" of classical statistics: upon "observing" x, one selects a decision from D according to the measure b(x,dy). If the measure b(x,·) is a point mass, b is called a *nonrandomized procedure*.

Let L be a nonnegative function defined on $\Theta \times D$; such a function is called a *loss* function. The triplet (E,D,L) is called a *statistical decision problem*. In all the applications of these notes $y \to L(\theta;y)$ will be lower semicontinuous. With this the case, define the *risk function* ρ of the procedure b by

$$(1.2) \qquad \rho(\theta;b) = \int_S \int_D L(\theta;y) b(x;dy) P_\theta(dx)$$

When there are several experiments under discussion, write $\rho(\theta;b;E)$ for $\rho(\theta;b)$.

The collection K_0 of (Markov kernel) procedures is convex, but K_0 lacks desirable compactness properties. The first task, therefore, will be to compactify K_0 in an appropriate space.

Define M to be the Banach space consisting of all finite signed measures on $(S,\mathcal{S})$, with the variation norm: $\|\mu\| = \sup \sum |\mu(A_i)|$, where the supremum is taken over all finite (measurable) partitions of S into sets $\{A_i\}$. Define V_0 to be the collection of all finite linear combinations of the form $\sum a_i \mu_i$, a_i real, where μ_i is, for each i, a finite signed measure absolutely continuous with respect to some P_{θ_i}, $\theta_i \in \Theta$. Define $V = V(E)$ to be the closure of V_0 in M. Of course, V_0 is a vector space, and V a Banach space.

(1.3) EXAMPLE (*Dominated experiments*). Given an experiment $E = \{P_\theta, \theta \in \Theta\}$, assume that there is a sigma finite measure ν_1 such that each P_θ is absolutely continuous with respect to ν_1. Then there is a probability ν such that $V\{E\}$ is isometrically isomorphic to $L^1(S,\mathcal{S},\nu)$. This may be proved using the well known fact that under the conditions here, there is a measure $\nu = \sum 2^{-i} P_{\theta_i}$, for appropriate $\theta_i \in \Theta$, such that $P_\theta << \nu$, all θ.

Let $C(D)$ be the Banach space of bounded continuous real functions on D, with supremum norm. A real-valued mapping b defined on $V \times C(D)$ will be called a *generalized procedure* if

(1.4) (a) b is bilinear

(b) b is positive ($b(\mu,c) \geq 0$ if $\mu \geq 0$, $c \geq 0$)

(c) $|b(\mu,c)| \leq \|\mu\|\|c\|$, $\mu \in V$, $c \in C(D)$

(d) $b(\mu,1) = \|\mu\|$ if $\mu \geq 0$

If $b(x,dy)$ is a Markov kernel procedure, it is also a generalized procedure if we define $b(\mu,c) = \iint c(y)b(x,dy)\mu(dx)$. To define the risk function of a generalized procedure b, suppose first that for each θ $L(\theta_i,\cdot) \in C(D)$. Then one naturally defines

$$\rho(\theta;b) = b(P_\theta;L(\theta;\cdot)) \tag{1.5}$$

If L is positive, but otherwise arbitrary, define

$$\dot{\rho}(\theta;b) = \inf b(P_\theta,c) \tag{1.6}$$

where the inf is computed over all continuous bounded functions c such that $c(y) \leq L(\theta;y)$. Let

$$K = \text{collection of all generalized procedures}$$

Specify a topology on K by: a net b_α in K converges to b if for each $\mu \in V$, $c \in C(D)$, $\lim b_\alpha(\mu,c) = b(\mu,c)$ (topology of pointwise convergence). Alternatively, neighborhoods of a fixed $b_0 \in K$ could be defined by $\{b: |b(\mu_i,c_i)-b_0(\mu_i,c_i)| < \varepsilon,$ for all $i = 1,\ldots,k\}$ where the μ_i, c_i are fixed elements of V, $C(D)$.

(1.7) THEOREM. (a) *The collection of generalized procedures* K *is compact and convex.*

(b) The mapping $b \to \rho(\theta;b)$ *is, for each* θ, *lower semicontinuous.*

PROOF. Compactness may be proved from the Tychonov theorem via the method used to prove Alaoglu's theorem in functional analysis. To prove (b), let b_α be a net converging in the topology of K to b_0. Then

$$\begin{aligned} \rho(\theta,b_0) &= \sup_{c \leq L(\theta;\cdot)} b_0(P_\theta,c) \\ &= \sup_c \lim_\alpha b_\alpha(P_\theta,c) \\ &\leq \underline{\lim} \sup_c b_\alpha(P_\theta,c) \\ &= \underline{\lim}\, \rho(\theta,b_\alpha) \end{aligned}$$

The next result shows that in passing to generalized procedures, we have really not added too much more to the Markov kernel procedures. Define a *simple measure* to be a measure supported by a finite collection of points. A *simple procedure* is a Markov kernel of the form $b(x,dy) = \sum_i I_{A_i}(x)\mu_i(dy)$ where $\{A_i\}$ is a finite partition of S and μ_i is a simple measure on D.

(1.8) THEOREM. *Assume* D *is separable and metric. Then the simple procedures are dense in* K.

It may be shown that, typically, there exist generalized procedures that are not given by the Markov kernel recipe. On the other hand, in special cases all procedures are Markov kernels.

(1.9) THEOREM. *Suppose the experiment* E *is dominated and that* D *is compact. Then all procedures are given by Markov kernels.*

2. Convergence of experiments

Let Θ be a *finite* parameter set: $\Theta = \{\theta_1,\ldots,\theta_d\}$. For a statistical experiment $E = \{P_\theta, \theta\in\Theta\}$, define $\mu = \sum_\theta P_\theta$. Let μ_0 be the distribution under μ of the vector $\{dP_\theta/d\mu: \theta\in\Theta\}$: for any Borel subset A of the unit simplex of $\mathbb{R}^d$,

$$\mu_0(A) = \mu\{\omega: (dP_{\theta_1}/d\mu(\omega),\ldots,dP_{\theta_d}/d\mu(\omega))\in A\}$$

The measure μ_0 is called the *canonical measure* of the experiment E. Two experiments E, F with the same (finite) Θ are called *equivalent* if they have the same canonical measure. The notion of equivalence just defined has several characterizations with great statistical importance. For example, as explained in more detail in section II.3, a famous theorem of Blackwell asserts that two experiments are equivalent iff each one is a "randomization" of the other; another characterization asserts that, for any decision space, the risk functions available in one experiment are the same as those available in the other.

Assume still that Θ is finite and let $E^n = \{P^n_\theta; (S^n,\mathcal{S}^n); \theta\in\Theta\}$ be a sequence of experiments. Let μ^n_0 be the canonical measure of E^n.

(2.1) DEFINITION. E^n *converges* to E if μ^n_0 converges weakly to μ_0; i.e., for every bounded continuous function f on $\mathbb{R}^d$

$$\lim \int f(x)\mu^n_0(dx) = \int f(x)\mu_0(dx)$$

This definition is purely analytic and, as such, is convenient for use in asymptotic theory. The characterizations mentioned in the preceding paragraph suggest that this mode of convergence entails certain decision theoretic consequences--in particular, that a decision problem in experiment E^n could be solved in E without losing too much. That this is indeed the case is discussed in the following section.

Let now $E^n = \{P_\theta, \theta \in \Theta\}$ be a sequence of experiments, Θ *arbitrary*.

(2.2) DEFINITION. E^n converges to E if, for every *finite* subset Θ_0 of Θ, the experiments $E_0^n = \{P_\theta^n, \theta \in \Theta_0\}$ converge to $E_0 = \{P_\theta, \theta \in \Theta_0\}$ in the sense of definition (2.2).

Checking convergence of experiments by computing canonical measures can be painful. The following simpler method is used frequently in subsequent chapters.

(2.3) PROPOSITION. *Let* E^n, E *be experiments with parameter set* Θ. *Suppose there exists* $\theta_0 \in \Theta$ *such that each* P_θ *is absolutely continuous with respect to* P_{θ_0}, *and that* P_θ^n *is contiguous to* $P_{\theta_0}^n$. *For a finite subset* Θ_0 *of* Θ *let* μ_{00} *be the distribution of the vector* $\{dP_\theta/dP_{\theta_0} : \theta \in \Theta_0\}$ *under* P_{θ_0}, *and define* μ_{00}^n *similarly for* E^n. *If, for each* Θ_0, μ_{00}^n *converges weakly to* μ_{00} *then* E^n *converges to* E.

(2.4) REMARK. Many applications proceed by showing that the distribution of $\log dP_\theta^n/dP_{\theta_0}^n$ converges to that of $\log dP_\theta/dP_{\theta_0}$.

To see heuristically why the proposition holds, let $\mu^n = \sum P_\theta^n$, $\mu = \sum P_\theta$. Then, presumably $\frac{dP_\theta^n}{d\mu} = \frac{dP_\theta^n/dP_{\theta_0}^n}{d\mu/dP_{\theta_0}^n}$ and the convergence of the canonical measures should follow from the convergence of $\{dP_\theta^n/dP_{\theta_0}^n\}$. Contiguity can be used to make such an argument rigorous. Here are two simple examples of convergence of experiments.

(2.5) EXAMPLE. Let $f(\lambda;x)$ be the exponential density with parameter λ $f(\lambda;x) = \lambda e^{-\lambda x}$, $x \geq 0$. Let $\Theta = \mathbb{R}^1$ and define P_θ^n to be the measure on $\mathbb{R}^n$ with density $\Pi f(1+\theta n^{-1/2};x_i)$. Let P_θ be $N(\theta,1)$, the normal distribution on the line with mean θ, variance 1. Then $\{P_\theta^n, \theta \in \Theta\}$ converges to $\{P_\theta, \theta \in \Theta\}$. This may be checked by direct use of (2.4). Chapter IV contains an important generalization of this one.

(2.6) EXAMPLE. Let $f(\lambda;x)$ be the uniform density on $[0,\lambda]$. Take $\Theta = \mathbb{R}^1$ and for $\theta \in \Theta$ let P_θ^n be the measure on $\mathbb{R}^n$ with density $\Pi f(1+\theta n^{-1};x_i)$. Let P_θ be the measure on the line with density $\exp\{-(x-\theta)\}$, $x > \theta$. Then $\{P_\theta^n\}$ converges to $\{P_\theta\}$. This can be proved by fairly straightforward computations; notice that (2.3) cannot be used as it stands.

3. The decision theoretic nature of convergence of experiments

This section explains the decision theoretic significance of the definition of convergence (2.2); the main result is due to Le Cam (1964). The entire development of these notes (except certain aspects of the convolution theorem) can be read without mastering (or even reading) the technical facts summarized below. However, if one wishes to work with freedom in this particular field of asymptotics, he will sooner or later have to come to grips with this theory of convergence.

Let $E = \{P_\theta, (S,\mathcal{S}), \theta \in \Theta\}$ be a statistical experiment. Let $(S_2,\mathcal{S}_2)$ be another measure space; denote by $L_\infty(S_2,\mathcal{S}_2)$ the Banach space of real, bounded $\mathcal{S}_2$-measurable functions on S_2. A *randomization* for E is a bilinear functional K on $V(E) \times L_\infty(S_2,\mathcal{S}_2)$ satisfying properties (1.4)(a)-(d) with L_∞ replacing $C(D)$ there. If K is a randomization, denote by KP_θ the element of L_∞^* (dual of L_∞) given by $f \to KP_\theta f \equiv K(P_\theta, f)$. Typically, KP_θ is then a finitely additive probability on S_2 and $KP_\theta f$ is the integral of f. Important examples of randomization are provided by the Markov kernels $K(x,dy)$ of $(S,\mathcal{S})/(S_2,\mathcal{S}_2)$ (cf. Definition (1.1)). In this case $KP_\theta f = \iint f(y)K(x,dy)P_\theta(dx)$, $f \in L_\infty$, which is the classical notion of randomization. As in the case of decision procedures, the general notion of randomization defined here merely compactifies the collection of randomizations given by Markov kernels.

(3.1) THEOREM. *Fix an experiment* $E = \{P_\theta, \theta \in \Theta\}$ *and a measure space* $(S_2,\mathcal{S}_2)$. *Let* $\mathbb{R}$ *denote the collection of randomizations of* E *(relative to* $(S_2,\mathcal{S}_2)$*).*

(a) $\mathbb{R}$ *is convex and compact for the topology of pointwise convergence.*

(b) *The randomizations given by simple Markov kernels are dense in* $\mathbb{R}$.

(c) *If* $(S,\mathcal{S})$, $(S_2,\mathcal{S}_2)$ *are locally compact, separable metric, then all transitions are given by Markov kernels..*

Next we shall build a metric that tells us how well a given experiment can be approached by randomizing some other experiment. To do this, let $E = \{P_\theta, (S,\mathcal{S}), \theta \in \Theta\}$, $F = \{Q_\theta, (S_2,\mathcal{S}_2), \theta \in \Theta\}$ be two experiments.

(3.2) DEFINITION. The *deficiency* of E to F is the number

$$\inf_K \sup_\theta \|Q_\theta - KP_\theta\|$$

Here the inf is over all randomizations K and the norm is that of $L^*_\infty(S_2)$. Of course $V(F)$ is a subspace of $L^*_\infty(S_2)$; using a projection argument based on facts about Riesz spaces (cf. Shafer, Ch. V) one may show that this inf may be computed over K such that $KP_\theta \in V(F)$. This replacement will be made throughout; then the norm that appears in the definition becomes the *variation norm* for signed measures on $\mathcal{S}_2$.

(3.3) DEFINITION. $\Delta(E,F) = \max\{\delta(E,F),\delta(F,E)\}$

Clearly $\Delta(E,F)$ is supposed to measure how much you can reproduce E by randomizing F, and conversely.

(3.4) THEOREM. (a) *Δ is a metric.*

(b) *There exists* $K_0 \in \mathbb{R}$ *such that* $\delta(E,F) = \sup_\theta \|Q_\theta - K_0 P_\theta\|$.

(c) $\Delta(E,F) = \sup \Delta(E(\Theta_0),F(\Theta_0))$

Here Θ_0 is a finite subset of Θ, $E(\Theta_0) = \{P_\theta, \theta\in\Theta_0\}$ and the sup is over all finite subsets Θ_0 of Θ.

The usefulness of (c) is that one can compute $\Delta(E,F)$ by first replacing Θ by a *countable* subset; this means that, for many purposes, one can without loss of generality assume that the experiments E, F are dominated (cf. (1.3)). Here are two examples.

(3.5) EXAMPLE. Let $E = \{P_\theta, \theta\in\Theta, (S,\mathcal{S})\}$ be an experiment and $(S_2,\mathcal{S}_2)$ a measure space. Suppose $T: S \to S_2$ is measurable. Define Q_θ, probability on $\mathcal{S}_2$, by $Q_\theta(A) = P_\theta\{T^{-1}A\}$, $A \in \mathcal{S}_2$. Let $F = \{Q_\theta, \theta\in\Theta\}$. Then $\delta(E,F) = 0$. Assume in addition that $\mathcal{S} = T^{-1}\mathcal{S}_2$. Then $\Delta(E,F) = 0$. In general the K such that $P_\theta = KQ_\theta$ will not be given by a Markov kernel.

(3.6) EXAMPLE. Let $E = \{P_\theta, \theta \in \Theta\}$ where Θ is *finite*. Let μ_0 be the canonical measure of E, defined as usual on the unit simplex of $\mathbb{R}^\Theta$. Let Q_θ be the measure on $\mathbb{R}^\Theta$, $Q_\theta(dx) = x_\theta \mu(dx)$, $x = (x_\theta) \in \mathbb{R}$. Let $F = \{Q_\theta, \theta \in \Theta\}$. Then $\Delta(E,F) = 0$. This result is due to Blackwell (1953) and gives one statistical interpretation to the notion of equivalence defined in II.2.

This example suggests the following result, proved by Le Cam in 1964.

(3.7) THEOREM. *Let* $E^n = \{P^n_\theta, \theta \in \Theta\}$, $E = \{P_\theta, \theta \in \Theta\}$ *be experiments with* Θ *finite; let* μ^n, μ *be their canonical measures. The following are equivalent:*

(a) $\lim_n \Delta(E^n, E) = 0$

(b) μ^n *converges weakly to* μ.

In short, E^n converges to E in the sense of section II.2 iff E^n (resp. E) can be arbitrarily closely reproduced from E (resp. E^n) via randomizations. Further, statistical interpretation is possible, thanks to the following further characterizations of Le Cam.

(3.8) THEOREM. *Let* $E = \{P_\theta, \theta \in \Theta\}$, $F = \{Q_\theta, \theta \in \Theta\}$ *be experiments,* Θ *arbitrary. The following are equivalent:*

(a) $\delta(E,F) \leq \varepsilon$

(b) *For every probability* μ *on* Θ *having finite support and for every decision space* D *and every loss function* L, $0 \leq L \leq 1$, *the Bayes risks* $\rho_E(\mu) = \inf_b \int \rho_E(\theta,b)\mu(d\theta)$ *satisfy*

$$\rho_E(\mu) \leq \rho_F(\mu) + \varepsilon$$

(c) *For every finite decision space* D *and every loss function* $L(\theta,\cdot)$, $0 \leq L(\theta,\cdot) \leq 1$ *and every procedure* b_F *of* (F,D,L) *there exists a procedure* b_E *of* (E,D,L) *such that*

$$\rho_E(\theta,b_E) \leq \rho_F(\theta,b_F) + \varepsilon$$

III. Two asymptotic optimality theorems

This chapter states and proves the asymptotic minimax theorem and the convolution theorem; these are the fundamental abstract asymptotic optimality results used throughout the remaining chapters. The minimax theorem is treated in section 1; the convolution theorem is in section 2.

1. Asymptotic minimax theorem

Fix a sequence of experiments $E^n = \{P_\theta^n,\ \theta\in\Theta\}$, $E = \{P_\theta,\ \theta\in\Theta\}$. Let D be a fixed decision space and L a loss function. Let $\rho_n(b,\theta)$ (respectively $\rho(b,\theta)$) be the risk function of a procedure b in the decision theoretic structure (E^n,D,L) (resp. (E,D,L)).

(1.1) THEOREM (Hajek-Le Cam asymptotic minimax theorem). *Suppose* E^n *converges to* E *(cf. II.2.2). Then*

$$\varliminf_{n\to\infty} \inf_b \sup_\theta \rho_n(b,\theta) \geq \inf_b \sup_\theta \rho(b,\theta) .$$

Notice the great generality here: there are *no* hypotheses on Θ at all. Hajek's version (1972) assumed $\{P_\theta\}$ a Gaussian shift family on $\mathbb{R}^n$ and was proved with delicate calculations; Le Cam's version (1972) is the form given here. We shall give a simple proof using only a minimal amount of the apparatus of Chapter II; a proof can also be based on the difficult theorem II.3.8. In nearly all of the applications of this paper, the hypothesis that E^n converge to E is checked by showing (II.2.3). In (1.1), the infimum is computed over *all* procedures b; for problems of hypotheses testing, this is usually inconvenient.

Here is a variant of (1.1) to cover such situations (cf. Millar, 1982c).

(1.2) THEOREM. *Suppose* E^n *converges to* E. *Let* $T(\alpha)$ *be the collection of all level* α *tests of* Θ_0 *versus* $\Theta-\Theta_0$ *(so* D *is the usual two point decision space with* 0-1 *loss function). Then*

$$\underline{\lim}_n \inf_{b\in T(\alpha)} \sup_\theta \rho_n(b,\theta) \geq \inf_{b\in T(\alpha)} \sup_\theta \rho(b,\theta)$$

We turn now to the proof of (1.1); the starting point is a suitable version of the ordinary (non-asymptotic) minimax theorem.

(1.3) MINIMAX THEOREM. *Let* X, Y *be subsets of two linear spaces. Assume* X *compact, convex and* Y *convex. Let* f *be a function on* $X \times Y$ *with values in* $[0,\infty]$ *such that*

(a) $f(\cdot,y)$ *is convex for each* $y \in Y$;

(b) $f(x,\cdot)$ *is concave for each* $x \in X$;

(c) $f(\cdot,y)$ *is lower semicontinuous for each* y.

Then

$$\inf_{x\in X} \sup_{y\in Y} f(x,y) = \sup_{y\in Y} \inf_{x\in X} f(x,y) .$$

A proof of this result is given by Kneser (C.R. Acad. Sci. Paris, 1952).

(1.4) DEFINITION. $M(\Theta)$ is the set of all probability measures μ on Θ such that μ has finite support. Define also for any procedure b with risk $\rho(\theta,b)$ and $\mu \in M$

$$\rho(b,\mu) = \int \rho(b,\theta)\mu(d\theta)$$

$$\rho(\mu) = \inf_b \rho(b,\mu)$$

Notice that $M(\Theta)$ is a convex subset of a linear space; K, the collection of all (generalized) procedures, is a compact convex subset of a linear space, and that the mapping $(b,\mu) \to \rho(b,\mu)$ satisfies the hypotheses of (1.3). The lower semicontinuity of $b \to \rho(b,\mu)$ follows from (II.1.7) and the fact that μ has finite support. The following result is then immediate from (1.3):

(1.5) THEOREM. *Fix a statistical experiment* $E = \{P_\theta, \theta \in \Theta\}$, *decision space* D, *loss* L. *Then*

$$\inf_b \sup_\theta \rho(b,\theta) = \sup_{\mu \in M(\Theta)} \inf_b \rho(b,\mu) .$$

(1.6) REMARK. Theorem 1.5 continues to hold if the infimum is taken over all $b \in K_0$ where K_0 can be any convex, closed subset of K (for example, K could be all level α tests).

PROOF OF (1.1). It suffices to show that for any $\mu \in M \equiv M(\Theta)$,

$$\underline{\lim_n}\, \rho_n(\mu) \geq \rho(\mu) \tag{1.7}$$

To see this, let

$$m \equiv \inf_b \sup_\theta \rho(b,\theta) = \sup_\mu \rho(\mu) \tag{1.8}$$

where we have used (2.5). Pick $\varepsilon > 0$ and take $\mu_0 \in M$ so that

$$\rho(\mu_0) \geq m - \varepsilon$$

Then, if (1.7) holds,

$$\sup_\mu \rho_n(\mu) \geq \rho_n(\mu_0) \geq m - 2\varepsilon$$

for all sufficiently large n. But by (1.5) again,

$$\sup_{\mu} \rho_n(\mu) = \inf_b \sup_{\theta} \rho_n(b,\theta) ,$$

proving (1.1).

Therefore, to prove (1.1), it is necessary only to show (1.7). For this, fix $\mu \in M$; let Θ_0 be the support of μ and set $\mu_\theta = \mu(\{\theta\})$, $\theta \in \Theta$, If $x \in \mathbb{R}^{\Theta_0}$, let x_θ be the θth coordinate of x. Define measures Q, Q^n by $Q = \sum_{\theta\in\Theta_0} P_\theta$, $Q^n = \sum P_\theta^n$. The vector process $\{dP_\theta/dQ: \theta \in \Theta_0\}$ takes values in the unit simplex of $\mathbb{R}^{\Theta_0}$: $\{x: x_\theta \geq 0, \sum x_\theta = 1\}$. Let Q_0, Q_0^n be the distribution of this process under Q, Q^n respectively. Because E^n converges to E, Q_0^n converges weakly to Q_0.

Next, notice that (since with no loss of generality, we may assume the loss function L continuous)

$$\rho(\mu) = \inf_b \rho(b,\mu) = \inf_{b\in K_0} \rho(b,\mu)$$

where K_0 is the collection of procedures of Markov kernel form; a proof of this may be based on (II.1.8). Therefore since $\{dP_\theta/dQ, \theta \in \Theta\}$ is a sufficient statistic (cf. II.3.6),

$$\begin{aligned}
\rho(\mu) &= \inf_b \sum_\theta \iint L(\theta,y)b(x,dy)P_\theta(dx)\mu_\theta \\
&= \inf_b \sum_\theta \mu_\theta \iint L(\theta,y)b(x,dy)x_\theta Q_0(dx) \qquad x = \{x_\theta\} \\
&= \inf_b \iint (\sum_\theta \mu_\theta L(\theta,y)x_\theta)b(x,dy)Q_0(dx) \\
&= \int \inf_{y\in D} \sum_\theta \mu_\theta L(\theta;y)x_\theta Q_0(dx) \\
&= \int h(x)Q_0(dx)
\end{aligned}$$

where $h(x) = \inf_{y\in D} \sum_\theta \mu_\theta L(\theta,y)x_\theta$. Similarly $\rho_n(\mu) = \int h(x)Q_0^n(dx)$. The function $x \to \sum_\theta \mu_\theta L(\theta,y)x_\theta$ is, for each y, concave and continuous over the simplex, so h is continuous. Therefore

$$\underline{\lim}_n \rho_n(\mu) = \underline{\lim} \int h(x)Q_0^n(dx)$$
$$= \int h(x)Q_0(dx)$$
$$= \rho(\mu)$$

proving the result.

2. Convolution theorem

We turn now to an abstract presentation of the convolution theorem. Unlike the discussion of III.1, this section makes heavy use of the concepts of II.3. Several approaches, much less technically advanced, are available in special situations. We prefer the present abstract approach because (a) the added flexibility is important for nonparametric applications (cf. Ch. IX and Millar (1982a)) and (b) the proof of the abstract version (2.10) can be given in a purely structural form, devoid of the messy calculations of the special cases.

To begin, let $E = \{P_\theta, (S,\mathcal{S}), \theta\in\Theta\}$ be an experiment. Let D be a decision space, assumed separable metric, and let $\mathcal{D}$ be the Borel sigma field on D.

(2.1) DEFINITION. A D-*statistic* T is a measurable map from S to D.

Denote by P_θ^T the measure on $\mathcal{D}$ given by

(2.2) $$P_\theta^T(A) = P_\theta\{s\in S: T(s)\in A\}, \quad A\in\mathcal{D}.$$

Denote by E^T the experiment

(2.3) $$E^T = \{P_\theta^T, \theta\in\Theta\}$$

Next, let $E^n = \{P^n_\theta, (S_n, \mathcal{S}_n), \theta \in \Theta\}$ be a sequence of experiments and T_n a D-statistic for E^n.

(2.4) DEFINITION. $\{T_n\}$ *converges in distribution* to T if for every bounded continuous function f on D and every θ:

$$\lim_n \int f(T_n)dP^n_\theta = \int f(T)dP_\theta .$$

Suppose that

$$(2.5) \qquad \begin{array}{l} E^n \text{ converges to } E, \text{ and} \\ T^n \text{ converges in distribution to } T . \end{array}$$

It is of interest to compare the experiments E and E^T. It may be shown, using II (3.8), (3.1(b)) for example that

$$(2.6) \qquad \delta(E,E^T) = 0$$

Simple examples point out, however, that it is easily possible that $\delta(E^T,E) > 0$ even if the Δ-distance between E^n and $E^{n,T_n} = \{P^n_\theta\{T_n \in dx\}, \theta \in \Theta\}$ is zero for every n! Informally, the sequence of statistics $\{T_n\}$ can somehow "lose information" in the limit. This leads to the following definition:

(2.7) DEFINITION. Assuming (2.5), call the sequence of statistics $\{T_n\}$ *distinguished* if $\Delta(E,E^T) = 0$.

(2.8) EXAMPLE. Bring in the experiments E^n, defined in II.2.5; so for each $\theta \in \mathbb{R}^1$, P^n_θ is a measure on $\mathbb{R}^n$ making $(x_1,\ldots,x_n) \in \mathbb{R}^n$, independent identically distributed, exponential with parameter $(1+\theta n^{-1/2})$. Let $T_n = n^{-1/2}\sum_{i=1}^n (x_i-1)$. Simple computations show that the P^n_θ

distribution of T_n converges to $N(\theta,1)$. It follows from the results of II.2.6, concerning convergence of E^n, that $\{T_n\}$ is distinguished. This result will be generalized considerably in Chapter VII.

(2.9) EXAMPLE. Let E^n be the experiment of II.2.6; so for $\theta \in \mathbb{R}^1$, P_θ^n is a measure on $\mathbb{R}^n$ making $(x_1,\ldots,x_n)$ independent identically distributed, uniform $[0,1+\theta n^{-1}]$. Let $T_n = n[1-\max x_i]$. Since the P_θ^n distributions of T_n converge to P_θ, defined in (II.2.6), the sequence $\{T_n\}$ is distinguished.

Here now is the basic convolution theorem in a form similar to that given by Le Cam (1972); the original version is due to Hajek (1968).

(2.10) THEOREM. *Let* $\Theta = \mathbb{R}^d$ *and* $E^n = \{P_\theta^n, \theta \in \Theta\}$, $E = \{P_\theta, \theta \in \Theta\}$ *statistical experiments. Assume* E^n *converges to* E. *Let* $D = \mathbb{R}^d$ *and suppose* T_n, R_n *are D-statistics for* E^n. *Assume further*

(i) *there are families of probabilities* $\{F_\theta, \theta \in \Theta\}$, $\{G_\theta, \theta \in \Theta\}$ *on* $\mathbb{R}^d$ *such that for each* θ

$$P_\theta^n\{T_n \in dy\} \Rightarrow F_\theta(dy)$$
$$P_\theta^n\{R_n \in dy\} \Rightarrow G_\theta(dy)$$

(ii) $\{T_n\}$ *is distinguished*

(iii) $F_\theta(A) = F_0(A-\theta)$, $G_\theta(A) = G_0(A-\theta)$ *for all* θ *and all Borel sets* A

(iv) F *is absolutely continuous with respect to Lebesgue measure.*

Then there is a probability μ *on* $\mathbb{R}^d$ *such that*

$$G_0 = F_0 * \mu$$

where $*$ *denotes convolution of probability measures.*

Applications of this result to the problem of efficient parametric estimation are given in VII.7; an extension, with applications, is given in VIII.3. We now discuss briefly one possible proof.

Let $E^T = \{F_\theta\}$, $E^R = \{G_\theta\}$. Because of (2.6) and (2.10(ii)) we see $\Delta(E,E^T) = 0$, $\delta(E,E^R) = 0$ and so $\delta(E^T,E^R) = 0$ by (II.3.4(a)). Therefore there exists a randomization K_0 such that $G_\theta = K_0 F_\theta$ for all θ, by II.3.4(b). By II.3.1(c), K_0 is a Markov kernel. Let $\mathcal{K}_0$ be the collection of all such randomizations (= Markov kernels) such that $KF_\theta = G_\theta$ for all θ. $\mathcal{K}_0$ is convex and closed in the topology given in II.3. Let $\mathcal{K}$ denote all (Markov kernel) transitions. The *group* $\mathbb{R}^d$ *acts on* $\mathcal{K}$ via the recipe

$$(gK)(x,A) = K(x+g,A+g) , \quad g \in \mathbb{R}^k, \quad K \in \mathcal{K}$$

Using (2.10(iii)-(iv)) we see that

$$g(\mathcal{K}_0) \subset \mathcal{K}_0$$

$$g(aK_1 + bK_2) = agK_1 + bgK_2 \qquad K_i \in \mathcal{K}$$

(i.e. g is *linear*). Because of this, the Markov-Kahutani fixed point theorem now applies (cf. Dunford, Schwarz, vol. I): there exists $K_1 \in \mathcal{K}_0$ such that

$$gK_1 = K_1$$

(as elements of $\mathcal{K}$). More explicitly, for every $Q \in V(E)$, every Borel set A and every $g \in \mathbb{R}^d$

$$\int K_1(x+g,A+g)Q(dx) = \int K_1(x,A)Q(dx)$$

Since $V(E) = L^1(dx)$ here (II.1.3), this leads to: for each $g \in \mathbb{R}^d$

$$K_1(x,A) = K_1(x+g,A+g) \text{ a.e. } x$$

where the exceptional set depends on g but not on A. This exceptional set can be shown independent of g by a variety of technical arguments; one can invoke a messy accounting of the null sets (Bickel, Berk,) or one can use a fairly standard argument involving lifting theorems. Here the invariance of Lebesgue measure under the group and local compactness are used. Assuming this done, it is then possible to define the measure $\mu(A)$ by

$$\mu(A) = K_1(0,A)$$

so that for almost all x

$$K_1(x,A) = \mu(A-x)$$

Since $K_1 \in K_0$, this proves the result.

IV. Some asymptotic expansions

Let Θ be an open subset of $\mathbb{R}^d$; fix $\theta_0 \in \Theta$. Let $\{P_\theta^n\} \equiv E^n$ be a sequence of experiments. This chapter provides an "asymptotic expansion" of $\log dP_\theta^n/dP_{\theta_0}^n$, under a hypothesis of quadratic mean differentiability (defined in (1.5) below). In conjunction with Proposition II.2.3, this result will give a large and useful collection of instances where the experiments E^n *converge*, in the sense of II.2.2, to a limit experiment E consisting of Gaussian measures. Even though Θ is assumed finite dimensional, the theory here will yield such convergence results even in infinite dimensional statistical problems, as shown in Chapter VI.

1. Quadratic mean differentiability

Throughout this chapter, Θ will be an open subset of $\mathbb{R}^d$. Let $\{P_\theta, \theta \in \Theta\}$ be a family of probabilities on the line, indexed by Θ. *Assume* each P_θ is absolutely continuous with respect to a sigma finite measure M. Denote by

(1.1) $\langle\cdot,\cdot\rangle$, $|\cdot|$ inner product and norm for $\mathbb{R}^d$

(1.2) $\langle\cdot,\cdot\rangle_M$, $|\cdot|_M$ inner product and norm for $L^2(M)$

(1.3) $$(dP_\theta)^{1/2} = (dP_\theta/dM)^{1/2}$$

(1.4) $$f(\theta,x) = (dP_\theta/dM)(x)$$

When context is clear, the subscript M will be omitted from $|\cdot|_M$.

Fix $\theta_0 \in \Theta$.

(1.5) DEFINITION. $\{P_\theta\}$ is quadratic mean differentiable (*qmd*) at θ_0 if there exists $\eta = \eta(\theta_0) = (\eta_1, \eta_2, \ldots, \eta_d)$, $\eta_i \in L^2(M)$ such that

$$|(dP_{\theta_0+\theta})^{1/2} - (dP_{\theta_0})^{1/2} - \langle\eta,\theta\rangle|_M = o(|\theta|)$$

for $\theta \in \mathbb{R}^d$, $|\theta| \to 0$.

(1.6) REMARK. There are variants of this definition which remove its dependence on the dominating measure M; these definitions have a certain elegance, but as a practical matter, in most applications there is a natural dominating measure M. Note, in particular, that η as defined here depends on M.

The definition (1.5) asserts that the map $\theta \to (dP_\theta)^{1/2}$ of Θ to $L^2(M)$ is Frêchet differentiable at θ_0.

(1.7) DEFINITION. If P, Q are two probabilities on the same measure space dominated by some sigma finite measure M, the *Hellinger distance* $\zeta(P,Q)$ is defined by

$$\zeta(P,Q) = |(dP)^{1/2} - (dQ)^{1/2}|_M$$

It is easy to see that this distance is independent of M. The qmd property of (1.5) implies in particular that the Hellinger distance between $P_{\theta_0+\theta}$ and P_{θ_0} is approximately a multiple of $|\theta|$ whenever θ is small. A simple, but important, additional geometric fact is provided by the following result.

(1.8) PROPOSITION. *Let* $\{P_\theta\}$ *be qmd at* θ_0 *with derivative* η, *as in* (1.5). *Then for any* $\theta \in \mathbb{R}^d$,

$$\langle\eta,\theta\rangle \textit{ is orthogonal to } (dP_{\theta_0})^{1/2}$$

in $L^2(M)$. *In particular, if*

$$\eta_0 = \eta/f^{1/2}(\theta_0)$$

then

$$\langle \eta_0, \theta\rangle \text{ is a random variable with mean } 0$$

under P_{θ_0}.

PROOF.
$$\begin{aligned}\int f^{1/2}(\theta_0)\langle\eta,\theta\rangle d\mu &= \lim_n n^{1/2} \int f^{1/2}(\theta_0)[f^{1/2}(\theta_0+\theta n^{-1/2}) - f^{1/2}(\theta_0)]d\mu \\ &= \lim_n n^{1/2} \int [f(\theta_0)^{1/2} f^{1/2}(\theta_0+\theta n^{-1/2}) - 1]dM \\ &= -\frac{1}{2}\lim n^{1/2} \int [f^{1/2}(\theta_0) - f(\theta_0+\theta n^{-1/2})^{1/2}]^2 dM \\ &= 0\end{aligned}$$

Here are some examples of qmd families.

(1.9) EXAMPLE. If Θ is an open subset of the real line, presumably one should be able to compute the qmd of P_θ by computing the pointwise derivative

$$\frac{d}{d\theta} f^{1/2}(\theta;x) = \frac{1}{2}\dot{f}(\theta,x)/f^{1/2}(\theta,x)$$

where $\dot{f}(\theta,x) = \frac{\partial}{\partial\theta} f(\theta,x)$. Under regularity assumptions, this is possible. Suppose Θ is a subinterval of $\mathbb{R}^1$, M = Lebesgue measure, and that

(i) $\theta \to f(\theta;y)$ is absolutely continuous for each y in a neighborhood of θ_0

(ii) $\dot{f}(\theta;y)$ exists for a.e. y, θ near θ_0

(iii) $I(\theta) \equiv \int [\dot{f}(\theta;y)]^2/f(\theta;y)dy$ exists, continuous at θ_0 and $I(\theta_0) > 0$.

Then (Hajek (1972)), $\{P_\theta\}$ is qmd at θ_0 and the qm derivative η is $\frac{1}{2}\dot{f}(\theta_0;y)/f^{1/2}(\theta_0;y)$. Further characterization of qmd in terms of "ordinary" differentiations can be found in Le Cam (1970).

(1.10) EXAMPLE (Exponential families in natural form). For θ in a subset of $\mathbb{R}^d$, let P_θ be a measure absolutely continuous with respect to a sigma finite measure M, satisfying

$$dP_\theta/dM = a(\theta)\exp\langle\theta,T\rangle$$

where $a(\theta)$ is a normalizing constant, and $T = (T_1,\ldots,T_d)$ is a Borel function on the measure space where M is defined. Here $\Theta = \text{int}\{\theta \in \mathbb{R}^d: 0 < a(\theta) < \infty\}$. If $\theta \in \Theta$, such a family is qmd and the derivative is easily calculated. See Johanssen (1979) for the theory of such families. These particular families are of considerable importance in applied statistics, because they form a reasonably broad but relatively easily analysed collection of distributions that can be used in statistical modelling.

(1.11) EXAMPLE. Let P_θ be the uniform distribution on $[0,\theta]$. This family is not qmd at any $\theta_0 > 0$. As another example, let P_θ have density $f(\theta,x) = c(\alpha)\exp\{-|y-\theta|^\alpha\}$ (with respect to Lebesgue measure) where $c(\alpha)$ is a normalizing constant. Then this family is qmd iff $\alpha > \frac{1}{2}$.

(1.12) EXAMPLE. This example and the next are important for nonparametric problems. Fix a density f with respect to a sigma finite measure M. Let $h \in L^2(M)$ be orthogonal to $f^{1/2}$. Define a density $f(\theta,x)$ by means of the recipe

$$f^{1/2}(\theta,x) = (1-\theta^2|h|_M^2)^{1/2} f^{1/2}(x) + \theta h , \quad \theta \text{ real.}$$

This definition indeed yields a density if θ is sufficiently small. The family of densities so defined is then qmd; the derivative at $\theta_0 = 0$ (relative to M), for example, is h. It is trivial to extend this definition to $\theta \in \mathbb{R}^d$ (h is replaced by a vector with components in $L^2(M)$, and so forth).

(1.13) EXAMPLE. Again fix a probability density f and let h be a function in $L^2(f(M))$ such that $\int fh = 0$. Define $f(\theta,x) = f(x)(1+\theta h(x))$. Assume for all small θ that $f(\theta,x) \geq 0$. Then $f(\theta,x)$ is a probability density. If h is assumed bounded, for example, then this family is qmd; the derivative at $\theta = 0$ is $\frac{1}{2}hf^{1/2}$.

2. Asymptotic expansions for i.i.d. observations from a qmd family

We shall now state the basic asymptotic expansion for qmd families in the "i.i.d. case". Fix a qmd family P_θ and a point θ_0. Let $\eta = (\eta_1,\ldots,\eta_d)$ be the qmd derivative at θ_0, $f(\theta,\ )$ the density of P_θ with respect to M.

(2.1) DEFINITION. $\Gamma = \Gamma(\theta_0)$ is the matrix with entries Γ_{ij} given by

$$\Gamma_{ij} = \langle \eta_i,\eta_j \rangle_M$$

It is then immediate that

$$|\langle \theta,\eta\rangle|_M^2 = \langle \theta,\Gamma\theta\rangle \tag{2.2}$$

(2.3) REMARKS. (a) It is easy to see that Γ does not depend on the dominating measure M.

(b) If η is given as in Example (1.9), then $\Gamma = \frac{1}{4}\int \frac{\dot{f}(\theta_0,x)^2}{f(x)}dx = \frac{1}{4}I(\theta_0)$, where $I(\theta_0)$ is the familiar *Fisher information number*.

If η is the qm derivative at θ, define $\eta_0(\theta,x)$ by

$$\eta_0(\theta,x) = \eta(\theta;x)/f^{1/2}(\theta;x) \tag{2.4}$$

Define product measures $P^n(\theta,dx)$ on $(\mathbb{R}^d)^n$ by

(2.5) $$P^n(\theta,dx) = \prod_1^n f(\theta,x_1)M(dx_i) \qquad x = (x_1,\ldots,x_n)$$

(2.6) THEOREM. *Under the measures* $P^n(\theta_0,dx)$, *for each* $\theta \in \mathbb{R}^d$,

$$\sum \log\{f(\theta_0+\theta n^{-1/2};x_i)/f(\theta_0,x_i)\} - 2n^{-1/2}\sum\langle\eta_0(\theta_0,x_i),\theta\rangle + 2\langle\theta,\Gamma\theta\rangle$$

converges to 0 *in probability as* $n\to\infty$.

(2.7) COROLLARY. *Under* $P^n(\theta_0,dx)$, $\sum \log\{f(\theta_0+\theta n^{-1/2},x_i)/f(\theta_0,x_i)\}$ *is asymptotically normal* $N(-2\langle\theta,\Gamma\theta\rangle,4\langle\theta,\Gamma\theta\rangle)$.

The corollary is immediate from (2.6); the proof of (2.6) will be given shortly. First, however, we interpret the result in terms of convergence of experiments as defined in Chapter II. Fix θ_0, let $\{P_\theta\}$ be qmd at θ_0, and define experiments

(2.8) $$E^n = \{P^n(\theta_0+\theta n^{-1/2},dx)\colon \theta\in\mathbb{R}^d\}$$

Since $\theta_0 \in \Theta$ and Θ is open, $\theta_0+\theta n^{-1/2} \in \Theta$ for large n, so $P^n(\theta_0+\theta n^{-1/2})$ is defined eventually for any $\theta \in \mathbb{R}^d$.

(2.9) THEOREM. *Suppose* $\Gamma = \Gamma(\theta_0)$ *is non-singular. Let* Q_θ *be the normal distribution on* $\mathbb{R}^d$ *with mean* θ *and covariance* $\frac{1}{4}\Gamma^{-1}$. *Then the experiments* E^n *converge to* $\{Q_\theta,\ \theta\in\mathbb{R}^d\}$.

PROOF OF (2.9). Because of Proposition II.2.3 and the fact that the expansion (2.6) is *quadratic in* θ, we need only show that for *each* θ $\log dP^n(\theta_0+\theta n^{-1/2})/dP^n(\theta_0)$ converges in distribution to $\log dQ_\theta/dQ_0$. This in turn is immediate from the following lemma, proved by simple computations with the multivariate normal density:

(2.10) LEMMA. *Let* $N(\theta,\Sigma)$ *be the normal distribution on* $\mathbb{R}^d$ *with mean vector* θ, *covariance* Σ. *Then the* $N(0,\Sigma)$*-distribution of* $\log dN(\theta,\Sigma)/dN(0,\Sigma)$ *is* $N(-\frac{1}{2}\langle\theta,\Sigma^{-1}\theta\rangle,\langle\theta,\Sigma^{-1}\theta\rangle)$.

The proof of (2.10) will be omitted. We turn now to the

PROOF OF (2.6). This proof will be accomplished through a series of lemmas. Recall θ_0 is fixed; for simplicity denote $\eta(\theta_0,x)$ by η and $\eta_0(\theta_0,x)$ by η_0.

(2.11) LEMMA. *Let* e *be a real function on the line such that* $e(0) = 0$ *and*

$$\lim_{x\to 0} x^{-2}[e(x)-ax-bx^2] = 0$$

Let $Y_{n,k}$, $k = 1,2,\ldots,n$, *be a triangular array of independent random variables such that* $\sum_k Y_{nk}$ *converges in distribution to* $N(\mu,\sigma^2)$ *and* $\lim_n \max_k P\{|Y_{nk}| > \varepsilon\} = 0$ *for every* $\varepsilon > 0$. *Then*

$$\sum_k e(Y_{nk}) - a\sum Y_{nk} - b\sigma^2 \longrightarrow 0$$

PROOF. By the central limit problem (cf. Loève), $\sum Y_{nk}^2 \to \sigma^2$ and $\max_k |Y_{nk}| \to 0$ (convergence in probability). If $e(x) = ax + bx^2 + x^2R(x)$ where $R(x) \to 0$ as $x \to 0$, then the difference given in the lemma is bounded by $|\sum Y_{nk}^2 R(Y_{nk})| \le \max_k R(Y_{nk}) \sum_k Y_{nk}^2 \to 0$. This proves (2.11).

Next define

$$Y_{nk} = [f^{1/2}(\theta_0+\theta n^{-1/2},x_k)/f^{1/2}(\theta_0,x_k)] - 1 \tag{2.12}$$

Notice that under $P^n(\theta_0)$, $\{Y_{nk}, k \ge 1\}$ are i.i.d. Take e in Lemma (2.11) to be $e(x) = 2\log(1+x) = 2x - x^2 + o(x^2)$, so that

$$\log f(\theta_0+n^{-1/2}\theta)/f(\theta_0) = e(Y_{nk})$$

By (2.11), if we show $\sum_k Y_{nk} \Rightarrow N(\mu,\sigma^2)$, then $\sum \log f(\theta_0+\theta n^{-1/2})/f(\theta_0) - 2\sum Y_{nk} + \sigma^2 \to 0$. Therefore, to establish the expansion (2.6), we shall analyze $\sum_k Y_{nk}$.

(2.13) LEMMA. $\lim_n \sum_j EY_{nj} = -\frac{1}{2}\langle\theta,\Gamma\theta\rangle$ *where* E *denotes expectation under* $P^n(\theta_0;dx)$.

PROOF.
$$E\,Y_{nj} = \int\left[\frac{f^{1/2}(\theta_0+\theta n^{-1/2})}{f^{1/2}(\theta_0)} - 1\right]f(\theta_0)M(dx)$$
$$= \int[f^{1/2}(\theta_0+\theta n^{-1/2})f^{1/2}(\theta_0) - (f^{1/2}(\theta_0))^2]$$
$$= -\frac{1}{2}\int[f^{1/2}(\theta_0+n^{-1/2}\theta) - f^{1/2}(\theta_0)]^2 dM$$

so

$$\sum EY_{nj} = -\frac{1}{2}n\int[f^{1/2}(\theta_0+n^{-1/2}\theta) - f^{1/2}(\theta_0)]^2 \to -\frac{1}{2}\int\langle\theta,\eta\rangle^2 dM$$

and the result follows from (2.2).

(2.14) LEMMA. $\sum_j Y_{nj} - EY_{nj} = n^{-1/2}\sum\langle\eta_0(x_i),\theta\rangle + o_p(1)$; *the latter converges in distribution to* $N(0,\langle\theta,\Gamma\theta\rangle)$.

PROOF. $Y_{nj} = f^{1/2}(\theta_0+\theta n^{-1/2},x_j)/f^{1/2}(\theta_0,x_j)^{-1}$ and $f^{1/2}(\theta_0+n^{-1/2}\theta,x_j) = f^{1/2}(\theta_0,x_j) + n^{-1/2}\langle\theta,\eta(x_j)\rangle + n^{-1/2}R_n(x_j)$ where $R_n(\cdot)\to 0$ in $L^2(M)$. Making this approximation in Y_{nj}, together with a similar one in EY_n, yields

$$Y_{nj} - EY_{nj} = n^{-1/2}\langle\eta_0,\theta\rangle + n^{-1/2}R_n/f^{1/2}(\theta_0) - n^{1/2}\,R_n f^{1/2}(\theta_0)\ .$$

An easy application of the central limit theorem shows that the sum of the last two terms goes to 0 (it is a triangular array of mean 0 random variables plus the variance of each row $\to 0$ since $R_n \to 0$ in

$L^2(M)$). This yields the expansion (2.14). Since $\langle \eta_0(x_i),\theta\rangle$ are i.i.d., mean 0 (by (1.8)) and have variance $\langle \theta,\Gamma\theta\rangle$ under $P^n(\theta_0)$, the asymptotic normality is evident.

(2.15) LEMMA. (a) $\lim_n \max_k P^n\{|Y_{nk}| > \varepsilon\} = 0$ *for every* $\varepsilon > 0$

(b) $\sum_j Y_{nj}$ *converges in distribution* $N(-\frac{1}{2}\langle h,\Gamma h\rangle, \langle h,\Gamma h\rangle)$

PROOF. (b) follows from (2.13), (2.14); (a) is easy.

The expansion (2.6) is now immediate on applying the foregoing lemmas.

V. Gaussian shift experiments

Chapter IV produced a large collection of experiments $E^n = \{P^n_\theta, \theta \in \Theta\}$ $\Theta \subset \mathbb{R}^d$ which converged in the sense of Chapter II to a limit experiment $E = \{P_\theta\}$ where P_θ is a Gaussian distribution on $\mathbb{R}^d$ with mean vector θ. In order to investigate nonparametric asymptotic theory, the class of such limit experiments must be enlarged: it is necessary to introduce experiments $\{P_h, h \in H\}$ where H is a separable Hilbert space and $\{P_h\}$ is a family of Gaussian measures parametrized by H. This chapter presents the basic facts for such a development; chapter VI, which is next, gives examples of experiments E^n which converge to such $\{P_h, h \in H\}$. Applications of Gaussian families $\{P_h\}$ with infinite dimensional parameter set H can be found in Chapters VIII, IX, X, XI and XII.

1. Cylinder measures

Let B be a Banach space (assumed separable for these notes, but that is not really necessary); let B^* be its dual--the collection of all continuous, real linear functionals on B. Denote by $\langle\ ,\ \rangle_B$ the duality relationship of B, B^*:

(1.1) $$\langle x,m\rangle_B = m(x)\ ,\quad x \in B,\ m \in B^*\ .$$

When context is clear, the subscript B will be omitted.

(1.2) DEFINITION. A *cylinder set* in B, based on $m_1,\ldots,m_k \in B^*$, is a set of the form

$$\{x \in B\colon (\langle x,m_1\rangle,\ldots,\langle x,m_k\rangle) \in A\}$$

where A is a Borel set of $\mathbb{R}^k$. A *cylinder function* is a real function on B of the form $x \to f((\langle x,m_1\rangle,\ldots,\langle x,m_k\rangle))$ where f is a real Borel function on $\mathbf{R}^k$. Denote by $C_{m_1,\ldots,m_k}$ the sigma field of all cylinder sets based on $m_1,\ldots,m_k$. Let $C_0 = \bigcup_{m_1,\ldots,m_k} C_{m_1,\ldots,m_k}$ be the field (*not* sigma field) consisting of all cylinder sets. Let C be the sigma field generated by C_0.

It is easy to see in the present case that

(1.3) C is the Borel sigma field on B

i.e., C is the sigma field generated by the open sets. Evidently C is also the smallest sigma field making all the maps $x \to \langle x,m\rangle$ measurable.

(1.4) DEFINITION. A *cylinder probability* on B is a finitely additive measure on C_0 which is countably additive on each $C_{m_1,\ldots,m_k}$.

We shall give examples of cylinder measures at the end of this section. In general, they cannot be extended to be countably additive on C, the Borel sigma field--an inconvenience that we shall deal with shortly. Nevertheless certain operations on cylinder measures are possible. The first of these is

(a) *Simple integration.* If Q is a cylinder probability on B and if f is a bounded cylinder function, then

$$\int f(x)Q(dx)$$

makes sense, since Q is countably additive on each $C_{m_1,\ldots,m_k}$.

This leads to the *second* operation:

(b) *Characteristic functional.* If Q is a cylinder measure, the characteristic functional ϕ of Q is the complex valued function defined on B^* by

$$\phi(m) = \int \exp\{i\langle x,m\rangle\}Q(dx)$$

The following simple *uniqueness theorem* is extremely useful:

(1.5) PROPOSITION. *Let* Q_1, Q_2 *be two cylinder probabilities on* B *having the same characteristic functional* ϕ. *Then* $Q_1 = Q_2$.

To see this, it is enough to show that $Q_1(A) = Q_2(A)$ for all A in some fixed $C_{m_1,\ldots,m_k}$. A typical such set A is of the form $\{x: (\langle x,m_1\rangle,\ldots,\langle x,m_k\rangle)\in A_0\}$, A_0 a Borel subset of $\mathbb{R}^k$. It is therefore enough to show that the joint distribution of the random variables $x \to \langle x,m_i\rangle$, $1 \leq i \leq k$, is the same under Q_1 and Q_2. The (multivariate) characteristic function of these under Q_j is

$$\begin{aligned}\psi_j(t) &= \int \exp\{i \sum_r t_r\langle x,m_r\rangle\}Q_j(dx) \\ &= \int \exp\{i\langle x,\sum t_r m_r\rangle\}Q_j(dx) \\ &= \phi(\sum t_r m_r) , \quad t \in \mathbb{R}^k\end{aligned}$$

and this last, by hypothesis, does not depend on j. The agreement of Q_1, Q_2 on $C_{m_1,\ldots,m_k}$ is then immediate from the uniqueness theorem for characteristic functions on Euclidean space.

The *third* operation on cylinder measures is:

(c) *Convolution*. Let Q_1, Q_2 be two cylinder measures on B. Define $Q_1 * Q_2$ to be the cylinder measures on B with characteristic function $\phi(m) = \phi_1(m)\cdot\phi_2(m)$ where ϕ_i is the characteristic functional of Q_i. It is easy to see that such a cylinder measure exists. Indeed, considerations given in the argument for (1.5) show how this measure may be defined on $C_{m_1,\ldots,m_k}$. If $C_{m_1,\ldots,m_k} \subset C_{m_1,\ldots,m_k,m_{k+1}}$ then this measure is consistently defined (in the usual sense) and so must exist on C_0.

The last operation on cylinder measures is extremely important for our applications.

(d) *Image of a cylinder measure.* Let Q be a cylinder probability on a Banach space B. Let τ be a continuous linear map from B to another Banach space B_1. The image of Q under τ, to be denoted by Q_τ, is the cylinder measure on B_1 given on the cylinder set $A = \{x \in B_1: (\langle x,m_1\rangle,\ldots,\langle x,m_k\rangle) \in A^0\}$, $A^0 \subset \mathbb{R}^k$, $m_i \in B_1^*$, by

$$(1.6) \qquad Q_\tau(A) = Q\{x \in B: (\langle \tau x,m_1\rangle,\ldots,\langle \tau x,m_k\rangle) \in A^0\}$$

Roughly speaking, Q_τ is the "distribution" of the B_1-valued random variable τ, under Q. If τ^* is the adjoint of τ: $\tau^*: B_1^* \to B^*$, then $\langle \tau x,m\rangle_{B_1} = \langle x,\tau^* m\rangle_B$, so the right side of (1.6) makes sense. It is immediate from (1.6) that the characteristic function of Q_τ is ϕ_τ given by

$$(1.7) \quad \phi_\tau(m) = \int \exp\{i\langle x,m\rangle_{B_1}\}Q_\tau(dx) = \int \exp\{i\langle x,\tau^* m\rangle_B\}Q(dx) = \phi(\tau^* m)$$

where ϕ is the characteristic function of Q.

We turn now to some examples. As usual, we do not strive here for utmost generality.

(1.8) EXAMPLE (Cylinder measures on $C[0,1]$). Let $\{X_t,\ 0 \le t \le 1\}$ be a real stochastic process such that $EX_t = 0$ and, for $0 \le t_1 < t_2 < \cdots < t_k \le 1$, the vector $(X_{t_1},\ldots,X_{t_k})$ has a Gaussian distribution. As usual in stochastic process theory, we write X_t, $X(t)$, $X(t,\omega)$ interchangeably. Let $R(s,t) = EX_sX_t$. Assuming that R is bounded on $[0,1]^2$, we can construct a cylinder measure on the Banach space $B = C[0,1]$, the continuous bounded functions with supremum norm. In order to do this, it is clear from (e.g.) the considerations discussed under item (b) above that it is enough to specify *consistently* the joint distribution of the random variables (defined on B)

$$x \longrightarrow (\langle x,m_1\rangle,\ldots,\langle x,m_k\rangle) \tag{1.9}$$

where $m_1,\ldots,m_k$ belong to $B^* = \{\text{finite signed measures on } [0,1]\}$ and where, for $x \in B$, $m \in B^*$,

$$\langle x,m\rangle = \int_0^1 x(t)m(dt) .$$

Since $\{X(t)\}$ is Gaussian and linear combinations of Gaussian random variables are still Gaussian, it is plausible that $\int X(t)m(dt)$ is Gaussian for any $m \in B^*$; i.e. the vector in (1.9) should be Gaussian. Since each X_i has mean 0, presumably

$$E\int X(t)m(dt) = \int EX(t)m(dt) = 0 \tag{1.10}$$

Therefore to specify the distribution of (1.9) we should only need to specify the covariance:

$$\begin{aligned} E\{\int X(t)dm_i(t)\int X(s)dm_j(s)\} &= \iint EX(t)X(s)dm_i(t)dm_j(s) \\ &= \iint R(s,t)dm_i(s)dm_j(t) \end{aligned} \tag{1.11}$$

This leads us to *define* the cylinder measure on B by specifying that the random vector on B, given by (1.9), should be normal with mean vector 0 and covariance given by the last line of (1.11). It is easy to see that this defines a cylinder measure on $B = C[0,1]$ (one need only check consistency of the assigned distributions). In this generality, the cylinder measure just constructed *cannot* always be extended to be countably additive on the Borel sets; this is connected with the problem of whether the process $\{X_t\}$ can be constructed to have continuous paths (an aspect we do not explore in these notes). The *characteristic function* ϕ *of such a cylinder measure* is easily calculated; since the real random variable $x \to \langle x,m \rangle$ is Gaussian, mean 0, variance $\iint R(s,t)dm(s)dm(t)$, it is immediate that

$$(1.12) \qquad \phi(m) = \int \exp\{i\langle x,m\rangle\} = \exp\{-\frac{1}{2}\iint R(s,t)m(ds)m(dt)\}$$

One could also define the cylinder measure by specifying its characteristic function via (1.12), a method which has the virtue of economic expression.

There are important examples of such cylinder measures; we mention only two.

(1.13) BROWNIAN MOTION. Brownian motion $\{X_t\}$ on $[0,1]$ is the Gaussian stochastic process with $EX_t = 0$ or $EX_tX_s = R(s,t) = \min\{s,t\}$. In *these* notes, this particular process is involved only in the applications concerning the estimation of the spectral function of a stationary process; nevertheless it is an extremely important example. The cylinder measure defined from this process via the recipes above can indeed be extended to the Borel sets of $C[0,1]$ (a well known fact, which we discuss in section 2). From a probabilistic point of view, this approach to Brownian motion is clumsy at best; however, insight into certain statistical applications is easier this way.

(1.14) BROWNIAN BRIDGE. In this example, the process $\{X_t\}$ on $[0,1]$ is Gaussian with $EX_t = 0$ and $R(s,t) = \min\{s,t\} - st$. Again, the resulting *cylinder* measure on $C[0,1]$ is well known to be countably additive on the Borel sets. Indeed, since $\{X_t\}$ yields the same cylinder measure on B as the process $\{W_t - tW_1\}$ where $\{W_t\}$ is Brownian motion on $[0,1]$, the countable additivity follows from that of $\{W_t\}$.

(1.15) EXAMPLE (Cylinder measures on $L^2[0,1]$). In the example (1.8), the cylinder measures there could just as well have been constructed on $L^2([0,1],\mu)$ where μ is (say) a finite measure. In this case, if $B = L^2[0,1]$, $B^* = B$, so that if $m \in B^*$, $\langle x,m\rangle = \int x(t)m(t)\mu(dt)$. With this minor change the rationale given in (1.8) suggests defining the cylinder measure by requiring that the vector (1.9) be Gaussian with mean 0, covariance $\iint R(s,t)m_i(s)m_j(t)\mu(ds)\mu(dt)$. Of course, Brownian motion, Brownian Bridge are again cylinder measures in this set-up too and are countably additive on the Borel sets of B. There are, therefore, sometimes several spaces B on which to construct a given cylinder measure; in the applications some choices are much more convenient than others.

2. Abstract Wiener space

Let H be a separable Hilbert space. Denote by $\langle\ \rangle_H$, $|\ |_H$ its inner product and norm. When context is clear, we shall omit the subscripts.

(2.1) DEFINITION. The *canonical normal cylinder measure* Q on H is the cylinder measures with characteristic function

$$\phi(h) = \exp\{-\tfrac{1}{2}|h|^2\}, \quad h \in H^* = H .$$

If $h_1,\ldots,h_k \in H$, then proceeding as in (1.5), it is easy to see that the joint characteristic function of the real random variables $x \to \langle h_i,x\rangle$

is, under Q,

$$\begin{aligned}\psi(t) &= \int \exp\{i \sum t_i \langle h_i, x\rangle\} Q(dx) \\ &= \exp\{-\tfrac{1}{2}|\sum t_i h_i|^2\} \\ &= \exp\{-\tfrac{1}{2}\langle t, \Gamma t\rangle\} \qquad t = (t_1, \dots, t_k) \in \mathbb{R}^k\end{aligned}$$

where Γ is the $k \times k$ matrix with entries $\langle h_i, h_j\rangle$. From this and basic facts about the multivariate normal distribution we obtain the

(2.2) PROPOSITION. *Under* Q, *the distribution of the random variables* $x \to \langle h_i, x\rangle$ *is Normal, mean* 0, *covariance* Γ.

In particular, if $e_1, \dots, e_k$ are orthonormal elements of H

(2.3) $\{\langle e_i, \cdot\rangle\}$ are i.i.d. $N(0,1)$ random variables.

It is easy to see that the cylinder measure Q exists, (2.2) permitting the necessary consistency conditions to be verified. Q will *not* be countably additive, however, when H is infinite dimensional. To see why, note that if $\{e_i, 1 \le i < \infty\}$ is an orthonormal basis for H, then $|x|^2 = \sum |\langle x, e_i\rangle|^2$ for any $x \in H$. Because of (2.3), if Q is countably additive, this series must equal $+\infty$ a.e. Q; so $Q\{x: |x|^2 = +\infty\} = 1$. On the other hand, $Q(H) = 1$ and $|x| < \infty$ if $x \in H$ --a contradiction.

To sidestep the fact that Q is not countably additive, we resort to the following device: Hunt for a separable Banach space B and a mapping $\tau: H \to B$ such that

(2.4) (a) τ is continuous, linear

(b) τ is one-to-one

(c) τH is dense in B

Let B^* be the dual of B, $\langle\ ,\ \rangle_B$ the usual duality relation (1.1) and let P_0 be the *image of* Q *under* τ as defined in V.1.

(2.5) DEFINITION. The triple (τ,H,B) is an *abstract Wiener space* if P_0 is countably additive on the Borel sets of B.

In short, the proposal is to make the standard normal of H countably additive by putting it on another space. Alternatively, one can view B as the "completion" of H under the norm $|h|_\tau = |\tau h|_B$ and the proposal is to make Q countably additive by "enriching" the space H. Important examples of abstract Wiener spaces will be given shortly. First let us note an important property.

(2.6) THEOREM. *Let* τ *satisfy (2.4). Then* (τ,H,B) *is an abstract Wiener space if and only if the distribution of* $\sum_{i=1}^{n} \langle e_i;\cdot\rangle\tau e_i \equiv Z_n$ *converges on* B *to* P_0.

(2.7) REMARK. Z_n has the same distribution as $\sum_1^n X_i \tau e_i$, where $X_1,\ldots,X_n$ are i.i.d. $N(0,1)$. Specializations of (2.6) therefore give, e.g., well-known representations of Brownian motion as the sum of a random series.

PROOF. We show only that if (τ,H,B) is abstract Wiener, then Z_n converges; the converse is simple. Let $m \in B^*$. Then

$$
\begin{aligned}
E\exp\{i\langle Z_n,m\rangle_B\} &= \exp\{-\tfrac{1}{2}\sum_1^n |\langle \tau e_i,m\rangle_B|^2\} \qquad (2.8)\\
&= \exp\{-\tfrac{1}{2}\sum_1^n \langle e_i,\tau^* m\rangle_H^2\}\\
&\to \exp\{-\tfrac{1}{2}|\tau^* m|_H^2\}
\end{aligned}
$$

which is the characteristic function of P_0. If H were finite dimensional, this would complete the proof, because of the theorem of Lévy relating convergence of characteristic functions to convergence of measures. This theorem, however, fails for general H. If we show Z_n has a subsequence converging in distribution, then the limit must be P_0 because of (2.8) and the uniqueness result for characteristic functions (VI.5). Therefore, it is necessary to show only that the distributions of $\{Z_n\}$ are tight. Let Q_n be the distribution of Z_n, P_n the distribution of $\tau(\cdot - \sum_1^n \langle e_i, \cdot \rangle e_i)$. Then the calculation with characteristic functions just made shows

$$P_0 = Q_n * P_n$$

where $*$ denotes convolution. Therefore, for any fixed $\varepsilon > 0$, there is a compact $K \subset B$ such that

$$1 - \frac{\varepsilon}{2} < P_0(K) = \int Q\{Z_n \in K-y\} P_n(dy) \ , \quad \text{all } n.$$

Hence there exists a sequence y_n such that $Q\{Z_n + y_n \in K\} \geq 1 - \frac{\varepsilon}{2}$. But the distribution of Z_n is *symmetric* (i.e., Z_n has the same distribution as $-Z_n$) so $Q\{-Z_n + y_n \in K\} \geq 1 - \frac{\varepsilon}{2}$, so $Q\{2Z_n \in K-K\} = Q_n\{\frac{1}{2}(K-K)\} \geq 1 - \varepsilon$. Since $K-K$ is compact and ε arbitrary, the distributions of Z_n are tight. Q.E.D.

(2.9) EXAMPLE (Classical Wiener space). Let $H = L^2(\mu)$ where μ is Lebesgue measure on the unit interval. Let $B = C[0,1]$, the Banach space of real continuous functions on $[0,1]$, with supremum norm. Define $\tau \colon H \to B$ by

$$(\tau h)(t) = \int_0^t h(s)ds$$

The adjoint τ^* is

$$(\tau^*m)(s) = m\{[s,1]\} \qquad m \in B^* \quad (= \text{signed measures}).$$

The characteristic function of P_0 therefore is

$$\exp\{-\tfrac{1}{2}|\tau^*m|^2\} = \exp\{-\tfrac{1}{2}\iint t \wedge s \; m(ds)m(dt)\}$$

since

$$|\tau^*m|^2 = \int m\{[u,1]\}^2 du = \iint I_{[0,s]}(u)dm(s)\int I_{[0,t]}(u)m(dt)du$$
$$= \iiint I_{[0,s\wedge t]}(u)dudm(t)dm(s)$$

This, of course, is the characteristic function of Brownian motion (cf. V.1.14). It is well known that P_0 here is countably additive, so (τ,H,B) is a Wiener space. (It is easy to check that τ is 1-1, τH dense in $C[0,1]$.)

(2.10) EXAMPLE (Brownian Bridge). Fix a probability density f on the line (with respect to a sigma finite nonatomic measure ν); write ds for $\nu(ds)$ from now on. Let H consist of all real functions h such that support $h \subset$ support f, $\int h^2(s)(ds) < \infty$ and $\int h(s)f^{1/2}(s)(ds) = 0$. Define τ on H by

$$(2.10a) \qquad (\tau h)(t) = \int^t h(s)f^{1/2}(s)(ds) \qquad t \in \text{support } f$$

Notice that τ is continuous. Let B be the closed linear span of τH, using supremum norm. Since support $h \subset$ support f, τ is 1-1. To compute τ^*, let m be a measure with support contained in that of f. Denote by F the cdf of f: $F(t) = \int^t f(s)ds$. Then for any c

$$\langle\tau h,m\rangle = \int [I_{(-\infty,t]}(s)-c]h(s)f^{1/2}(s)m(dt)ds = \langle h,\tau^*m\rangle$$

from which we see

$$(\tau^*m)(s) = f^{1/2}(s)[m\{[s,\infty)\} - c]$$

where $c = \int f(s)m\{[s,\infty)\}ds = \int F(u)m(du)$ since τ^*m must be in H. A simple computation shows

$$|\tau^*m|^2 = \int [F(u \wedge v) - F(u)F(v)]m(du)m(dv)$$

Let

(2.11) $$W^0 = \{W^0(t),\ 0 \leq t \leq 1\}$$

be the standard Brownian Bridge stochastic process, i.e. the Gaussian process with covariance $s \wedge t - st$. It is well known that the distribution of W^0 is countably additive on the Borel sigma field of $C[0,1]$. The characteristic function of P_0 is $\exp\{-\frac{1}{2}|\tau^*m|^2\}$ by V.1.6 and the calculation of τ^* above together with Example V.1.8 shows that this is the same characteristic function as that associated with the process $\{W^0(F(t))\}$. That is, (τ,H,B) is an abstract Wiener space and P_0 is the distribution of $W^0 \circ F$. It may be shown, incidentally, that B here consists of all continuous bounded functions of the form $t \to x(F(t))$, $x \in C[0,1]$, $x(0) = x(1) = 0$.

(2.12) EXAMPLE (Brownian Bridge). Let f be a density on the line. Let H be the subspace of $L^2(f\,dx)$ consisting of all functions h with $\int hf = 0$, support h $\subset$ support f. Let τ be

$$(\tau h)(t) = \int_{-\infty}^{t} h(s)f(s)ds$$

and B the Banach space of continuous functions $\overline{\tau H}$, closure under supremum norm. If m is a signed measure, then as in Example (2.10) $\tau^*m(s) = m([s,\infty)) - \int m[s,\infty)f(s)ds$, and so forth. It is then easy to see that (τ,H,B) here is an abstract Wiener space and that P_0 is the same as in (2.10).

(2.13) EXAMPLE (L^2 spaces). In studying a triple (τ,H,B) with the hope of proving it an abstract Wiener space, a main task is identifying the cylinder measure P_0, image of the standard normal of H under the mapping τ. Calculation of τ^* is often painful; this example provides a situation--common in applications--where a different method works. Assume $H = L^2(E,d\mu)$, $B = L^2(F,d\nu)$ where μ, ν are sigma finite measures and E, F are (e.g.) complete separable metric. Quite often the mapping τ has the form $(\tau h)(t) = \langle k_t,h\rangle_H$ where for each $t \in E$, $k_t \in H$. This is true in all the examples used so far--(2.9), (2.10), (2.12). P_0 is the "distribution" of the "random variable" $x \to \tau x$ and to identify P_0 we need only find the distribution of $\langle \tau\cdot,g_1\rangle_B,\ldots,\langle \tau\cdot,g_k\rangle_B$, $g_i \in B$. These will be Gaussian, mean 0, and the covariance must be

$$\int_H \langle \tau x,g_i\rangle_B \langle \tau x,g_j\rangle_B Q(dx) \tag{2.14}$$

But $\langle \tau x,g_i\rangle = \int \langle k_t,x\rangle_H g_i(t)\nu(dt)$. Making these substitutions in (2.14), changing order of integration and remembering

$$\int \langle h,x\rangle\langle e,x\rangle Q(dx) = \langle h,e\rangle$$

shows that (2.14) is equal to

$$\iint R(s,t)g_1(s)g_2(t)\nu(ds)\nu(dt) \equiv \langle g_1,Rg_2\rangle_B \tag{2.15}$$

where $R(s,t) = \langle k_s,k_t\rangle_H$. In particular, the characteristic function of P_o must be

$$\phi(g) = \exp\{-\tfrac{1}{2}\langle g,Rg\rangle^2\}$$

Example V.1.15 then permits identification of P_0 as the "distribution" of some process.

(2.16) EXAMPLE (Hilbert Schmidt operators). Let H_1, H_2 be two separable Hilbert spaces. A continuous linear mapping $\tau\colon H_1 \to H_2$ is called Hilbert-Schmidt (H-S hereafter) if, for some orthonormal basis $\{e_k\}$ of H_1

(2.17) $$\sum_1^\infty |\tau e_k|^2 < \infty$$

It may be shown that if property (2.17) holds for one orthonormal basis, then it holds for all others. It is a well known fact about H-S operators that they have a convenient representation that can be described as follows. Let $\{e_k\}$, $\{\psi_k\}$ be orthonormal bases of H_1, H_2 respectively. Then if τ is H-S, there exist scalars a_k, $\sum a_k^2 < \infty$ such that

(2.18) $$\tau h = \sum a_k \langle h,e_k\rangle_1 \psi_k$$

The adjoint τ^* is then easily seen to be

(2.19) $$\tau^* = \sum a_k \langle \psi_k,\cdot\rangle_2 e_k$$

Here $\langle\ ,\ \rangle_i$ denotes inner product for H_i.

For many applications, the following characterization of H-S operators is convenient. Let E_1, E_2 be subsets of $\mathbb{R}^k$; put Lebesgue measure on E_1, E_2 and let H_i be the collection of real functions h on E_i such that $\int_{E_i} h^2(s)ds < \infty$. Then $\tau\colon H_1 \to H_2$ is H-S if and only if there exists R, a non-negative measurable function on $E_1 \times E_2$ such that $\int_{E_1\times E_2} R(s,t)^2 dsdt < \infty$ and

$$(2.20) \qquad (\tau h)(t) = \int_{E_1} R(s,t)f(s)ds$$

cf. Kato. Here is an important illustration of this. If $H_1 = H_2 = L^2([0,1],ds)$, ds = Lebesgue measure, then $(\tau h)(t) = \int_0^t h(s)ds$ is H-S, with $R(s,t)$ of (2.20) given by $R(s,t) = I_{[0,t]}(s)$. For another example, take H_1 to be the Hilbert space H of Example (2.10), and define τ as in (2.10a). Let μ be a sigma finite measure on the line and take $H_2 = L_2(\mu)$. By the Schwarz inequality $|\tau h(t)|^2 \leq F(t) \wedge [1-F(t)]|h|$. Then τ maps H_1 into H_2 if $\int F(t) \wedge [1-F(t)]\mu(dt) < \infty$. It may be shown that τ is then H-S.

The importance of Hilbert-Schmidt operators for abstract Wiener spaces is demonstrated by the following result.

(2.21) PROPOSITION. *Let* H, B *be Hilbert spaces and let* $\tau: H \to B$ *be a continuous linear mapping, 1-1, and such that* τH *is dense in* B. *Then* P_0, *the image of the standard normal of* H, *is countably additive if and only if* τ *is H-S.*

PROOF. We prove only that if τ is H-S, then P_0 is countably additive. Because of (2.19), the characteristic function of P_0 is $\xi(h)$.

$$(2.22) \qquad \exp\{-\tfrac{1}{2}\textstyle\sum a_k^2 \langle h,\psi_k\rangle^2\} = \xi(h) \ , \quad h \in B$$

Let $X_1, X_2, \ldots$ be i.i.d., $N(0,a_i^2)$, random variables on some probability space (Ω,P). Set $Y_n(\omega) = \sum_1^n X_i(\omega)\psi_i$. Then Y_n is B-valued and $|Y_n|_B^2 = \sum X_i^2$ and $\sum X_i^2 < \infty$ a.e. since $\sum EX_i^2 = \sum a_i^2 < \infty$. Therefore $Y = \sum X_i(\omega)\psi_i$ is a B-valued random variable, defined on (Ω,P). Let Q_0 be the distribution of Y: $Q_0(A) = P\{\omega: Y(\omega) \in A\}$, A a Borel set of B. Q_0, of course, is countably additive since P is; its characteristic function is easily seen to be (2.22): $E \exp\{i\langle h,Y\rangle_B\} = \exp\{-\frac{1}{2}\sum\langle h,\psi_i\rangle^2 a_i^2\}$, since the X_i are independent, $N(0,a_i^2)$. By the uniqueness theorem for characteristic functions, $Q_0 = P_0$. Q.E.D.

3. Likelihood ratios

(3.1) DEFINITION. Let (τ,H,B) be an abstract Wiener space. Let P_0 be the image on B of the standard normal distribution on H. For $h \in H$, define the probability P_h by

$$P_h(A) = P_0(A-\tau h)$$

The experiment $\{P_h, h\in H\}$ will be called the *standard normal shift family* for (τ,H,B).

This family shares many similarities with the usual $N(\theta,I)$ experiments on $\mathbb{R}^k$. In this section we shall see that the P_h are mutually absolutely continuous, and shall compute the derivatives dP_h/dP_0.

To get a start on this define

(3.2) $$L_h(x) = \langle z,x\rangle_B\ , \quad x\in B, \quad z\in B^* \text{ if } h = \tau^*z$$

It is clear that for each fixed h, $h = \tau^*z$, that $L_h(\cdot)$ is a real random variable on B. Under P_0, it is immediate from (2.2) that

(3.3) the distribution of L_h is $N(0,|h|^2)$ and if $h_i = \tau^*z_i$, $1 \le i \le k$

(3.4) the distribution of $(L_{h_1},\ldots,L_{h_k})$ is multivariate normal with covariance $\{\langle h_i,h_j\rangle\}$

(3.5) $$h \rightarrow L_h \text{ is linear}$$

The definition of L_h will now be extended to *all* $h \in H$, not just $h \in \tau^*B^*$. For this purpose, note

(3.6) LEMMA. τ^*B^* *is dense in* $H^* = H$. (*Need* τ 1-1 here.)

PROOF. Suppose $\langle \tau^*z,h\rangle_H = 0$ for all $z \in B^*$. Then $\langle z,\tau h\rangle_B = 0$ for all z so $\tau h = 0$. But τ is 1-1, so $h = 0$. Since τ^*B^* is a subspace, its closure must be H by (e.g.) the projection theorem for Hilbert space. Q.E.D.

To define L_h for arbitrary $h \in H$, take z_n to be any sequence in B^* such that $\tau^*z_n \rightarrow h$ in H; such sequences exist by (3.6). Define

(3.7) $$L_h(\cdot) = \lim \langle z_n,\cdot\rangle_B$$

where the limit is in $L^2(B,dP_0)$. This limit exists as asserted because

(3.8) $$\int |\langle z_n,\cdot\rangle - \langle z_m,\cdot\rangle|^2 dP_0 = \int |\langle z_n - z_m,\cdot\rangle|^2 dP_0 = |\tau^*(z_n - z_m)|_H^2 \rightarrow 0$$

using (3.3) for the second equality. It is clear that the definition of L_h in (3.7) does not depend on the sequence chosen and that (3.3), (3.4) and (3.5) continue to hold.

(3.9) THEOREM. *Let* $\{P_h,\ h \in H\}$ *be the standard normal shift experiment for* (τ,H,B). *Then* P_h *is absolutely continuous with respect to* P_0 *and*

$$dP_h/dP_0 = \exp\{L_h - \tfrac{1}{2}|h|_H^2\}$$

(3.10) COROLLARY. $\log dP_n/dP_0 = L_h - \frac{1}{2}|h|^2$ *is* $N(-\frac{1}{2}|h|^2, |h|^2)$ *under* P_0.

PROOF. The corollary is immediate from (3.3). The theorem will be proved by showing that the characteristic functions of the measures P_h and $\exp\{L_h - \frac{1}{2}|h|^2\}dP_0$ are equal. The characteristic function of P_h is

$$(3.11)\qquad \int \exp\{i\langle z,y\rangle_B\}P_h(dy) = \int \exp\{i\langle z,y+\tau h\rangle_B\}P_0(dy)$$
$$= \exp\{i\langle z,\tau h\rangle_B - \frac{1}{2}|\tau^* z|^2\}$$

On the other hand, the characteristic function of $\exp\{L_h - \frac{1}{2}|h|^2\}dP_0$ is

$$(3.12)\qquad \int \exp\{i\langle z,x\rangle_B + L_h(x) - \frac{1}{2}|h|^2\}P_0(dx)$$
$$= \exp\{-\frac{1}{2}|h|^2\}\int \exp\{i\langle z,x\rangle_B + L_h(x)\}P_0(dx)$$

The last integral is of the form $E\exp\{iX_1 + X_2\}$ where (X_1, X_2) is a bivariate Normal rv with mean 0, covariance $\langle\tau^* z, h\rangle$ because of the definition of L_h. The equality of (3.11), (3.12) now follows by direct calculation of $E\exp\{iX_1 + X_2\}$. Q.E.D.

(3.13) EXAMPLE (Classical abxtract Wiener space). In this example L_h has traditionally been given a different representation. Referring to (V.2.9), let $h = \tau^* m$, $m \in B^*$, so $(\tau^* m)(t) = m\{[t,1]\}$. If m is a simple measure with mass a_i at t_i, $0 \le t_1 < \dots < t_k \le 1$, then $\tau^* m = h$ is a step function. In this case, for $x \in B = C[0,1]$, direct calculation yields

$$L_h(x) = \int x(t)m(dt)$$
$$= \int h(t)dx(t)$$

where, as usual, if h is any step function ($h(x) = b_i$ on $(s_i, s_{i+1}]$, $0 \le s_1 < \dots < s_k = 1$), $\int h(t)dx(t) = \sum h(s_i)[x(s_{i+1}) - x(s_i)]$. The representation of L_h just made for $h = \tau^*m$ (i.e. h a step function) can be extended to arbitrary m by taking a sequence m_n of simple measures that converges weakly to m (use the L^2 convergence argument given in the general construction of L_h). If $h = \tau^*m$, L_h so constructed is still denoted formally by

$$L_h(x) = \int h(t)dx(t)$$

the stochastic integral (Wiener integral). It is defined for all $h \in L^2[0,1]$ (even though $t \to x(t)$ is not absolutely continuous so $dx(t)$ does not have its usual meaning).

In terms of processes, the result (3.9) says the following in the present example. If $\{X_t\}$ is Brownian motion, then P_h is the distribution on $C[0,1]$ of the process $X_t + \int_0^t h(s)ds$. P_h is absolutely continuous with respect to P_0 and its derivative is

$$dP_h/dP_0 = \exp\{\int h(s)dX(s) - \tfrac{1}{2}|h|^2\} .$$

VI. Optimality theory for Gaussian shift experiments

This chapter provides examples of experiments $\{P_h^n, h \in H\}$, H a Hilbert space, which converge to the Gaussian shift experiment $\{P_h, h \in H\}$ of V.3. In order to apply the asymptotic minimax theorem to such convergent families, the minimax risk in the limit experiment needs to be computed; this is done in section 2 below. Section 3 presents a version of the convolution theorem (cf. III.2) appropriate to experiments $\{P_h^n\}$ converging to a Gaussian shift experiment.

1. Convergence to a Gaussian shift experiment

This section presents two examples of experiments converging to the Gaussian shift experiments; these examples will find application in Chapters VII, VIII, IX and XI.

(1.1) EXAMPLE. Let f be a density on the line; let H be the collection of functions h such that $\int h(s)f^{1/2}(s)ds = 0$, $\int h^2(s)ds < \infty$, support $h \subset$ support f, and define τ as in (V.2.10a) and B as in Example V.2.10. Define, for $h \in H$, the density $f(h,h;x)$ by

$$(1.2) \qquad f^{1/2}(n,h;x) = (1-|h|^2/4n)^{1/2}f^{1/2}(x) + h(x)/2n^{1/2} .$$

This definition gives a density if $|h|^2 \leq 4n$; define $f(n,h)$ arbitrarily if $|h| > 4n$. Define product measures P_h^n on $\mathbb{R}^n$ by

$$(1.3) \qquad P_h^n(dx) = \prod_1^n f(n,h;x_i)dx_i \ , \quad x = (x_1,\ldots,x_n)$$

By Example IV.1.12 and the asymptotic expansion for qmd families (Chapter IV), it is immediate that under P_0^n

$$(1.4) \qquad \log dP_h^n(x)/dP_0^n(x) = n^{-1/2}\sum[h(x_i)/f^{1/2}(x_i)] - \frac{1}{2}\int h^2(s)ds + o_p(1)$$

for each fixed h. Let $L_h^n = n^{-1/2}\sum_1^n h(x_i)/f^{1/2}(x_i)$. The central limit theorem and the fact that the expansion (1.4) is quadratic in h show that if $h_1,\ldots,h_k$ are fixed elements of H, then $(L_{h_1}^n,\ldots,L_{h_k}^n)$ converges under P_0^n to a multivariate normal distribution with mean 0, covariance $\langle h_i,h_j\rangle_H$. It follows from II.2.3, V.3.4, V.3.9 that *the experiments* $E^n = \{P_h^n, h\in H\}$ *converge to* $E = \{P_h, h\in H\}$.

(1.5) EXAMPLE. Fix a density f on the line. Let (τ,H,B) be the abstract Wiener space of Example V.2.12. For $h\in H$ satisfying $\sup_t |h(t)| < n$, define a density

$$f(n,h;x) = f(x)[1+n^{-1/2}h(x)] \tag{1.6}$$

and define product measures P_h^n on $\mathbb{R}^n$ by

$$P_h^n(dx) = \prod_1^n f(n,h;x_i)dx_i\ , \quad x=(x_1,\ldots,x_n)$$

By Example IV.1.13 and the asymptotic expansion of qmd families in Chapter IV, it is immediate that

$$\sum \log dP_h^n/dP_0^n = n^{-1/2}\sum h(x_i) - \frac{1}{2}\int h^2(x)f(x)dx + o_p(1) \tag{1.7}$$

under P_0^n. Let Λ be any subset of H such that if $h\in\Lambda$, then $\sup_t |h(t)| < \infty$. Let $E^n = \{P_h^n, h\in\Lambda\}$; let $E = \{P_h, h\in\Lambda\}$ where $\{P_h, h\in H\}$ is the standard normal shift experiment of (τ,H,B). As in Example VI.1.1, the central limit theorem, V.3.4, V.3.9 show that E^n *converges to* E.

2. Minimax risk for a Gaussian shift experiment

For this section, suppose given an abstract Wiener space (τ,H,B) and let $\{P_h, h\in H\}$ be its Gaussian shift experiment. The decision theoretic problem shall consist in taking B as decision space and if ℓ is a real function on B, the loss will be $\ell(x-\tau h)$ if P_h is 'true' and $x \in B$ is chosen from the decision space. In this framework we wish to compute the minimax risk. This is the abstract problem appropriate for applications to estimation; the analogous development for testing problems will not be discussed in these notes (see Millar (1982c)). The natural condition to put on the loss function ℓ just mentioned is the following.

(2.1) DEFINITION. A non-negative function ℓ on a separable Banach space B is *subconvex* if it satisfies

(i) $\ell(x) \geq 0$

(ii) $\ell(x) = \ell(-x)$

(iii) for each c, $\{x \in B: \ell(x) \leq c\}$ is closed, convex.

Of course, ℓ is then lower semi-continuous.

(2.2) EXAMPLES. If $| \ |_B$ is the norm of B, let g be any non-decreasing lower semi-continuous function on $[0,\infty)$. Define $\ell(x) = g(|x|_B)$. Then ℓ is subconvex. In particular the indicator of $\{x: |x| > c\}$ and $x \to |x|^\alpha$ $(\alpha > 0)$ are both subconvex.

It is easy to see that if ℓ is subconvex, so is $\min\{\ell, a\}$ for any non-negative number a. If ℓ_1, ℓ_2 are subconvex and if $\ell_1(x) \leq \ell_2(x)$, then $\ell_1 + \ell_2$ is subconvex; in general the sum of two subconvex functions will not be subconvex. If ℓ_n is subconvex for all n and if $\ell_n \uparrow \ell$, then ℓ is subconvex. A further useful technical fact is the following.

(2.3) PROPOSITION. *If* ℓ *is subconvex, there is a sequence* ℓ_n *of bounded, uniformly continuous, subconvex cylinder functions, such that* $\ell_n \uparrow \ell$.

A proof of this can be constructed as follows. For a sequence $a = \{a_0 = 0 < a_1 < a_2 < \cdots < a_n\}$, define $\ell_a(x) = a_k$ if $a_k < \ell(x) \le a_{k+1}$. Then $\ell_a(x) = \sum a_k I\{x: a_k < \ell(x) \le a_{k+1}\} = \sum (a_k - a_{k-1}) I\{x: \ell(x) > a_k\}$ is subconvex, $\ell_a \le \ell$. Since $I\{x: \ell(x) > a_k\}$ is the indicator of the complement of a closed convex set, it can be approximated from below by the indicator of a *cylinder* convex set, since any convex set in B is the intersection of hyperplanes containing it. Since the indicator of an open set A can be approximated from below by the uniformly continuous function $1 - \exp\{-nd(x, A^c)\}$, d being the metric of B, the result follows.

The minimax result for general H will be deduced from the special case $H = \mathbb{R}^d = B$.

(2.4) PROPOSITION. *Let* $H = \mathbb{R}^d = B$, τ = *identity, so* $\{P_h, h \in H\}$ *is the standard normal shift experiment of* $\mathbb{R}^d$. *If* ℓ *is subconvex on* B *and* $\int \ell(x) P_0(dx) < \infty$, *then*

$$\inf_b \sup_h \iint \ell(x-h) b(y,dx) P_h(dx) = \int \ell(x) P_0(dx)$$

REMARK. It is not hard to deduce that all procedures for this problem are given by Markov kernels. The minimax procedure is then $b(y;dx)$ = unit mass at $\{y\}$. If $X_1, \ldots, X_n$ are i.i.d. P_h and we wish to estimate $h \in \mathbb{R}^d$, then the theory of sufficient statistics lets us use only procedures based on $\bar{X}$, the sample mean; the result (2.4) then shows that $\bar{X}$ is minimax for any subconvex loss function. For less general loss functions, this result is in many common texts, e.g. those of Bickel-Doksum and Ferguson.

PROOF. The proof will be sketched only and will be based on the following well known decision theoretic fact:

(2.5) FACT. Suppose b_0 is any procedure whose risk function $\rho(\theta,b_0) \equiv c$ is identically constant. If μ_n is a sequence of probabilities on Θ and if b_n is μ_n-Bayes, with Bayes risk $\rho(\mu_n,b_0)$ converging to c, then b_0 is minimax.

See the texts of Bickel-Doksum, Ferguson for proofs. Now take μ_n in the present case to be $N(0,\sigma_n^2 I)$ where $\sigma_n^2 \uparrow \infty$. The posterior distribution of θ, given x, under this prior μ_n, is $N(\sigma_n^2 x(1+\sigma_n^2)^{-1}, \sigma_n^2(1+\sigma_n^2)^{-1} I)$. Let $n(x,\sigma_n^2;dy)$ denote the density of this posterior distribution. By the usual argument involving conditional expectation, the μ_n-Bayes procedure is the one that, when x is observed, picks $y \in \Theta = \mathbb{R}^d$ that minimizes the posterior risk; i.e., the function of y given by

$$(2.6) \qquad \int \ell(\theta-y)n(x,\sigma_n^2;d\theta) = \int \ell(\theta-(y-x\sigma_n^2(1+\sigma_n^2)^{-1}))n(0,\sigma_n^2;d\theta)$$

To find this y, use Anderson's lemma (PAMS 6, 1970-76, 1955):

(2.7) *If* f *is a probability density on* $\mathbb{R}^d$ *such that* $\{x: f(x) \geq a\}$ *is convex, symmetric for every* $a > 0$, *and if* c *is a convex, symmetric set in* $\mathbb{R}^d$, *then for every* y,

$$\int_c f(z+y)dz \leq \int_c f(z)dz$$

If P_n is the $N(0,\sigma_n^2(1+\sigma_n^2)^{-1} I)$ distribution, then (2.6) can be continued

$$= \int_0^\infty P_n\{\theta: \ell[\theta - (y - x\sigma_n^2(1+\sigma_n^2)^{-1})] > t\}dt$$

$$\geq \int_0^\infty P_n\{\theta: \ell(\theta) > t\} \quad \text{(Anderson's lemma)}$$

$$= E\,\ell(Z_n)$$

where Z_n has distribution P_n. It follows that the procedure $x \to x\sigma_n^2(1+\sigma_n^2)^{-1}$ is μ_n-Bayes. Its risk $E\ell(Z_n)$ is easily shown to converge to EZ, $Z \sim N(0,I)$; indeed Z_n converges in distribution to Z, so a proof can be based on the hypothesis that $\int \ell(x)P_0(dx) < \infty$ and the fact that the densities of P_n are bounded above by a multiple of P_0.

Here is the main result of this section.

(2.8) THEOREM. *Let* (τ,H,B) *be an abstract Wiener space,* $\{P_h, h \in H\}$ *its Gaussian shift experiment. Let the decision space* $D = B$; *let* ℓ *be subconvex on* B, $\int \ell(x)P_0(dx) < \infty$. *Then*

(2.9) $$\inf_b \sup_h \int \ell(x-\tau h)b(y,dx)P_n(dy) = \int \ell(x)P_0(dx)$$

(2.10) REMARK. In Example V.2.10, where P_0 is the distribution of $W^0 \circ F$, W^0 = Brownian Bridge, the last integral will often be written

$$\int \ell(x)P_0(dx) = E\ell(W^0 \circ F)$$

PROOF. Let m equal the left side of display (2.9). By taking $b_0(x,dy)$ to be unit mass at $\{x\}$, it is obvious that

$$m \leq \sup_h \int \ell(y-\tau h)b_0(x,dy)P_h(dx)$$

$$= \int \ell(x-\tau h)P_h(dx)$$

$$= \int \ell(x)P_0(dx)$$

Therefore, it is necessary only to prove $m \geq \int \ell(x)P_0(dx)$. Because of (2.3a) and the dominated convergence theorem, we can assume without loss of generality that ℓ is a cylinder subconvex function on B, based on $m_1,\ldots,m_{k_0}$ in B^*. Using the Gram-Schmidt orthogonalization procedure (and trivially changing the definition of ℓ, if necessary), we can choose the m_i so that $e_i = \tau^* m_i$, $1 \leq i \leq k_0$, is an orthonormal set in H. Complete $\{e_i\}$ to form an orthonormal basis of H. For $k \geq k_0$, let $H_k = \text{span}\{e_i, 1 \leq i \leq k\}$ and let A_k be the collection of all procedures with decision space τH_k (instead of $\overline{\tau H}$). Then for each $k \geq k_0$

(2.11)
$$m \geq \inf_b \sup_{h \in H_k} \iint \ell(x-\tau h) b(y,dx) P_h(dy)$$
$$\geq \inf_{b \in A_k} \sup_{h \in H_k} \iint \ell(x-\tau h) h(y,dx) P_n(dy)$$

Indeed, the first inequality in (2.11) holds because $H_k \subset H$. The second inequality holds because $k \geq k_0$ and ℓ is a cylinder function based on $m_1,\ldots,m_{k_0}$. To see this, suppose you choose $x \in \tau H_k^{\perp}$ ($\perp$ denoting orthocomplement) so $x = \tau q$, $q \perp e_1,\ldots,e_k$; then the loss is

$$\begin{aligned}\ell(\tau q-\tau h) &= \ell(\langle m_1,\tau q-\tau h\rangle,\ldots,\langle m_{k_0},\tau q-\tau h\rangle)\\ &= \ell(\langle e_1,q-h\rangle,\ldots,\langle e_{k_0},q-h\rangle)\\ &= \ell(\langle e_1,-h\rangle,\ldots,\langle e_{k_0},-h\rangle)\end{aligned}$$

That is, if $x \in \tau H_k^{\perp}$ is chosen from the decision space, then you lose exactly as much as you would if you chose 0, which is in τH_k; in particular, because of the nature of ℓ, there is no advantage in going outside the decision space τH_k. (A completely rigorous discussion of this last point makes use of II.1.8.)

Next, define a real valued function ℓ_0 on $\mathbb{R}^k$ by

$$\begin{aligned}(2.12)\qquad \ell_0(a) &= \ell_0((a_1,\ldots,a_{k_0}))\\ &= \ell(a_1\langle e_1,e_1\rangle,\ldots,a_{k_0}\langle e_{k_0},e_{k_0}\rangle)\end{aligned}$$

where $a = (a_1,\ldots,a_{k_0},\ldots,a_k)$; ℓ_0 is *subconvex* on $\mathbb{R}^k$. If $h = \sum_i^k e_i a_i$ and $a = (a_1,\ldots,a_k)$, it is clear that

$$(2.13)\qquad \ell(\tau h) = \ell_0(a)$$

Because of V.3, if $h = \sum_i^k a_i e_i$, then

$$(2.14)\qquad dP_h/dP_0 = \exp\{\textstyle\sum a_i Le_i - \frac{1}{2}\sum a_i^2\}$$

and since the e_i are orthonormal, $\{Le_i\}$ is an i.i.d. $N(0,1)$ sequence. Let Q_a be the $N(a,I)$ distribution on $\mathbb{R}^k$. Then dQ_a/dQ_0 has the same structure as (2.14). Finally, note that if b is a procedure in A_k, it chooses points $x \in B$ of the form $x = \sum_{i=1}^k a_i \tau e_i$; since τ is 1-1, the a_i are uniquely determined, and so such a procedure may be regarded as choosing points $a = (a_1,\ldots,a_k) \in \mathbb{R}^k$.

Taking the considerations of the preceding paragraph into account, it is immediate that

$$\inf_{b\in A_k}\ \sup_{h\in H_k} \int \ell(x-\tau h)b(y,dx)P_h(dy)$$

is precisely the same as the minimax risk in the estimation of the mean in the Gaussian shift family of $\mathbb{R}^k$:

$$\begin{aligned}&= \inf_b\ \sup_{a\in\mathbb{R}^k} \int \ell_0(x-a)b(y,dx)Q_a(dy)\\ &= \int \ell_0(x)Q_0(dx)\end{aligned}$$

using (2.4) for the last equality. Notice that we needed (2.4) with a *general* subconvex loss function to carry out the argument. Reference to the definition of ℓ_0 shows that this last quantity is

$$= E\,\ell\Big(\sum_1^k X_i \tau e_i\Big)$$

where the X_i are i.i.d., $N(0,1)$. Since $\sum X_i \tau e_i$ converges in distribution to P_0 (cf. V.2.6), the last display on letting $k \to \infty$ is equal to

$$\int \ell(x) P_0(dx) \qquad \text{Q.E.D.}$$

The following simple extension of (2.4) is needed in Chapter VII. Let P_θ be the normal distribution on $\mathbb{R}^d$ with mean vector θ and covariance matrix Σ; assume Σ nonsingular.

(2.14) PROPOSITION. *If* ℓ *is subconvex on* $\mathbb{R}^d$ *and* $\int \ell(x)P_0(dx) < \infty$, *then*

$$\inf_b \sup_\theta \iint \ell(x-\theta) b(y,dx) P_\theta(dx) = \int \ell(x) P_0(dx) = E\ell \circ \Sigma^{1/2}(Z)$$

where Z *is* $N(0,I)$.

PROOF. This may be proved using the method of (2.4). Alternatively, it may be derived from (2.4) as follows. Let Q_θ be the normal distribution with mean vector $\Sigma^{-1/2}\theta$, covariance the identity. Then $\{Q_\theta\}$, $\{P_\theta\}$ are equivalent (cf. II.3.5) and hence have the same minimax risk. Let Q'_θ be $N(\theta,1)$. Then using non-randomized procedures for notational simplicity, we find the minimax risk of $\{Q_\theta\}$ satisfies

$$\inf_T \sup_\theta \int \ell(T-\theta)dQ_\theta = \inf_T \sup_\theta \int \ell \circ \Sigma^{1/2}(T-\theta)dQ'_\theta$$

Since $\ell \circ \Sigma^{1/2}$ is subconvex on $\mathbb{R}^d$, the second expression in the last display is $\int \ell \circ \Sigma^{1/2}(x) dQ_0'(x)$ by (2.4) and this implies the desired result.

3. Convolution theorem for Hilbertian parameter set

Let (τ, H, B) be an abstract Wiener space. Let $E^n = \{P_h^n, (S^n, \mathcal{S}^n), h \in H\}$ be a sequence of experiments *converging* to $\{P_h, h \in H\}$, the standard Gaussian shift experiment for (τ, H, B). Let T_n, R_n be B-valued statistics for E^n. Denote by $P_h^n(T_n \in dy)$ the distribution of T_n (on B) when P_h^n is the measure on $(S_n, \mathcal{S}_n)$. Assume

(3.1) $P_h^n(T_n \in dy)$ converges weakly (on B) to $P_h(dy)$ for all h

(3.2) $P_h^n(R_n \in dy)$ converges weakly to $G_h(dy)$ for all h

where, for each $h \in H$, G_h is a probability on the Borel sets of B. Assume further

(3.3) $$G_h(A) = G_0(A - \tau h)$$

for every Borel set A of B. Under these assumptions, one may formulate an extension of the Hajek-Le Cam convolution theorem of Chapter III.

(3.4) THEOREM. *Assume (3.1)-(3.3). Then there is a probability* μ *on* B *such that*

$$G_0 = \mu * P_0$$

where $*$ *denotes convolution of measures.*

(3.5) REMARK. Roughly speaking, the theorem addresses this situation: you are supposed to estimate something in B and you are given two estimators (after n "observations") T_n and R_n; and the question is:

which estimator should you use? The statistical content of the theorem is: since convolution "spreads out mass", R_n is asymptotically more "sloppy" than T_n, so you should use T_n which is more precise. This thought is expressed more explicitly in the contexts of estimating a cdf (Chapter VIII) and of estimating a $\theta \in \mathbb{R}^d$ for some qmd family (Chapter VII); for the original motivations, you should read the paper of Hajek (1968).

(3.6) REMARK. The proof of Theorem (3.4) will be given in outline only; further amplification can be found in the recent work of Millar (1982) which contains many more applications. In particular, it is important for applications that the Hilbert space H above can be replaced by a dense subset of H_0; if the hypotheses are reformulated for H_0 instead of H, then the results continue to hold. The utility of the aforementioned technical modifications are detailed in Millar (1982). Further refinement would have T_n converging under P_h^n to a shift family F_h, T_h distinguished, and $\{F_h\}$ equivalent to $\{P_h\}$.

PROOF OF (3.4) (outline). The general Hajek-Le Cam theorem of III.2 can be reformulated for any locally compact group admitting invariant means; local compactness and existence of "Haar measure" is essential in the argument. Unfortunately an infinite dimensional Hilbert space is not locally compact (nor is there Haar measure), so one cannot routinely apply the arguments outlined in III.2.

Let $\{e_i\}$ be an orthonormal basis of H where $e_i = \tau^* m_i$ for some $m_i \in B^*$; since $\tau^* B^*$ is dense in H, such a basis exists. Let C_k be the cylinder sets based on $m_1, \ldots, m_k$, so $A \in C_k$ implies there is a Borel set $A_0 \in \mathbb{R}^k$ such that

$$\{x: (\langle m_1,x\rangle,\ldots,\langle m_k,x\rangle)\in A_0\} = A$$

For such a k fixed, let $h = \sum_1^k a_i e_i$; set $a = (a_1,\ldots,a_k)$ and let $Q_a^{(k)}$ denote the normal distribution on $\mathbb{R}^k$ with mean vector a, covariance the identity. A simple computation (which uses the fact that $e_i = \tau^* m_i$) shows that

$$P_h(A) = Q_a^{(k)}(A_0)$$

Similarly, let $G_h^{(k)}$ denote the probability on $\mathbb{R}^k$ given by $G_h^{(k)}(A_0) = G_h(A)$, with A, A_0 as above; of course, $G_h^{(k)}(A_0) = G_0^{(k)}(A_0-a)$ if $h = \sum_1^k a_i e_i$, $a = (a_1,\ldots,a_k)$. Finally, define the random variable $T_{n,k}$ with values in $\mathbb{R}^k$ by

$$(\langle m_1,T_n(m)\rangle,\ldots,\langle m_k,T_n(m)\rangle) ;$$

similarly define $R_{n,k}$. Then for each fixed k, the system $\{R_{n,k};T_{n,k};G_a^{(k)}Q_a^{(k)}\}$ satisfies the conditions of the classical Hajek convolution theorem (III.2.10) for measures on $\mathbb{R}^k$. Hence there exists a probability μ_k on $\mathbb{R}^k$ such that

$$G_0^{(k)} = \mu_k * Q_0^{(k)}$$

Of course μ_k may be defined on C_k by defining the measure of $A\in C_k$ to be $\mu_k(A_0)$ if $A = \{(\langle m_1,x\rangle,\ldots,\langle m_k,x\rangle)\in A_0\}$. In short, for $m_1,\ldots,m_k$ fixed, we see that for each $A\in C_k$ there is a measure μ_k on C_k such that

$$G_0(A) = \mu_k * P_0(A)$$

It is easy to see that μ_k is consistently defined on $C_0 \equiv \cup C_k$ so there is a cylinder measure μ on C_0 such that $G_0(A) = \mu * P_0(A)$, all $A\in C_0$.

The sigma field generated by C_0 is the Borel sigma field of B. On the other hand, G_0, P_0 are both countably additive on the Borel sigma field; it follows by a standard argument that μ must also be. For details, see Millar (1982).

VII. Classical parametric estimation

For this chapter, Θ will be an open subset of $\mathbb{R}^d$ and $\{P_\theta, \theta\in\Theta\}$ a quadratic mean differentiable family (see IV.1 for the definition). Let $X_1,\ldots,X_n$ be i.i.d. observations--each X_i having P_θ as distribution, but θ is "unknown". The statistical problem is now to estimate θ "efficiently". This chapter shows first of all (VII.1) that there are no estimators that converge to θ at rate faster than $n^{1/2}$; the argument given is elementary, while a more general approach is described in Chapter XII. Next, section 2 describes two statistical notions of efficient estimation: the LAM property and a concept depending on the notion of convolution. Section 3 introduces the 1-step maximum likelihood estimator and shows that it is efficient in the LAM sense, defined in section 2. In section 3, the same estimator is shown to be efficient in the "convolution" sense; there, to show the additional possibilities, an efficiency result is also given for a common parametric family *not* satisfying the qmd property.

1. Estimation at rate $n^{1/2}$

This section shows that in a qmd family the parameter θ cannot be estimated at a rate faster than $n^{1/2}$. More precisely, let $\{P_\theta, \theta\in\Theta\}$ be a qmd family, described in the first paragraph of this section; let P_θ^n denote the n-fold product measure of P_θ. That is, P_θ^n is a probability on $(\mathbb{R}^d)^n$ such that $(x_1,\ldots,x_n) \in (\mathbb{R}^d)^n$ is a random vector with each component independent, distributed according to P_θ. For each n, let T_n be an *estimator* of θ: T_n is a measurable function from $(\mathbb{R}^d)^n$ to Θ.

(1.1) DEFINITION. A sequence of estimators $\{T_n\}$ is $n^{1/2}$-*consistent* if, for each $c > 0$ and each θ_0,

$$\lim_{a\uparrow\infty} \lim_{n} \sup_{\theta\in A(n,c)} P_\theta^n\{|T_n-\theta| > a\} = 0$$

where $A(n,c) = \{\theta\in\Theta: n^{1/2}|\theta-\theta_0| \le c\}$.

Roughly speaking, a $n^{1/2}$-consistent sequence of estimators can "pin down" the unknown θ with an accuracy of $n^{-1/2}$, approximately. Chapter X provides a general theorem that shows that, in qmd families, it is quite easy to produce a sequence of $n^{1/2}$-consistent estimators. To put this in perspective, it is necessary to know that one cannot find a sequence $\{T_n\}$ that estimates θ "faster" than $n^{1/2}$:

(1.2) PROPOSITION. *Let* $\{P_\theta, \theta\in\Theta\}$ *be a qmd family,* Θ *an open subset of* $\mathbb{R}^d$. *Fix* $\theta_0 \in \Theta$. *It is impossible that there exists a sequence of estimators* T_n *such that*

$$\lim_{n} \sup_{|\theta-\theta_0|\le n^{-1/2}} P_\theta^n\{n^{1/2}|T_n-\theta| > \varepsilon\} = 0$$

for every $\varepsilon > 0$.

PROOF. We give a simple direct argument; a general argument based on the asymptotic minimax theorem will be given in Chapter XII.

If $\zeta(P,Q)$ is the Hellinger distance between probabilities P, Q (cf. IV.1.7) and if P^n, Q^n are products of P, Q, then $\zeta^2(P^n,Q^n) = 2[1-(1-\frac{\zeta^2(P,Q)}{2})^n]$. Take $\theta_n = \theta_0 + \theta n^{-1/2}$; then $\zeta^2(P_{\theta_n},P_{\theta_0})$ is of order n^{-1} because of qmd. It follows from the result of the previous sentence with $P = P_{\theta_0}$, $Q = P_{\theta_n}$ that $\zeta(P_{\theta_n}^n,P_{\theta_0}^2)$ remains bounded away from 0 as n increases. For any probabilities P, Q, the L_1 distance $|P-Q|_1 = \sup_{\phi:\ -1\le\phi\le 1} |\int \phi d(P-Q)| = \int |\frac{dP}{d\mu} - \frac{dQ}{d\mu}| d\mu$, $\mu = \frac{P+Q}{2}$ satisfies

$$\zeta^2(P,Q) \le |P-Q|_1 \le 2\zeta(P,Q)$$

so the L^1 distance between $P^n_{\theta_n}$, $P^n_{\theta_0}$ stays bounded away from 0. Now consider testing between $P^n_{\theta_0}$ and $P^n_{\theta_n}$; if ϕ is any test, the sum of the error probabilities is

$$\int \phi dP^n_{\theta_0} + \int (1-\phi) dP^n_{\theta_n} = 1 - \int \phi d(P^n_{\theta_n} - P^n_{\theta_0})$$

so the smallest the sum of the errors can ever be is

$$1 - \frac{1}{2}|P^n_{\theta_n} - P^n_{\theta_0}|_1 = \inf_\phi \{\int \phi dP^n_{\theta_0} + \int (1-\phi) dP^n_{\theta_n}\}$$

Because of the foregoing estimates on $|P^n_{\theta_n} - P^n_{\theta_0}|_1$, it follows that the sum of errors in testing $P^n_{\theta_0}$ vs. $P^n_{\theta_n}$ can never go to zero as $n \to \infty$. On the other hand, for a sequence of estimators T_n, consider the tests ϕ_n which accept θ_0 iff T_n is closer to θ_0 than to θ_n; if T_n satisfies the condition of the proposition, then the sum of errors for these tests go to zero as $n \to \infty$. This contradiction completes the proof.

2. Efficiency and the LAM property

Again fix a qmd family $\{P_\theta, \theta \in \Theta\}$, let P^n_θ be the product measure of P_θ, and suppose the problem is to estimate θ. In many such problems, it happens that commonly employed estimators T_n satisfy

(2.1) $n^{1/2}(T_n - \theta)$ converges in distribution, under P^n_θ, to a multivariate normal distribution $N(0, K(\theta))$

For ease of exposition, suppose now that θ is real. A naive claim (due to Fisher), that could be based on considerations involving the information inequality, asserts that *all* asymptotically normal estimators of θ should satisfy $K(\theta) \geq I(\theta)^{-1}$, where $I(\theta)$ is the Fisher information (cf. IV.2.3). If this claim were reasonable, then one could naturally define a sequence of estimators $\{T_n\}$ to be *efficient* if the asymptotic variance $K(\theta)$ was as small as possible, i.e., $K(\theta) = I(\theta)^{-1}$.

Unfortunately there are difficulties with this classical definition of efficiency. First of all, the definition excludes from consideration all estimators that are not asymptotically normal. There is no convincing reason for doing this. In the second place, Fisher's conjecture is false: there are asymptotically normal estimators whose asymptotic variance is less than $I(\theta)^{-1}$.

(2.2) HODGES EXAMPLE. Let P_θ^n be the n-fold product measure of $N(\theta,1)$. Set $\bar{X}_n = (X_1+\cdots+X_n)/n$ and define $T_n = \bar{X}_n$ if $|\bar{X}_n| > n^{-1/4}$, $T_n = 0$ if $|\bar{X}_n| \leq n^{-1/4}$. The asymptotic variance of $n^{1/2}(T_n-\theta)$ is 1 if $\theta \neq 0$ and is 0 if $\theta = 0$. This asymptotic variance is less than $I(\theta)^{-1} \equiv 1$ everywhere and strictly so at $\theta = 0$.

Of course no one would really use an estimator like T_n in Hodges example. There are two basic ways to exclude such unreasonable estimators and thereby rescue the intent of the faulty definition of efficiency proposed above.

One method is to note that, in Hodges example, the convergence of $n^{1/2}(T_n-\theta)$ to its asymptotic limit is rather erratic in θ; if the "rate" at which an estimator converges to its limit varies drastically with θ (which is unknown), then one cannot use asymptotic theory to help make inference on θ. Therefore, one proposal is to insist that all estimators under consideration enjoy a certain uniformity in their convergence to the limit. This leads to the notion of a "regular" estimator and the convolution theorem as a tool for defining efficiency. We discuss this in section 4.

Another approach to the difficulties arising from Hodges example is to insist that one use only estimators that enjoy a small amount of decision theoretic optimality--in particular, a local asymptotic minimax (LAM) property. As we shall see this also excludes unreasonable estimators like the Hodges example, but it does so indirectly. Experience suggests that very few--if any--estimators judged good by practicing statisticians fail to have a certain LAM property. We turn now to define this concept for the present problem of estimating θ in a qmd family.

Let L be a subconvex function on $\mathbb{R}^d$. Because of (1.2), it is reasonable to specify the loss, when estimator of θ, T_n is used by

(2.3) $$L(n^{1/2}(T_n-\theta))$$

if P_θ^n is the true distribution of the data $x_1,\ldots,x_n$. This leads to the following definition.

(2.4) DEFINITION. A sequence T_n' of estimators is *locally asymptotically minimax* (LAM) if for each $\theta_0 \in \Theta$

$$\lim_{c\uparrow\infty} \liminf_{n} \inf_{T_n} \sup_{\theta\in A(n;c;\theta_0)} \int L(n^{1/2}(T_n-\theta))dP_\theta^n$$

$$= \lim_{c} \lim_{n} \sup_{\theta\in A(n;c;\theta_0)} \int L(n^{1/2}(T_n'-\theta))dP_\theta^n$$

Here $A(n;c;\theta_0) = \{\theta\in\Theta: |\theta-\theta_0| \le cn^{-1/2}\}$ and the infimum is computed over all the estimators T_n available in the experiment $\{P_\theta^n, \theta\in\Theta\}$.

Technically, this definition depends on L; as we shall see if T_n' is LAM for one L it is for most others (e.g. all bounded L).

We may now apply the results of Chapters V, VI to get a basic asymptotic minimax lower bound. Let

(2.5) $\Gamma(\theta)$ be the matrix defined in IV.2.1

(2.6) THEOREM. *Assume* $\Gamma(\theta_0)$ *is nonsingular. If* L *is any subconvex function on* $\mathbb{R}^d$*, then*

$$\lim_{c\uparrow\infty} \lim_n \inf_{T_n} \sup_{\theta\in A(n;c;\theta_0)} \int L(n^{1/2}(T_n-\theta))dP_\theta^n \ge \int \ell(x)dP_0(x)$$

where P_0 *is* $N(0,\frac{1}{4}\Gamma(\theta_0)^{-1})$.

(2.7) REMARKS. The usual way to show that a sequence T_n' of estimators is LAM is the following. Let $\theta_n = \theta_0 + \theta_{1n}/n^{-1/2}$ where θ_{1n} is any sequence satisfying $|\theta_{1n}| \le c$. One then shows that under $P_{\theta_n}^n$, $n^{1/2}(T_n'-\theta_n)$ converges in distribution to P_0. This will indeed prove T_n' LAM (at θ_0) since for fixed c, one can choose θ_n to achieve $\sup_{\theta\in A(n;c;\theta_0)} \int L(n^{1/2}(T_n'-\theta))dP_\theta^n$. If Θ is real (say) and $\ell(x) = x^2$, then $\int \ell(x)dP_0(x)$ is the asymptotic variance of $n^{1/2}(T_n'-\theta_0)$ (if T_n' is LAM); so we see in this particular case that a LAM estimator will have the smallest asymptotic variance. In addition the LAM estimator will enjoy mild but desirable theoretic optimality properties--the Hodges example will not be LAM, as is easy to see. Motivated by these remarks, one is led to the very reasonable thought of *defining efficiency* as the LAM property (2.4).

PROOF. It will suffice to establish the result for bounded L; if L is not bounded, replace it by $\min\{L,a\}$ (which is still subconvex) and then let $a \uparrow +\infty$. Define experiments $E_c^n = \{Q_\theta^n: |\theta| \leq c\}$ where $Q_\theta^n = P^n_{\theta_0+\theta n^{-1/2}}$. These converge to $E = \{P_\theta, |\theta| \leq c\}$ where P_θ is $N(\theta, \frac{1}{4}\Gamma(\theta_0)^{-1})$, as discussed in VI.1. If $\theta \in A(n;c;\theta_0)$, $(T_n-\theta)n^{1/2} = \tilde{T}_n - \tilde{\theta}$ where $\tilde{T}_n$ is the estimator $\tilde{T}_n = (T_n-\theta_0)n^{1/2}$ and $\tilde{\theta} = (\theta-\theta_0)n^{-1/2}$. This being so, the asymptotic minimax theorem and a change of notation yield

$$\begin{aligned} &\liminf_n \inf_{T_n} \sup_{\theta\in A(n;c;\theta_0)} \int L(n^{1/2}(T_n-\theta))dP_\theta^n \\ &= \liminf_n \inf_{T_n} \sup_{|\theta|\leq c} \int L(T_n-\theta)dQ_\theta^n \\ &\geq \inf_b \sup_{|\theta|\leq c} \int L(x-\theta)b(y,dx)P_\theta(dy) \end{aligned}$$

Next, let $c \uparrow \infty$. Presumably, this should replace the $|\theta| \leq c$ set in the last display by $\theta \in \mathbb{R}^d$; appeal to VI.2.14 would then complete the proof. To carry this out rigorously, let b be any (generalized) procedure for the experiment $\{P_\theta\}$ and $\rho(\theta,b)$ its risk. Then using III.1.5

$$\begin{aligned} \inf_b \sup_{|\theta|\leq c} \int L(x-\theta)b(y,dx)P_\theta(dy) &= \inf_b \sup_{|\theta|\leq c} \rho(\theta,b) \\ &= \sup_{\mu\in M_c} \inf_b \rho(\mu,b) \end{aligned}$$

where the supremum is taken over M_c, the set of all simple probabilities with finite support in the set $\{\theta: |\theta| \leq c\}$; since M_c is convex, III.1.5 indeed applies. Of course, $M_c \uparrow M = \{\text{all simple probabilities}\}$ as $c \uparrow \infty$, so

$$\sup_{\mu\in M_c}\inf_b \rho(\mu,b) + \sup_{\mu\in M}\inf_b \rho(\mu,b) = \inf_b \sup_{\mu\in M} \rho(\mu,b)$$

by yet another application of (III.1.5). Q.E.D.

3. One-step MLE

This section provides an example of a LAM estimator of θ when $\{P_\theta\}$ is a qmd family. This estimator, which is called the one-step maximum likehood estimator, can be motivated by the following heuristics. All notations not explicitly defined in this section can be found in Chapter IV.

Suppose given a family $\{P_\theta\}$ with densities $f(\theta,x)$. Define for $x = (x_1,\ldots,x_n)$ the likehood function

$$L(\theta,x) = \prod_1^n f(\theta;x_i)$$

For the purposes of these heuristics, suppose Θ real. The maximum likelihood estimator of θ is the function $\hat{\theta}: x \to \Theta$ satisfying

$$L(\hat{\theta}(x),x) = \sup_\theta L(\theta,x) .$$

Such a $\hat{\theta}$ may not exist; if it does, it may depend on the particular versions of the densities and it may not be unique. Moreover, unless extremely stringent regularity assumptions are satisfied, it will typically behave in a miserable fashion. It is therefore remarkable that a sample variant on this recipe yields an estimator with excellent asymptotic properties, under quite reasonable conditions. To describe this heuristically, notice first that in computing the MLE, it would be enough to find θ which maximizes $\sum \log f(\theta,x_i)/f(\theta_0,x_i)$ for any fixed θ_0. If θ_0 were chosen within $c/\sqrt{n}$ of θ, one could use the expansion IV.2.6 to approximate this sum and then compute a modified MLE by maximizing this approximation. But since we don't know θ, how can we pick the θ_0? We shall prove in Chapter X that there exist estimators of θ which are $\sqrt{n}$ consistent in the sense of Definition VII.1.1. Let $\tilde{\theta}$ be one such. Let $\eta_0(\theta)$ be the modified qm derivative given in IV.2.4. Then presumably IV.2.6 applies and

$$\sum \log f(\theta;x_i)/f(\tilde{\theta},x_i) = 2\sum \eta_0(\tilde{\theta})(\tilde{\theta}-\theta) - 2\Gamma(\tilde{\theta}-\theta)^2 n$$

The right side being a simple quadratic expression is easily maximized in θ, giving the following estimator of θ:

$$\tilde{\theta}_n + \frac{1}{2n\Gamma}\sum \eta_0(\tilde{\theta})$$

The estimators given next are simple variants of this recipe. They are called one-step MLE's, because, as a glance at the foregoing heuristics show, they arise by applying the Newton-Raphson maximizing technique, but carrying out one step of the iteration only.

We turn now to the technical definition of the 1-step MLE. The following hypotheses shall be in effect for the rest of this section.

(3.1) $\{P_\theta, \theta \in \Theta\}$ is a qmd family with q.m. derivative $\eta(\theta)$ at each $\theta \in \Theta$: Θ is an open subset of $\mathbb{R}^d$; $f(\theta,x)$ will denote the density of P_θ, as in Chapter IV.

(3.2) the matrix $\Gamma(\theta)$, defined in IV.2.1, is nonsingular for each θ

(3.3) if θ_n is a sequence in Θ such that P_{θ_n} converges weakly to P_θ, then $\theta_n \to \theta$

The "identifiability" hypothesis (3.3) can be varied--see later on. Under (3.1), (3.2) there are asymptotic expansions of the form

$$(3.4)\quad \sum_1^n \log f(\theta_0+\theta n^{-1/2};x)/f(\theta_0,x) - \Delta_n(\theta_0,x) + 2\langle\theta,\Gamma(\theta_0)\theta\rangle \to 0$$

in probability, under $P^n_{\theta_0}$. According to IV.2.6 $\Delta_n(\theta,x)$ may be chosen to satisfy

$$(3.5)\qquad \Delta_n(\theta,x) = 2n^{-1/2}\sum_1^n \eta_0(\theta;x_i)\ ,\quad x=(x_1,\ldots,x_n)$$

but other forms of Δ_n are sometimes more convenient. Here is the basic recipe for the one-step MLE.

Bring in $\tilde\theta_n$, a $n^{1/2}$-consistent estimator of θ. That such exists (and is given by natural recipes) is shown in Chapter X. Discretize $\tilde\theta_n$ as follows: partition Θ into abutting cubes of side $n^{-1/2}$; for any $v \in \Theta$, let v^* be the centre of the cube containing v. Define the discretization $\hat\theta_n$ of $\tilde\theta_n$ by

$$(3.6)\qquad \hat\theta(x) = v^* \quad \text{if} \quad \tilde\theta(x) = v$$

Evidently $\hat\theta$ is still $n^{1/2}$-consistent.

Let $\Gamma_n(\theta)$ be any matrix such that if $\theta_n = \theta + h_n n^{-1/2}$, $|h_n| \le c$, then

$$(3.7)\qquad \Gamma_n(\theta_n) \to \Gamma(\theta)$$

That such matrices always exist is shown at the end of this section. Of course, if $\theta \to \Gamma(\theta)$ is continuous (a condition satisfied in all of the common examples), then $\Gamma_n(\theta)$ may be taken to be Γ.

Finally, choose $\Delta_n(\theta)$ to satisfy

$$(3.8)\quad \Delta_n(\theta+h_n n^{-1/2}) - \Delta_n(\theta+g_n n^{-1/2}) - 4\Gamma_n(\theta+g_n n^{-1/2})(h_n-g_n) \to 0$$

under $\{P^n_{\theta+g_n n^{-1/2}}\}$, whenever $h_n \to h$, $g_n \to g$ in $\mathbb{R}^n$. At the end of this section, this will be shown to be possible always; one can retain the special form (3.5) of Δ_n provided $\theta \to \eta_0(\theta)$ enjoys modest smoothness properties.

Define the one step estimator of θ by

$$T_n = \hat{\theta}_n + (\tfrac{1}{4})n^{-1/2}\Gamma_n^{-1}(\hat{\theta}_n)\Delta_n(\hat{\theta}_n) \tag{3.9}$$

(3.10) THEOREM. *Under hypotheses (3.1)-(3.3),* $\{T_n\}$ *is LAM, for any bounded subconvex loss* L.

The hypothesis (3.3) is there only to ensure existence of a $n^{1/2}$-consistent estimator; in the literature, the existence of such an estimator is often taken as a hypothesis.

PROOF. *Step 1.* Let $h_n \in \mathbb{R}^d$, $|h_n| \le c$, $h_n \to h$; set $\theta_n = \theta_0 + h_n/n^{1/2}$. It suffices to prove that $\sqrt{n}(T_n - \theta_n)$ converges in distribution to $N(0, \frac{1}{4}\Gamma^{-1}(\theta_0))$ under $P^n_{\theta_n}$. Indeed, if this is shown one can take h_n so that

$$\sup_{\theta \in A(n;c;\theta_0)} \int L(n^{1/2}(T_n - \theta))dP^n_\theta = \int L(n^{1/2}(T_n - \theta_n))dP^n_{\theta_n}$$

(or comes within n^{-2} of achieving this sup). Then for every c

$$\lim_n \sup_{\theta \in A(n;c;\theta_0)} \int L(n^{1/2}(T_n - \theta))dP^n_{\theta_n} = \int L(x)dP_0$$

where P_0 is $N(0, \frac{1}{4}\Gamma^{-1}(\theta_0))$.

Step 2. We now prove $n^{1/2}(T_n - \theta_n)$ is asymptotically normal, as required in Step 1. If $\hat{\theta}_n$ is the discretized $n^{1/2}$-consistent estimate, write $\hat{\theta}_n = \theta_n + g_n n^{-1/2}$; g_n is random, of course. If g_n were not random, then we would have

$$(3.11)\qquad \Delta_n(\tilde{\theta}_n) - \Delta_n(\theta_n) + 4\Gamma(\theta_n)(\hat{\theta}_n - \theta_n)n^{1/2} \to 0$$

under $P^n_{\theta_n}$, because of (3.8). However, because $\hat{\theta}_n$ is discretized, (3.11) still holds. To see this, note that if b is a fixed positive number and if $|\hat{\theta}_n - \theta_0| < bn^{-1/2}$, then the number of cubes covering the ball of radius $n^{-1/2}b$ about θ_0 is a finite number N_b --independent of n (N_b depends on d, the dimension of Θ). If $\{\theta_{ni}\}$ is a list of the centers of the cubes that intersect the aforementioned ball, then (3.11) holds for each i, with θ_{ni} replacing $\hat{\theta}_n$. Therefore

$$\max_{1 \le i \le N_b} [\Delta_n(\theta_{ni}) - \Delta_n(\theta_n) + 4\Gamma(\theta_n)(\theta_{ni} - \theta_n)n^{1/2}] \to 0$$

on the set $|\hat{\theta}_n - \theta_0| \le bn^{-1/2}$; i.e., since b is arbitrary

$$(3.12)\qquad \Delta_n(\theta_n) = \Delta_n(\hat{\theta}_n) + 4\Gamma(\theta_n)(\hat{\theta}_n - \theta_n)n^{1/2} + \xi_n$$

where $\xi_n \to 0$ under $P^n_{\theta_n}$, as was to be shown. For similar reasons $\Gamma_n(\hat{\theta}_n) \to \Gamma(\theta_0)$.

By definition of T_n

$$\Delta_n(\hat{\theta}_n) = 4\Gamma_n(\hat{\theta}_n)(T_n - \hat{\theta}_n)n^{1/2}$$

so by (3.12)

$$(3.13)\qquad \begin{aligned} \Delta_n(\theta_n) &= 4\Gamma_n(\hat{\theta}_n)(T_n - \hat{\theta}_n)n^{1/2} + 4\Gamma_n(\theta_n)(\hat{\theta}_n - \theta_n)n^{1/2} + \xi_n \\ &= 4\Gamma_n(\hat{\theta}_n)(T_n - \theta_n)n^{1/2} + 4[\Gamma_n(\theta_n) - \Gamma_n(\hat{\theta}_n)]n^{1/2}(\hat{\theta}_n - \theta_n) + \xi_n \end{aligned}$$

Because of (3.7), the discretization and $n^{1/2}$-consistency of $\hat{\theta}_n$ the term $4[\Gamma_n(\theta_n) - \Gamma_n(\hat{\theta}_n)]n^{1/2}(\hat{\theta}_n - \theta_n)$ goes to 0. It follows that

$$\Delta_n(\theta_n) = 4\Gamma_n(\hat{\theta}_n)(T_n-\theta_n)n^{1/2}$$

is asymptotically $N(0,4\Gamma(\theta_0))$ since $\Delta_n(\theta_n)$ is. Because of (3.7), we see that $4\Gamma(\theta_0)(T_n-\theta_0)n^{1/2}$ is asymptotically $N(0,4\Gamma(\theta_0))$, completing the proof of step 2.

Step 3. In this step is provided an example of matrices $\Gamma_n(\theta)$ satisfying (3.7). Let $e_1,\ldots,e_d$ be the usual basis for $\mathbb{R}^d$: e_i is a vector of 0's except for a 1 in the ith component. Let the i-j entry of $\Gamma_n(\theta)$ be defined by

$$\int v(n;i;\theta)v(n;j;\theta)f_\theta d\mu$$

where

$$v(n;i;\theta) = [f^{1/2}(\theta+e_i n^{-1/2})/f^{1/2}(\theta)]-1$$

Using the qmd property, it is not difficult to check that Γ_n satisfies the required condition.

Step 4. We check in this step that Δ_n can always be chosen to satisfy (3.8). Since, $P^n_{\theta_0+h_n n^{-1/2}}$ is contiguous to $P^n_{\theta_0}$, it suffices to establish the convergence in (3.8) under the probabilities $P^n_{\theta_0}$. Because of (3.7) we see that we really need only arrange that Δ_n satisfy

$$(3.14) \qquad \Delta_n(\theta_0+h_n n^{-1/2}) - \Delta_n(\theta_0) - 4\Gamma(\theta_0)h_n \rightarrow 0$$

under $P^n_{\theta_0}$. Write $\theta_n = \theta_0 + h_n n^{-1/2}$. There are several natural ways to arrange for (3.14). For example, motivated by the idea in Step 3, one could define

$$\xi_n(\theta;y) = \sum_1^d n^{1/2}[f(\theta+n^{-1/2}e_i;y)/f(\theta;y) - 1]e_i$$

and then take

$$\Delta_n(\theta) = 2n^{-1/2}\sum_1^n\{\xi_n(\theta;x_i)-E_\theta\xi_n(\theta)\}$$

4. Efficiency and the convolution theorem

Sections 2 and 3 presented an approach to asymptotically efficient estimation based on the LAM criterion. In this section a second approach to the problem is discussed: it is based on the notion of "regular estimate" and the convolution theorem. Two examples will be given: (a) estimation of θ in a qmd family and (b) estimation of the parameter in the uniform $[0,\theta]$ model. The uniform case illustrates the fact that the approach to efficiency via convolution theorem applies easily to many cases where the limit experiment is non-Gaussian.

The present approach to efficiency has an interesting feature: the formulation of the efficiency concept is free of decision theoretic concepts--e.g., no loss functions appear. However, as is apparent from the development of III.2, decision theory is intimately involved--indeed, the notion of convergence of experiments involved is inherently decision theoretic (cf. II.3), even though a purely analytic definition of it can be given.

(a) *Convolution theorem for 1-step MLE*

Let $\{P_\theta, \theta \in \Theta\}$ be a qmd family as in section 3; the problem again is to take n iid observations and then estimate θ. All further notations will follow those of section 3 and Chapter IV. A sequence of estimators $\{V_n\}$ is defined to be *regular* at $\theta_0 \in \Theta$ if

(4.1) $n^{1/2}(V_n - \theta_n)$ converges in distribution under $P^n_{\theta_n}$ to a distribution G that depends only on θ_0.

Here, as in section 3, θ_n is any sequence of the form $\theta_0 + \theta n^{-1/2}$.

Evidently this definition formulates a certain amount of "uniformity" in the way V_n converges to its asymptotic limit; see discussion following (2.2). It is easy to see that the Hodges estimator will not be regular in this sense.

(4.2) THEOREM. *Assume (3.1) and (3.2). If* V_n *is any estimator of* θ, *regular at* θ_0 *(cf. (4.1)), then*

$$G = \mu * P_0$$

where μ *is a probability on* $\mathbb{R}^d$ *and* P_0 *is the normal distribution with mean* 0, *covariance* $(\frac{1}{4})\Gamma(\theta_0)^{-1}$.

(4.3) REMARK. Since convolution "spreads out mass", it is natural to define a sequence of regular estimators $\{T_n\}$ to be *efficient* if the corresponding measure μ is unit mass at $\{0\}$. If this be granted, then the 1-step MLE is efficient in this sense too, by the argument given in step 2 of the proof of VII.3.10.

PROOF. If Q_θ^n denotes the measure $P_{\theta_n}^n$, $\theta_n = \theta_0 + \theta n^{-1/2}$, then $\{Q_\theta^n, \theta \in \Theta\}$ converges to $\{P_\theta\}$ where P_θ is $N(\theta, \frac{1}{4}\Gamma(\theta_0)^{-1})$. Next note that if $R_n = n^{1/2}[T_n - \theta_0]$, then since $n^{1/2}(V_n - \theta_n) \Rightarrow G$ we have

$$R_n \Rightarrow G_\theta \quad \text{under} \quad Q_\theta^n \tag{4.4}$$

where $G_\theta(A) = G(A-\theta)$. If M_n is the 1-step MLE then

$$n^{1/2}(M_n - \theta_n) \Rightarrow P_0 \quad \text{under} \quad Q_\theta^n \tag{4.5}$$

so if we define $T_n = n^{1/2}(M_n - \theta_0)$, then

$$T_n \Rightarrow P_\theta \quad \text{under} \quad Q_\theta^n \tag{4.6}$$

It is immediate that $\{T_n\}$ is distinguished relative to $\{Q_\theta^n\}$, $\{P_\theta\}$. Theorem III.2.10 now applies, showing that $G = \mu * P_0$ for some μ. Q.E.D.

(b) *Uniform families*

Let P_θ be the uniform distribution on $[0,\theta]$. This family is not qmd. Let P_θ^n denote the product of n copies of P_θ. The statistical problem now is to estimate θ. Define an estimator V_n to be *regular* at $\theta_0 > 0$ if

(4.7) $n[\theta_n - V_n]$ converges weakly under $P_{\theta_n}^n$ to a distribution G which depends only on θ_0

Here θ_n is any sequence of the form $\theta_n = \theta_0 - n^{-1}\theta$, for some $\theta \in \mathbb{R}$. The usual estimator of θ, namely $M_n = \max_{1 \leq i \leq n} \{X_i\}$, is regular in this sense. To simplify notation slightly, we assume from now on that $\theta_0 = 1$. Define the probability F to be the measure on the line with density

$$F(dx) = e^{-x}dx \,, \quad x > 0 \,. \tag{4.8}$$

(4.9) THEOREM. *Let* V_n *be any sequence of estimators regular at* $\theta_0 = 1$. *Then there exists a probability* μ *such that*

$$G = \mu * F$$

(4.10) REMARK. As in remark (4.3), Theorem (4.9) has a natural interpretation as an efficiency result. In particular, M_n is then efficient.

PROOF. Let $Q_\theta^n = P_{\theta_n}^n$. Let F_θ be the shift of F: $F_\theta(A) = F(A-\theta)$. Then by II.2.6 $\{Q_\theta^n\}$ converges to $\{F_\theta\}$. Since $n(\theta_n - V_n) \Rightarrow G$ it follows that

$$R_n \Rightarrow G_\theta \quad \text{under} \quad Q_\theta^n \tag{4.11}$$

where $R_n = n^{-1}[\theta_0 - V_n]$, $G_\theta(A) = G(A-\theta)$. By III.2.9, if $T_n = n[1 - \max X_i]$, T_n is distinguished relative to $\{Q_\theta^n\}$, $\{F_\theta\}$ and in fact

$$T_n \Rightarrow F_\theta \quad (Q_\theta^n) \tag{4.12}$$

It is now immediate from III.2.10 that $G = \mu * F$. Q.E.D.

VIII. Optimality properties of the empirical distribution function

This chapter begins the first substantial nonparametric applications of the theory of Chapters II-VI.

1. Asymptotic minimax character of the empirical cdf

This section proves by purely abstract methods the Dvoretsky-Kiefer-Wolfowitz theorem, which asserts that the empirical distribution function is asymptotically minimax. Previous proofs have involved painful calculation, usually involving multinomial distributions.

Let $X_1, X_2, \ldots, X_n$ be i.i.d. real observations. The empirical cdf $\hat{F}_n$ is defined by

(1.1) $$\hat{F}_n(t) = n^{-1} \sum_1^n I_{(-\infty,t]}(X_i) \; ;$$

here I_A denotes the indicator function of the set A. Variants of this definition make $\hat{F}_n$ continuous and piecewise linear. We shall assume this done when technical niceties clearly require it. Let g be an increasing function on $[0,\infty)$ and set

(1.2) $$\ell(x) = g(|x|_k)$$

where x is a real function on the line and

(1.3) $$|x|_k = \sup_t |x(t)|$$

The function g is assumed to satisfy

(1.4) $$g(u) \leq c \exp\{u^{2-\varepsilon}\} \quad \text{for all} \quad u > 0$$

for some $c > 0$ and some ε, $0 < \varepsilon \leq 2$. It is clear that ℓ is *subconvex* on the space (say) of real bounded continuous functions (sup norm). For a given distribution $F(dx)$ denote by

(1.5) $$F(t)\ ,\ \text{the cdf of } F$$

(1.6) $$F^n(dx) = \prod_1^n F(dx_i)\ ,\quad x = (x_1,\ldots,x_n)$$

It is well known (cf. Billingsley (1968); Parthasarathy (1968)) that, under F^n,

(1.7) $n^{1/2}|\hat{F}_n - F|_k$ has a distribution that does not depend on F,

and

(1.8) if F is continuous, $n^{1/2}(F_n - F)$ converges in distribution to P_0, the distribution of the process $\{W^0 \circ F(\cdot)\}$

Here W^0 is the Brownian Bridge (cf. Chapters V, VI) and $W^0 \circ F$ is the mean 0 Gaussian stochastic process with covariance $EW^0 \circ F(t) W^0 \circ F(s) = F(s \wedge t) - F(s)F(t)$.

We may now formulate our basic asymptotic decision theoretic problem. The parameter set Θ is to consist of all continuous distributions F; the probabilities P_θ^n at time n are given by (1.6). That is, we observe n iid observations from some continuous distribution F and the task is to estimate F. So Θ is again the decision space. Because of (1.8), it is reasonable to introduce the following loss at time n: if F is true and x is our estimate, the loss will be

(1.9) $$\ell(n^{1/2}(x-F))$$

This can be justified exactly as in the case of classical parametric estimation--see VII.2, VII.3.

(1.10) THEOREM.

(a) $\lim_n \inf_b \sup_F \iint \ell(n^{1/2}(x-F))b(y,dx)F^n(dy) = E\ell(W^0)$

(b) $\lim_n \sup_F \int \ell(n^{1/2}(\hat{F}_n-F))dF^n = E\ell(W^0)$

This result asserts that $\hat{F}_n$ is asymptotically minimax. The inf in (a), (b) is taken over all procedures (= estimators of F) and the sup is over all continuous distributions F.

PROOF. From a result of Kiefer and Wolfowitz (TAMS (1958)). There exist constants c, b, independent of n, such that

$$F^n\{n^{1/2}|\hat{F}_n-F|_k > t\} \le c\exp\{-bt^2\}$$

It follows that $\ell(n^{1/2}(\hat{F}_n-F))$ is uniformly integrable under F^n. Since $n^{1/2}(\hat{F}_n-F)$ converges in distribution on the space of continuous functions to $W^0 \circ F$ and since $n^{1/2}|\hat{F}_n-F|_k$ is distribution free, the $\sup_F$ in (b) is superfluous and

$$\lim_n \int \ell(n^{1/2}(\hat{F}_n-F))dF^n = E\ell(W^0)$$

since g has at most a countable number of discontinuities while $|W^0|_k$ has a continuous distribution. This proves part (b).

To prove (a) it suffices to prove that the left side of (a) is at least as big as $E\ell(W^0)$ since we know (b). For this we can assume without loss of generality that ℓ is bounded (if ℓ is not bounded, replace it by $\min\{\ell,a\}$, which is still subconvex and then let $a \uparrow \infty$). Bring in f_0, the uniform density on [0,1]; let

$$(1.11) \qquad f(n,h;\cdot) = f_0(1+n^{-1/2}h)$$

where $\int_0^1 hf_0 dx = 0$. Let $H = \{h: \int hf_0 = 0, \int_0^1 h^2 < \infty\}$ and let

$$(1.12) \qquad H_0 = \{h \in H: |h|_k < \infty\}$$

Then for each $h \in H_0$, $f(n,h,\ )$ is a probability density as long as $n > |h|_k$. Note that H_0 is dense in H. Let $F(n,h;\ )$ be the cdf of $f(n,h)$ and let $F^n(h;\ )$ be the n-fold product measure of $F(n,h)$. If Λ is any subset of H_0, then the experiments $\{F^n(h),\ h \in \Lambda\}$ *converge* to $\{P_h,\ h \in \Lambda\}$ where

$$(1.13) \qquad \{P_h,\ h \in H\} \text{ is the Gaussian shift experiment of } (\tau,H,B)$$

where (τ,H,B) is given in V.2.12 with f there equal to f_0. In particular $B = C[0,1]$, P_0 is the distribution of W^0, and $(\tau h)(t) = \int_0^t h(s)f_0(s)ds$. Define

$$(1.14) \qquad \Lambda_m = \{h \in H: |h|_k \leq m\}$$

Then if $n \geq m$

$$(1.15) \qquad \inf_b \sup_F \iint \ell(n^{1/2}(x-F))b(y,dx)F^n(dy)$$
$$\geq \inf_b \sup_{h \in \Lambda_m} \iint \ell(n^{1/2}(x-F(n,h)))b(y,dx)F^n(h;dy)$$

Since $F(n,h) = F_0 + \tau h n^{-1/2}$, F_0 = cdf of f_0, it follows that

$$n^{1/2}(x-F(n,h)) = x' - \tau h\ , \quad x' = n^{1/2}[x-F_0]$$

Therefore, if we enlarge the decision space to all continuous bounded functions (and hence enlarge the set of all possible decision procedures b), (1.15) can be continued

(1.16) $$\geq \inf_b \sup_{h\in\Lambda_m} \iint \ell(x-\tau h)b(y,dx)F^n(h;dy)$$

since the inf is computed over a larger set than before. By the asymptotic minimax theorem, the limit of the expression in (1.16) is at least

(1.17) $$\inf_b \sup_{h\in\Lambda_m} \iint \ell(x-\tau h)b(y,dx)P_n(dy) \equiv \inf_b \sup_{h\in\Lambda_m} \rho(h,b)$$

where ρ has the evident definition (cf. II.1.2). The left side of (1.10a) is therefore at least as big as the expression in (1.17) for all m. Presumably we may now let $m\uparrow\infty$ and replace $\inf_b \sup_{h\in\Lambda_m} \rho(h,b)$ by $\inf_b \sup_{h\in H_0} \rho(h,b)$; if this is permissible, then the continuity of $h\rightarrow P_h$ would allow replacement of $\inf_b \sup_{h\in H_0}$ by $\inf_b \sup_{h\in H}$. The result would then follow immediately from VI.2.8. The argument just outlined can be made rigorous by exactly the same method used in the proof of VII.2.6. Q.E.D.

(1.18) REMARK (Other loss functions). It is important to note that other common loss functions can also be used, provided the argument for Theorem (1.10) is only slightly extended. The most common alternative to the loss (1.9) is when F is true and x is our estimate:

$$g(n\int(x(t)-F(t))^2F(dt)) = \ell(n;x,F)$$

Here g is an increasing function on $[0,\infty)$. This is the Cramer-von Mises loss function. It is known that $n\int[\hat{F}_n(t)-F(t)]^2F(dt)$ has a distribution that does not depend on F, if F is continuous. To carry through the argument for asymptotic minimaxity with this loss, one must reduce $\ell(n;x,F)$, when $F = F(n,h)$ to an expression like $\ell(x'-\tau h)$. If g were *uniformly continuous* then

$$(1.19)\quad g(n\int[x(t)-F(n,h;t)]^2F(n,h;dt)) = g(\int(x'(t)-\tau h(t))^2F_0(dt)) + o(1)$$

where $o(1)$ does not depend on $h \in \Lambda_m$; here we use again the representation $F(n,h) = F_0 + \tau h$ and $x' = \sqrt{n}[x-F_0]$. The function $\ell_0(x) = g(\int x(t)^2F_0(dt))$ is subconvex; therefore, provided g is uniformly continuous, the same argument will now apply. To treat general g, it is necessary only to produce $g_n \uparrow g$, g_n uniformly continuous. From this we see that the Theorem (1.10) continues to hold with asymptotic limit $E\ell_0(W^0)$, $\ell_0(x) = g[\int_0^1 x^2(t)dt]$.

2. A generalization

The proof given in section 1 can be adapted to a variety of interesting situations. Here is one possible formulation.

Let C be a collection of continuous cdf's on the line. At time n one observes $X_1,\dots,X_n$ iid F, $F \in C$. The problem is to estimate F, *knowing* that it belongs to C; one is permitted to use any cdf estimator at all, even estimators that do not pick elements of C. Let ℓ be a subconvex function on the usual Banach space of continuous functions. The parameter set is C, the decision space is all continuous cdf's and loss at time n will be given by $\ell(n^{1/2}(x-F))$ if F is 'true' and x is chosen from the decision space.

Basically, the empirical cdf will, in this situation, still be the best estimator provided C is reasonably 'fat'. Such constrained estimation problems arise in several fields of application, e.g. reliability theory. Here is a collection of hypotheses on C under which $\hat{F}_n$ will still be asymptotically minimax.

Suppose C contains a distinguished cdf F_0 with density f_0 (with respect to some measure μ). Denote by H the Hilbert space $\{h \in L^2(\mu): \int hf_0 d\mu = 0, \text{ support } h \subset \text{support } f_0\}$. Assume

(2.1) there are convex sets $H_n \subset H$ such that if $h \in H_n$, then $f_0(1+k^{-1/2}h)$ is a probability density in C for all $k \geq n$

(2.2) $H_n \subset H_{n+1}$, $\cup H_n$ dense in H

(2.3) $\int \ell(n^{1/2}(\hat{F}_n - F))dF^n$ is distribution free over C and converges to $E\ell(W^0 \circ F_0)$ as $n \to \infty$

In the case of section 1, $H_n = \Lambda_n$. Under the hypotheses above, the following theorem holds, showing that $\hat{F}_n$ is again asymptotically minimax.

(2.4) THEOREM.
$$\liminf_n \inf_b \sup_{F \in C} \iint \ell(n^{1/2}(x-F))b(y,dx)F^n(dy)$$
$$= \lim_n \sup_{F \in C} \int \ell(n^{1/2}(\hat{F}_n - F))dF^n$$
$$= E\ell(W^0 \circ F_0)$$

Here F^n is, as usual, the product measure of F. The proof is a simple modification of that in section 1.

Here is a simple illustration of its use; more examples can be found in Millar (1979), including many collections C important for reliability theory.

(2.5) EXAMPLE (Decreasing densities). Let C be the class of all continuous cdf's on $[0,\infty)$ having a decreasing density with respect to Lebesgue measure. Then the empirical cdf is asymptotically minimax in C.

To prove this, it suffices to pick F_0; it is easy to see that the uniform distribution, although belonging to C, is inconvenient since it is "on the bounding of C"--i.e. (2.1) is impossible to satisfy. Therefore let F_0 be the probability with density $f_0(x) = e^{-x}$, $x \geq 0$. Then $(1+hn^{-1/2})f_0$ with $\int h(x)e^{-x}dx = 0$ will be a density if $\sup_x |h(x)n^{-1/2}| < 1$; such a density will belong to C if it has a negative derivative:

$$f_0'(1+hn^{-1/2}) + f_0 h' n^{-1/2} \leq 0$$

Therefore one is led to consider H_n of the form

$$H_n = \{h: \int h(x)e^{-x}dx = 0,\ |h| < \tfrac{1}{2}n^{1/2},\ h'(x) \leq (\tfrac{1}{2})n^{1/2}\}$$

Evidently $\{H_n\}$ satisfies (2.1)-(2.3), so $\hat{F}_n$ is asymptotically minimax for this choice of C.

REMARK. The result just established proves that the empirical cdf is asymptotically minimax among all distributions that have concave cdf on $[0,\infty)$, since this latter class is larger than the one just treated.

3. An efficiency property of $\hat{F}_n$

This section again addresses the problem of trying to estimate a continuous cdf F, after trying n iid observations. Just as in the situation of classical parametric estimation, there are very perverse estimators similar to the Hodges estimator VII.2.2: the asymptotic limit looks excellent, but there are still unsettling aspects about the estimator:

(3.1) EXAMPLE. Fix G a cdf and let $\hat{F}_n$ be the empirical cdf. Define a new estimator $\tilde{F}_n$ by

$$\tilde{F}_n(\cdot) = G(\cdot) \quad \text{if} \quad |\hat{F}_n - G|_k \leq n^{-1/4}$$
$$= \hat{F}_n \quad \text{otherwise}$$

If $X_1,\ldots,X_n$ are iid G, then $\lim_n P\{\tilde{F}_n \equiv G\} = 1$; if $X_1,\ldots,X_n$ are iid F, then $n^{1/2}(\hat{F}_n - F) \Rightarrow W^0 \circ F$. That is, if the data has distribution G, then $\tilde{F}_n$ is better than $\hat{F}_n$, while if the data has any other distribution, $\tilde{F}_n$ appears no worse than $\hat{F}_n$, judging from its asymptotic limit.

Just as for the Hodges estimator, however, such an estimator as that of (3.1) converges to its limit in an extremely erratic manner. Therefore we shall again introduce a notion of "regular estimator" that specifies that all estimators under consideration enjoy a modest amount of uniformity in their convergence to the limit.

To this end fix F a continuous cdf having density f with respect to a measure μ. Let H be the Hilbert space consisting of all real functions $h \in L^2(\mu)$ such that $\int hf^{1/2} d\mu = 0$ and support $h \subset$ support f. Define densities $f(n,h;x)$ by the recipe

$$f^{1/2}(n,h;x) = (1 - |h|^2/4n)^{1/2} f^{1/2}(x) + (\tfrac{1}{2}) h(x) n^{-1/2}$$

where $|\ |$ is the norm of H. Such $f(n,h)$ will indeed be a density as long as $|h|^2 \leq 4n$. Let $F(n,h)$ be the cdf of $f(n,h)$ and let $F^n(h;dx)$ be the n-fold product measure of F: $F^n(h,dx) = \prod_1^n f(n,h;x_i)$, $x = (x_1,\ldots,x_n)$. Then the experiments $\{F^n(h);\ h \in H\}$ converge to $\{P_h,\ h \in H\}$ where $\{P_h\}$ is the standard Gaussian shift experiment of (τ,H,B) given in V.2.10. In particular P_0 is the distribution of

$W^0 \circ F$, $(\tau h)(t) = \int^t h(s) f^{1/2}(s)\mu(ds)$ and B is a Banach space of continuous functions.

(3.2) DEFINITION. A sequence $\{V_n\}$ of cdf estimators is *regular* at F if, for each $h \in H$, the distribution of $n^{1/2}(V_n - F(n,h))$ converges, under $F^n(h)$, to a measure G that depends only on F. Here convergence is on the usual Banach space of continuous functions, although a number of other spaces could be used--see Chapter X for some ideas.

It is clear that this definition is an exact analogue of the classical Hajek definition of regularity for classical parametric estimation (VII.4.1). It ensures a certain uniformity in the convergence to the asymptotic limit. It is not hard to see that the estimator in example (3.1) is not regular in this sense. Simple arguments show that the empirical cdf *is* regular in the sense of (3.2). The following result was proved by Beran (1977) using a different method.

(3.3) THEOREM. *Let* V_n *be a cdf estimator, regular at* F. *Then there exists a measure* μ *such that*

$$G = \mu * P_0$$

Here G is given in (3.2) and P_0 is the distribution of $W^0 \circ F$.

(3.4) REMARK. As in the classical case, this is an efficiency result: no regular estimator can have an asymptotic distribution that is less spread out than P_0. It is natural to define a sequence V_n to be *efficient* if for each F the corresponding μ is unit mass at 0. The empirical cdf is efficient in this sense.

PROOF OF (3.3). Note first that $F(n,h)$, the cdf of the density $f(n,h)$, satisfies $F(n,h) = F + n^{-1/2}\tau h$. Therefore, since $n^{1/2}[V_n - F(n,h)] \Rightarrow G$, it follows that

$$R_n \Rightarrow G_h \quad \text{under} \quad F^n(h)$$

where $R_n = n^{1/2}[V_n - F]$ and $G_h(A) = G(A - \tau h)$. If $\hat{F}_n$ is the empirical cdf, it is a regular estimator and so if $T_n = n^{1/2}(\hat{F}_n - F)$ then

$$T_n \Rightarrow P_h \quad \text{under} \quad F^n(h)$$

where $P_h(A) = P_0(A - \tau h)$. It is now immediate from VI.3.4 that there exists μ and $G = \mu * P_0$. Q.E.D.

4. Some variants

Sections 1-3 described the optimality of the empirical cdf as an estimator of a cdf F, known to belong to a collection C. The optimality of $\hat{F}_n$ depended upon finding F_0 and a kind of "neighborhood" around it so that this "neighborhood" was entirely on C. This asserts that C needs to be fairly "fat". On the other hand, if C is rather smaller, $\hat{F}_n$ will not be asymptotically minimax. To illustrate the general idea we begin with a typical example.

(4.1) EXAMPLE (Symmetric densities). Let C consist of all distributions on $[0,1]$ having a density f with respect to Lebesgue measure that is symmetric about $\frac{1}{2}$: $f(x) = f(1-x)$, $0 < x < 1$. The problem is to take n iid observations and then estimate F. Instead of $\hat{F}_n$, it would be natural to try its symmetrization: $\tilde{F}_n(t) = [\hat{F}_n(t) + 1 - \hat{F}_n(1-t)]/2$. We now show that $\tilde{F}_n$ is asymptotically minimax. To do this note that if f is a density of $F \in C$ then by orthogonal decomposition in $L^2[0,1]$

$$f^{1/2} = (1-|h|^2)^{1/2} f_0^{1/2} + h \ , \quad h \perp f_0^{1/2} \tag{4.2}$$

where f_0 is the uniform density on $[0,1]$. Since f, f_0 are symmetric, h must be also. Therefore bring in the Hilbert space $H = \{h \in L^2[0,1], h \perp f_0^{1/2}, h \text{ symmetric}\}$. For $h \in H$, define $f(n,h)$ by

$$f^{1/2}(n,h;x) = (1-|h|^2/4n)^{1/2} f_0^{1/2} + (h/2)n^{-1/2} \tag{4.3}$$

if $|h| \leq 4n$ ($|h|$ is the L^2 norm of h) and define $f(n,h)$ to be an arbitrary density in C if $|h| > 4n$. Let $F(n,h)$ be the cdf of $f(n,h)$; $F^n(h)$ the product measure. The key observation is that

$$\text{for every } n, \quad \{F(n,h): h \in H\} = C \tag{4.4}$$

This leads to the asymptotic minimax character of $\tilde{F}_n$.

(4.5) THEOREM.
$$\liminf_n \sup_b \sup_{F \in C} \iint \ell(n^{1/2}(x-F)) b(y,dx) F^n(dy)$$
$$= \lim_n \sup_{F \in C} \iint \ell(n^{1/2}(\tilde{F}_n - F)) dF^n$$
$$= E\ell(W')$$

where W' *is the distribution on* $C[0,1]$ *of the process* $[W^0(t) - W^0(1-t)]/2$ *and where* ℓ *is the loss function given in (1.2).*

(4.6) REMARK. It is easy to see that $n^{1/2}(\tilde{F}_n - F)$ converges in distribution to the distribution of $[W^0(F(t)) - W^0(1-F(t))]/2$.

PROOF. Assume for convenience that $\ell(x) = g(|x|_k)$ with g uniformly continuous; this hypothesis is easily removed. Then ignoring $o(1)$ terms and using the fact that $F(n,h) = F_0 + \tau h n^{-1/2} + o(n^{-1/2})$ where $(\tau h)(t) = \int_0^t h(s) f_0^{1/2}(s)ds$, we have

$$(4.7) \qquad \liminf_{n} \inf_{b} \sup_{F \in C} \iint \ell(n^{1/2}(x-F)) b(y,dx) F^n(dy)$$

$$= \liminf_{n} \inf_{b} \sup_{h \in H} \iint \ell(n^{1/2}(x-F)) b(y,dx) F^n(dy)$$

$$= \liminf_{n} \inf_{b} \sup_{h \in H} \iint \ell(x - \tau h) b(y,dx) F^n(h;dy)$$

Define π: map of $C[0,1] \to C[0,1]$ by $(\pi x)(t) = \frac{1}{2}[\pi(t) - \pi(1-t)]$. Then π is continuous, linear and

$$(4.8) \qquad |\pi x|_k \leq |x|_k$$

Since $h \in H$, $\pi\tau h = \tau h$; hence because of (4.8) any estimate x can be improved using πx instead. Therefore the last line in (4.7) is equal to

$$(4.9) \qquad \liminf_{n} \inf_{b} \sup_{h \in H} \iint \ell(\pi(x - \tau h)) b(y,dx) F^n(h;dy)$$

$$= \liminf_{n} \inf_{b} \sup_{h \in H_e} \iint \ell(\pi(x - \tau h)) b(y,dx) F^n(h;dy)$$

where in the last line of (4.9) $H_e = \{h \in L^2[0,1], \int h f_0^{1/2} = 0\}$ and the inf is over all procedures with values in $C[0,1]$. Since $\ell \circ \pi$ is subconvex on $C[0,1]$, the last line is at least as big as

$$E\ell \circ \pi(W^0) = E\ell(W')$$

using the argument in section 1. This proves that $E\ell(W')$ is a lower bound for the asymptotic minimax value. That $\tilde{F}_n$ satisfies the statement in the theorem is easily deduced using (4.6). Q.E.D.

(4.10) AN EXTENSION. Let C be a collection of distributions satisfying the following conditions:

(4.11) there is $F_0 \in C$ with density f_0 such that each $F \in C$ has density f satisfying

$$f^{1/2} = (1-|h|^2)f_0^{1/2} + h$$

for some $h \in H_0$, a subspace of $H = \{h: \int h^2 < \infty,\ h \perp f_0^{1/2}\}$

(4.12) the recipe in (4.11) gives a density in C for all h in a dense subset of H_0

(4.13) for h in a dense subspace of H_0, the density $f(n,h)$ defined via

$$f^{1/2}(n,h) = (1-|h|^2/n)f_0^{1/2} + h/n^{1/2}$$

is a density of C

Define $(\tau h)(t) = \int^t h(s)f_0^{1/2}(s)ds$. Assume also

(4.14) there is a continuous linear map $\pi\colon B = \overline{\tau H} \to \overline{\tau H_0}$ such that

$$|\pi x|_k \leq |x|_k\ , \quad \pi x = x \ \text{ if } \ x \in \overline{\tau H_0}$$

The assumption (4.14) means $\overline{\tau H_0}$ is a complemented subspace of the Banach space B. Assume ℓ, loss function, is as in (1.2).

With these assumptions, we have the following result, proved by the same methods as for (4.5).

(4.15) THEOREM. $F_0 + \pi(\hat{F}_n - F_0)$ *is an asymptotic minimax estimator of* $F \in C$ *and the asymptotic minimax value is* $\int \ell(\pi x) P_0(dx)$, P_0 *being the distribution of* W^0.

One can apply (4.15) to show that if C = all cdf's of some qmd family, then the empirical cdf is not asymptotically minimax. (It is even easier if one shows that $\hat{F}_n$ is not *locally* asymptotically minimax.) Alternatively, one could do this directly: if $F(\theta;t)$ is the cdf of P_θ, then $F(\hat{\theta}_n;t)$ should be a much better estimator than $\hat{F}_n$, where $\hat{\theta}_n$ is the 1-step MLE for example. It is easy to see $\hat{\theta}_n$ is better in the *local* asymptotic sense: see Chapter IX for the definition.

IX. Recent developments in the theory of asymptotically optimal nonparametric inference

The basic approach to proving asymptotic minimaxity of the empirical cdf, and a convolution theorem for it, can be extended to a variety of important nonparametric situations. This chapter presents a few of the possibilities, but gives only a brief description of the technical argument. All of the results here are specializations of a single (but slightly complicated to state) general result; see Millar (1982a,b) for a statement of it, complete proof, and many more examples. Minimum distance procedures can be based on all of the examples here; this is discussed in Chapter X.

1. Stationary Gaussian processes

Let $X_1,\dots,X_n$ be a real stationary Gaussian sequence with mean 0, and spectral function $F(t)$, $-\pi \leq t \leq \pi$, so that $EX_kX_{k+n} = \int_{-\pi}^{\pi} e^{int}F(dt)$. The basic problem is to construct an optimal estimate of F. Assume $F(0) = 0$; we shall consider estimating $F(t)$ for $0 \leq t \leq \pi$ (this reduction involves only a normalization). A common estimate of F is

$$(1.1) \qquad \hat{F}_n(t) = (2\pi n)^{-1}\int_0^t |\sum_{k=1}^{n} X_i e^{-iyk}|^2 dy$$

In this section, we (a) show $\hat{F}_n$ has a certain asymptotic minimax property and (b) give a convolution theorem for it. The minimaxity was first shown by Levit-Samarev although they did not explicitly work in an infinite dimensional framework. The convolution result (b) is new, and is developed in more detail in Millar (1982a). The development of this chapter illustrates the fact that the general approach of the preceding chapters can be applied to situations involving dependent observations.

We begin with the minimaxity; it will be a *local* asymptotic minimax result in the spirit of the classical result of VII.2. That is, we fix a point F in the parameter space, construct "neighborhoods" about F, that "shrink" at rate $n^{-1/2}$, and establish the minimaxity over these neighborhoods. See Chapter VII for some motivation for this optimality property in the case of classical estimation. As a glance at section VIII.1 shows, the argument for asymptotic minimaxity there is really a local argument (again the main development concerned "neighborhoods" shrinking at rate $n^{-1/2}$); it could be given a global formulation only because of certain "distribution free" aspects of the empirical distribution.

Fix a spectral function F; assume F has a density f with respect to Lebesgue measure, and that $\int_0^\pi f^2(t)dt < \infty$. Define $F(h;\cdot)$ spectral function by

$$F(h;dt) = (1 + h(t)f(t))f(t) \tag{1.2}$$

where h is a real function satisfying $\int_0^\pi h^2 f^2 < \infty$. Let H be the Hilbert space consisting of all h satisfying the condition in the preceding sentence. Fix α, $\frac{1}{2} < \alpha \leq 1$; let $\Lambda_c = \{h \in H: |h(s)-h(t)| \leq c|s-t|^\alpha, |h|_H \leq c\}$ and let $H_0 = \bigcup_{c>0} \Lambda_c$. Then H_0 is a dense linear subspace of H. Let W be standard Brownian motion on the line. Set

$$T(t) = 2\pi \int_0^t f^2(s)ds \tag{1.3}$$

$$W_F(t) = W(T(t)) , \quad 0 \leq t \leq \pi \tag{1.4}$$

Let P_h^n denote the distribution of the stationary Gaussian sequence $X_1,\ldots,X_n$ if the spectral function is $F(hn^{-1/2};\cdot)$. For the theorem below, $b(x,dy)$ denotes any procedure which chooses a spectral function, when x is observed.

(1.5) THEOREM (Local asymptotic minimax property of $\hat{F}_n$). *If* ℓ *is a bounded, subconvex loss function on* $C[0,\pi]$, *then*

$$\lim_{c\uparrow\infty}\lim_{n}\inf_{b}\sup_{h\in\Lambda_c}\iint \ell[n^{1/2}(y-F(n^{-1/2}h))]b(x,dy)P_h^n(dx)$$
$$= E\ell(W_F)$$
$$= \lim_{c\uparrow\infty}\lim_{n}\sup_{h\in\Lambda_c}\int \ell[n^{1/2}(\hat{F}_n-F(n^{-1/2}h))]dP_h^n$$

(1.6) REMARK. The hypothesis that ℓ be bounded is easily weakened.

PROOF (Sketch). Because of known expansions of $\log dP_h^n/dP_0^n$ (Davies, 1973), it may be shown that the experiments $\{P_h^n, h\in H_0\}$ converge to $\{P_n, h\in H_0\}$, where P_h is the standard Gaussian shift experiment of (τ,H,B) where $B = C[0,\pi]$ and $(\tau h)(t) = \int_0^t hf^2$. The form of H confirms the fact that P_0 is the distribution of W_F. As in the argument for asymptotic minimaxity of the empirical cdf

$$\ell[n^{1/2}(y-F(n^{-1/2}h))] = \ell(y'-\tau h)$$

$y' = (y-F)n^{1/2}$, so that by the asymptotic minimax theorem

$$\lim_{n}\inf_{b}\sup_{h\in\Lambda_c}\iint \ell[n^{1/2}(y-F(n^{-1/2}h))]b(x,dy)P_h^n(dy)$$
$$\geq \inf_{b}\sup_{h\in\Lambda_c}\iint \ell(y-\tau h)b(x,dy)P_h^n(dy)$$

Let $c\uparrow\infty$; since H_0 is dense in H the argument given in VIII.2 shows that

$$\lim_{c}\inf_{b}\sup_{h\in\Lambda_c}\iint \ell(y-\tau h)b(x,dy)P_h^n(dx)$$
$$= \inf_{b}\sup_{h\in H}\iint \ell(y-\tau h)b(x,dy)P_h^n(dx)$$
$$= E\ell(W_F)$$

the last equality following from the main result of VI.2. Therefore the first expression in (1.5) is at least as big as $E\ell(W_F)$. To show equality there, and also the second equality of (1.5), it suffices to take a sequence $h_n \in \Lambda_c$ and show that the $P^n_{h_n}$ distribution of $n^{1/2}[\hat{F}_n - F(n^{-1/2}h_n$ converges in $C[0,\pi]$ to W_F. For this one can employ arguments of Ibragimov (1963).

We turn now to a convolution theorem. As in the classical case VII.4 and the cdf case VIII.3 it is necessary to define a notion of *regular* estimate. One possibility is to call a sequence $\{V_n\}$ of spectral function estimates regular at F if, for any $h \in H_0$,

(1.7) the P^n_h distribution of $n^{1/2}[V_n - F(n^{-1/2}h)]$ converges to a distribution G that does not depend on H.

Here convergence is on the space of bounded continuous functions on $[0,\pi]$. It can be shown that $\hat{F}_n$ defined in (1.1) is regular in this sense. Since H_0 is dense in H, the extension of the convolution theorem mentioned in VI.3 yields the following result, via arguments similar to those in VIII.4.

(1.8) THEOREM. *If* $\{T_n\}$ *is regular at* F *then there is a probability* μ *on* $C[0,\pi]$ *such that*

$$G = \mu * P_0$$

where P_0 *is the distribution of* W_F.

2. Estimation of a quantile function

Let F be a cdf on the line. Define $\xi(F)$ to be the increasing, right continuous function on the interval $[0,1]$ given by

(2.1) $$\xi(F)(t) = \inf\{s: F(s) \geq t\}\ , \quad 0 < t < 1\ .$$

Suppose given $X_1,\ldots,X_n$ iid real random variables with unknown distribution F; the problem addressed in this section is the optimal estimation of $\xi(F)$, the *quantile function of* F. We shall present a result showing the local asymptotic minimax character of $\xi(\hat{F}_n)$, where $\hat{F}_n$ is the empirical cdf. Of course $\xi(\hat{F}_n)$ is the usual empirical quantile function. We shall also give a convolution theorem for this estimator. Both of these results are new and are proved in detail in Millar (1982a,b). Since the problem of this section is *much* more technically involved than the analogous problem for the empirical cdf, we give only a heuristic discussion, emphasizing the structural considerations that make the aforementioned optimality properties apparent.

One technicality that needs immediate attention involves the fact that, even if F is smooth, $\xi(F)$ can be miserable at the endpoints of $[0,1]$: $\xi(F)$ usually is not a *bounded* continuous function on $[0,1]$, even for nice F. Therefore, a problem arises as to what Banach space B is appropriate so that for all F of interest, $\xi(F) \in B$. Many choices are possible; for simplicity let us pick $0 < \alpha < \beta < 1$ and try to estimate $\xi(F)$ optimality only on $[\alpha,\beta]$. We shall work only with those continuous F such that $\xi(F)$ is also continuous. Therefore we shall regard each such $\xi(F)$ as a member of the Banach space $C(\alpha,\beta)$, ignoring $\xi(F)(t)$ for $t \leq \alpha$, $t \geq \beta$.

We turn first to the LAM property. Fix a cdf F_0. Let H be the Hilbert space consisting of all functions h on $[0,1]$ such that $\int h^2(s)ds < \infty$, $\int h(s)ds = 0$. Let F_h be the cdf with density

$$dF_h(t) = f(t)[1 + h(F_0(t))]$$

where f is the density of F_0 with respect to Lebesgue measure. Define τ on H by $(\tau h)(t) = \int^{F_0(t)} h(s)ds$, so that $F_h = F_0 + \tau h$. Let B be the closure of τH under sup norm. Then (τ, H, B) is an abstract Wiener space, and if $\{P_h, h \in H\}$ is its Gaussian shift family, it is easy to see that P_0 is the distribution of $W^0(F_0)$; indeed, this is a minor variant of the set-up of VIII.2, and could actually have been used there. If P_h^n is the product measure for $F_{n^{-1/2}h}$, then the experiments $\{P_h^n, h \in H\}$ converge to $\{P_h, h \in H\}$.

We now try to see heuristically that $\xi(\hat{F}_n)$ must be LAM. To do this the main job is to evaluate

$$\lim_{c \uparrow \infty} \liminf_{n} \inf_{T_n} \sup_{h \in N_c} \int \ell[n^{1/2}(T_n - \xi(F_{n^{-1/2}h}))]dP_h^n \tag{2.2}$$

where $N_c = \{h \in H: |h| \leq c\}$, T_n is an estimator of $\xi(F)$, and ℓ is subconvex on $C[\alpha,\beta]$. The key observation is that the *functional* ξ is *differentiable*: there is a continuous linear map $\xi': B \to C(\alpha,\beta)$ such that

$$\begin{aligned} \xi(F_{hn^{-1/2}}) &= \xi(F_0 + n^{-1/2}\tau h) \\ &= \xi(F_0) + n^{-1/2}\xi'(\tau h) + o(h^{-1/2}) \end{aligned} \tag{2.3}$$

Under modest regularity on F_0, τh, this derivative ξ' is given by

$$\xi'(x)(t) = -x(F_0^{-1}(t))/f(F_0^{-1}(t)) \tag{2.4}$$

where $x \in B$, F_0^{-1} is the same as $\xi(F_0)$, and f is the density of F_0. This can be seen heuristically by drawing a simple picture; see also XII.1.24. Because of (2.3),

$$\ell[n^{1/2}(T_n - \xi(F_{n^{-1/2}h}))] \doteq \ell[T_n' - \xi' \circ \tau h] \tag{2.5}$$

where $T_n' = n^{1/2}[T_n - \xi(F_0)]$ is just another "estimator". Applying (2.5) in (2.2) leads by the usual argument involving the asymptotic minimax theorem to this lower bound to (2.2):

$$\inf_T \sup_h \int \ell(T-\xi' \circ \tau h)dP_h \tag{2.6}$$

On the other hand, *suppose* ξ' is a 1-1 map. Then the experiment $\{Q_h, h \in H\}$ where Q_h is the P_h-distribution of the random variable $\xi': B \to C[\alpha,\beta]$, is *equivalent* to $\{P_h\}$ (cf. II.3.5). It is easy to see that $\{Q_h\}$ is the Gaussian shift experiment for the abstract Wiener space (τ_0, H, B_0) where $\tau_0 = \xi' \circ \tau$, $B_0 = \overline{\tau \circ B}$. Because of the equivalence, (2.6) is equal to

$$\inf_T \sup_h \int \ell(T-\tau_0 h)dQ_h = \int \ell(x) hQ_0(dx) \tag{2.7}$$

Using now, for the first time, the actual form of the derivative in (2.3), the second expression in (2.7) is

$$E\ell(Z) \tag{2.8}$$

where $Z = \{Z(t), \alpha < t < \beta\}$ is the stochastic process $Z(s) = W^0(s)/f(F_0^{-1}(s))$. Therefore (2.8) is the lower bound to the asymptotic minimax risk in (2.2). To see heuristically that $\xi(\hat{F}_n)$ should achieve this lower bound, it is necessary only to see if

$$n^{1/2}[\xi(\hat{F}_n) - \xi(F_{n^{-1/2}h_n})] \tag{2.9}$$

converges to Z whenever h_n is a sequence in N_c. Because of (2.3) it is plausible that the expression (2.6) is approximately

$$\xi'(n^{1/2}(\hat{F}_n - F_{n^{-1/2}h})) \tag{2.10}$$

Since $n^{1/2}(\hat{F}_n - F_{n^{-1/2}h}) \Rightarrow W^0 \circ F_0$, the limit of (2.7) should be

(2.11) $$\xi'(W^0 \circ F_0)$$

which has the same distribution as Z. So the empirical quantile should be LAM. The argument just given is not rigorous: ξ' is *not* 1-1, the way things are set up; and ξ will not be differentiable unless h is fairly smooth. These matters can be arranged by cutting down H somewhat, and working with a subspace that consists of nice h --as done in the case of estimating the spectral function (section 1). Arguments of Shorack can be used to shore up the heuristics of (2.9)-(2.11). This leads to the following result; see Millar (1982) for a detailed argument.

(2.12) THEOREM. *Let* F_0 *have density* f, *with an interval for its support. There is an infinite dimensional subspace* $H_0 \subset H$, *such that if* $N_c = \{h \in H_0 : |h| \leq c\}$, *then*

$$\begin{aligned}\lim_{c\uparrow\infty} \lim_n \inf_{T_n} \sup_{h\in N_c} \int \ell[n^{1/2}(T_n - \xi(F_{n^{-1/2}h}))]dP_h^n \\ = E(Z) \\ = \lim_{c\uparrow\infty} \lim_n \sup_{h\in N_c} \int \ell[n^{1/2}(\xi(\hat{F}_n) - \xi(F_{n^{-1/2}h}))]dP_h^n\end{aligned}$$

where Z *has the distribution in (2.8).*

Let us turn briefly to a convolution theorem for the present estimation problem; the discussion will be totally heuristic. Bring in the space H_0 mentioned in (2.12). A sequence V_n of quantile function estimators is regular at the cdf F_0 if for every $h \in H_0$

(2.13) $n^{1/2}[V_n - \xi(F(n^{-1/2}h))]$ converges, under P_h^n, to a distribution G_0 on $C[\alpha,\beta]$ that does not depend on h.

If $\{V_n\}$ is regular, set $R_n = n^{1/2}[V_n-\xi(F_0)]$. Then since ξ has derivative ξ' (cf. (2.3)), it follows from (2.13) that

(2.14) $$P_h^n\{R_n \in dy\} \Rightarrow G_h$$

where $$G_h(A) = G_0(A-\xi'\circ\tau h)$$

Similarly, if $T_n = n^{1/2}[\xi(\hat{F}_n)-\xi(F_0)]$, then, according to the heuristics of (2.9)-(2.11)

(2.15) $$P_h^n\{T_n \in dy\} \Rightarrow Q_h(dy)$$

where Q_h is the Gaussian shift experiment introduced after (2.6). Since $\{Q_h\}$ is equivalent to $\{P_h\}$, $\{T_n\}$ is distinguished. Therefore, R_n, T_n, G_h, Q_h appear to conform to the structure of the basic infinite dimensional convolution theorem (VI.3), suggesting the following result (see Millar (1982a) for detailed proof).

(2.16) THEOREM. *If* F_0 *satisfies the hypotheses of (2.12), and if* $\{V_n\}$ *is a sequence of estimators regular at* F_0, *then there is a measure* μ *on* $C[\alpha,\beta]$ *such that*

$$G_0 = \mu * Q_0$$

3. Censored data: the Kaplan Meier estimate

Let (Y_{1i},Y_{2i}), $1 \le i \le n$, be iid pairs of non-negative random variables, with cdf's G_1, G_2 respectively--both assumed continuous. Let $X_i = \max\{Y_{1i},Y_{2i}\}$ and $\delta_i = 1$ if $X_i = Y_{1i}$, $\delta_i = 2$ if $X_i = Y_{2i}$. Then (X_i,δ_i), $1 \le i \le n$, forms an iid sequence with values in

$S = R' \times \{1,2\}$. The problem is to estimate G_1 optimally using the observations $\{(X_i,\delta_i)\}$. The structure of this problem is very similar to that discussed in section 2, so we shall discuss only the structural similarity, and try to make it plain that a LAM result and a convolution theorem are expected in the present situation.

To set this up, let ν be a measure on R' such that G_i is absolutely continuous with respect to ν, with density g_i $(i=1,2)$. Let μ be a counting measure on $\{1,2\}$. Then on $S = R^+ \times \{1,2\}$ define a density with respect to $\nu \times \mu \equiv m$

$$(3.1) \qquad f(x,i) = [(1-G_2(x))g_1(x)]^{2-i}[(1-G_1(x))g_2(x)]^{i-1}$$

$x \in R^+$, $i=1,2$. Let F be the cdf of the density f (F has a 2-dimensional argument). Then the pairs (X_i,δ_i) are iid (F). If $F_i(x) = \int^x f(s,i)\nu(ds)$, $F_+ = F_1 + F_2$ then

$$(3.2) \qquad 1 - G_1(x) = \exp\{-\int^x f_1(s)[1 - F_+(s)]^{-1}\nu(ds)\} \equiv \xi(F)(x)$$

If ξ could be extended to all probability measures on $S = R^+ \times \{0,1\}$, then a natural estimate for $1-G_1$ would be

$$\xi(\hat{F}_n)$$

where $\hat{F}_n$ is the empirical measure of $\{(X_i,\delta_i), 1 \leq i \leq n\}$. Such an extension of ξ is fairly simple. Let M be any probability on S. Set $M_i(t) = M\{(-\infty,t] \times \{i\}\}$, $M_+(t) = M_1(t) + M_2(t)$ and define

$$\xi(M)(t) = \exp\{-\int_0^t [1-M_+(s)]^{-1} dM_1(s)\}$$

With this definition, $\xi(\hat{F}_n)$ is asymptotically equivalent to the well-known Kaplan-Meier estimate.

Proceeding exactly as in section 2, fix F, with density f, given by (3.1) and let H be the Hilbert space consisting of all real functions on $S = R^+ \times \{1,2\}$ such that $\int h^2 dm < \infty$, $\int hf^{1/2} dm = 0$, support $h \subset$ support f. Let $F(h,\cdot)$ be the cdf (on S) of the measure having density $f(h;\ )$ given by

$$f^{1/2}(h;x) = (1-|h|^2/4)^{1/2} f^{1/2} + h/2$$

Let P_h^n be the product measure of $F(hn^{-1/2})$. Let

$$\tau h(u,v) = \int_0^u \int_0^v h(s,t) f^{1/2}(s,t) m(ds,dt)$$

B = closure of τH in supremum norm. Using the development of Chapter VI, it is not difficult to check that $\{P_h^n,\ h \in H\}$ converges to $\{P_h,\ h \in H\}$, the Gaussian shift experiment for the abstract Wiener space (τ,H,B).

Just as for the case of the estimation of a quantile function, the map ξ is differentiable (cf. (2.3)): there is a continuous linear map $\xi': B \to C$, where C is a Banach space of continuous functions on the line such that

$$\xi(F(n^{-1/2}h)) - \xi(F) = n^{-1/2}\xi' \circ \tau h + o(n^{-1/2})$$

as $n \to \infty$. This derivative has been calculated explicitly by Wellner (1982); however, it is important to note that, in order to perceive that a LAM result should hold in the present situation, the explicit form of ξ' is not really needed. Presumably

$$\xi(\hat{F}_n) - \xi(F(n^{-1/2}h)) = \xi'(\hat{F}_n - F_{n^{-1/2}h}) \Rightarrow \xi' \circ Z$$

where Z is a B-valued random variable with distribution P_0. Assuming ξ' is one-to-one (something that can be arranged by reducing B appropriately), the arguments of (2.5)ff in the preceding section lead to the following result. Let $N_c = \{h \in H: |h| \leq c\}$.

(3.3) THEOREM. *If* ℓ *is bounded and subconvex on* B, *then*

$$\lim_{c\uparrow\infty} \liminf_{n} \inf_{b} \sup_{h\in N_c} \int \ell[n^{1/2}(x-\xi(F(n^{-1/2}h)))]b(y,dx)P_h^n(dy)$$

$$= \lim_{c\uparrow\infty} \lim_{n} \sup_{h\in N_c} \int \ell[n^{1/2}(\xi(\hat{F}_n)-\xi(F(n^{-1/2}h)))]dP_h^n$$

$$= \int \ell\circ\xi'(x)P_0(dx)$$

The reader is reminded that the foregoing discussion is heuristic. To carry through the argument just indicated, a certain number of technical niceties have to be dealt with; these require some effort, but basically are not really difficult. The point that we make is, aside from some technicalities, the presence of a LAM result here is *apparent from the abstract structure of the problem alone*--to see its form it is not even necessary to calculate the various distributions involved.

X. Minimum distance procedures

1. Introduction

Let us begin with what is perhaps the most commonly known minimum distance set up. Let Θ be an open subset of $\mathbb{R}^d$ and, for $\theta \in \Theta$, let P_θ be a probability on the line. The cdf of P_θ will be written $F_\theta(t) = P_\theta\{(-\infty,t]\}$. One can regard F_θ as an element of several different Banach spaces--e.g. $L^2(\mu)$ where μ is a finite measure, or a space of bounded continuous functions, or L_∞. Suppose we regard F_θ as an element of a Banach space B, with norm $|\ |_B$. Suppose further that $X_1, X_2, \ldots, X_n$ are independent, identically distributed random variables and based on these "observations", we wish to estimate θ: that is, find the measure P_θ, $\theta \in \Theta$, that "best" approximates the distribution of the $\{X_i\}$. Define an estimate $\hat{\theta}_n$ of θ by specifying $\hat{\theta}_n$ to be the point in Θ that satisfies

$$\inf_{\theta \in \Theta} |F_\theta - \hat{F}_n|_B = |F_{\hat{\theta}_n} - \hat{F}_n|_B$$

Assuming there are no problems about existence and uniqueness, $\hat{\theta}_n$ is called a minimum distance estimate of θ (based on $\hat{F}_n$). Obviously $\hat{\theta}_n$ depends on the Banach space chosen; so for any given statistical problem there will be a vast number of minimum distance estimators.

The basic throught behind minimum distance estimation has a long history. A profound pioneering effort was undertaken by Neyman (1948) in his study of minimum chi square procedures; we shall discuss this method in section 5. After Neyman, another early pioneer was Wolfowitz, who was the first to advertise the method as a general principle, and stimulate much subsequent investigation; to him is due the first paradigm localization argument. Since this time, many people have worked on minimum distance methods; W. Parr has a nice recent survey.

Our development shall begin from the assumption that the Banach space is, in fact, a separable Hilbert space. This is done largely for mathematical convenience--the theory is relatively simple and the procedures have a neat asymptotically normal representation. Minimum distance procedures in non-Hilbertian B are generally difficult and usually non-Gaussian asymptotically. Nothing really important seems to be lost by contenting ourselves with the Hilbertian framework: the minimum distance procedures developed in such framework work extremely well in practice. In fact, many common useful procedures can be regarded (at least asymptotically) as minimum distance procedures (see X.3 and XII.4) below). Minimum distance procedures are also robust--they do not deteriorate when the data is subjected to a certain amount of "contamination"--see Chapter XI for more discussion of this point.

Section 2 gives a basic asymptotic representation of minimum distance estimators; this is done again in section 7, even more abstractly. Section 3 discusses in detail the application of the basic result to estimators based on the empirical cdf. Further examples are given in sections 4-6; some of them illustrate the point that the basic abstract asymptotic result of section 2 works easily outside the iid framework. Section 8 proves the existence of $n^{1/2}$-consistent estimators for qmd families $\{P_\theta\}$, a fact needed in Chapter VII.

2. Asymptotic normality

This section develops a basic asymptotic normality theorem for minimum distance estimators in a Hilbertian framework. In order that the theory cover the wide variety of possible applications, it is essential to proceed with a fair amount of abstraction. An even more abstract version can be found in Section 7.

Let Θ be an open subset of $\mathbb{R}^d$. Let P_n, $n=1,2,\dots$ be a sequence of probabilities, P_n on the measure space $(S_n, \mathcal{S}_n)$. Let B be a separable Hilbert space with norm $|\ |_B$, inner product $\langle\ \rangle_B$; often the subscript B will be omitted when it is clear from context. For each $\theta \in \Theta$ and $n = 1,2,\dots$, suppose $\xi_n(\theta,\omega)$ is a B-valued random variable. Sections 3 et seq give many examples of the stochastic processes $\{\xi_n(\theta,\ ), \theta\in\Theta\}$. Define the Θ-valued random variables $\hat{\theta}_n$ by

$$\inf_{\theta\in\Theta} |\xi_n(\theta)|_B = |\xi_n(\hat{\theta}_n)|_B \tag{2.1}$$

That is, $\hat{\theta}_n$ any point in Θ which achieves $\inf_{\theta} |\xi_n(\theta)|_B$. Under the assumptions to be expounded below, $\hat{\theta}_n$ will exist, uniquely. Sections 3 et seq explain how various choices of $\xi_n(\theta)$ lead to $\hat{\theta}_n$'s that are well known minimum distance estimators.

Fix now θ_0, an arbitrary point of Θ. There are three basic assumptions that must hold at θ_0. The first of these is the *hypothesis of identifiability*:

(2.2) for every $\varepsilon > 0$, c, there exists $\delta > 0$ such that

$$P_h\{\omega: \inf_{\theta:|\theta-\theta_0|>c} |\xi_n(\theta,\omega)-\xi_n(\theta_0,\omega)| > \delta\} \geq 1-\varepsilon \quad \text{for all } n$$

In several very important applications, $\xi_n(\theta)-\xi_n(\theta_0)$ is free of ω (i.e., is non random) and independent of n; in this case the definition is much simpler. The requirement that (2.2) hold "for all n" can be reduced to "all sufficiently large n", but we do not dwell on this. The second assumption is the *hypothesis of boundedness*

(2.3) Under the measures P_n, the B-valued random variables $n^{1/2}\xi_n(\theta_0)$ are norm bounded in probability

This means that for every ε there exists $c > 0$ such that $P_n\{n^{1/2}|\xi_n(\theta_0)| > c\} < \varepsilon$ for all n. In many applications, an even stronger assumption holds: the *hypothesis of convergence*. Under this assumption

(2.4) the B-valued random variables $n^{1/2}\xi_n(\theta_0)$ converge in distribution to a B-valued random variable W

The third assumption is the *hypothesis of differentiability*: there exists $\eta = \eta(\theta_0) = (\eta_1,\ldots,\eta_d)$, $\eta_i \in B$ such that

(2.5) $$\xi_n(\theta,\omega) = \xi_n(\theta_0,\omega) + \langle\eta,\theta-\theta_0\rangle + |\theta-\theta_0|\varepsilon(|\theta-\theta_0|;\omega)$$

where $\langle\ ,\ \rangle$, $|\ |$ are inner product and norm for $\mathbb{R}^d$ and where, for each ω, $\varepsilon(\cdot,\omega)$ is an increasing function on the line such that, for any $\varepsilon > 0$, $c > 0$, there is a $\delta > 0$ such that

$$P_n\{\omega: \sup_{|\theta|<\delta} \varepsilon(|\theta|;\omega) \leq c\} \geq 1-\varepsilon\ , \quad \text{all } n$$

The vector η is non-random (i.e., free of ω) and is assumed to be *non-singular*: there exists $c_0 > 0$ such that

(2.6) $$|\langle\eta,\theta\rangle|_B \geq c_0|\theta|\ , \quad \text{all } \theta \in \mathbb{R}^d$$

Since $\mathbb{R}^d$ is finite dimensional, (2.5) is equivalent to the assumption that the η_i, $1 \leq i \leq d$, are linearly independent in B. Extensions of the theory to infinite dimensional Θ, however, require the hypothesis (2.5). In several important applications, $\xi_n(\theta)-\xi_n(\theta_0)$ is independent of n, ω; and in these cases (2.4) merely asserts *Frêchet differentiability* of certain B-valued functionals on Θ.

Let B_η be the subspace of B spanned by $\eta_1,\ldots,\eta_d$. Let

(2.7) $$\pi \text{ be projection of } B \text{ to } B_\eta$$

Since η is non-singular, the linear operator T, mapping $\mathbb{R}^d$ to B_η, via

(2.8) $$T(t) = \langle t,\eta\rangle$$

is continuous, one-to-one, onto and so the inverse T^{-1} exists, continuous and linear.

(2.9) THEOREM. *If the hypotheses of identifiability, boundedness, and differentiability hold and if* $\hat{\theta}_n$ *is the minimum norm estimate of (2.1), then, with probability approaching* 1 *as* $n\to\infty$, $\hat{\theta}_n$ *exists and is unique. Moreover,*

$$\xi_n(\hat{\theta}_n) = (1-\pi)\xi_n(\theta_0) + o_p(n^{-1/2})$$
$$\hat{\theta}_n - \theta_0 = -T^{-1}\circ\pi\circ\xi_n(\theta_0) + o_p(n^{-1/2})$$

If in addition the hypotheses of convergence holds,

(2.10) $$n^{1/2}[\hat{\xi}_n(\hat{\theta}_n)-\xi_n(\theta_0)] \Rightarrow -\pi\circ W \quad \textit{in } B$$

(2.11) $$n^{1/2}[\hat{\theta}_n-\theta_0] \Rightarrow -T^{-1}\circ\pi\circ W \quad \textit{in } \mathbb{R}^d$$

(2.12) COROLLARY. *If* W *is Gaussian, then* $n^{1/2}[\xi_n(\theta_n)-\xi_n(\theta_0)]$ *and* $n^{1/2}[\hat{\theta}_n-\theta_0]$ *are asymptotically normal.*

The corollary is an immediate consequence of the theorem because π, $T^{-1}\circ\pi$ are bounded linear operations on a Gaussian random element. A proof of (2.8) appears in section X.9. To see heuristically the form of the limit, note first that (2.2) and (2.3) suggest that the element $\hat{\theta}_n \in \Theta$ minimizing the norm $|\xi_n(\theta)|_B$ should be close to θ_0. This being so, if θ is close to θ_0, then because of differentiability

$$\begin{aligned}|\xi_n(\theta)|^2 &= |\xi_n(\theta)-\xi_n(\theta_0)+\xi_n(\theta_0)|^2 \doteq |\xi_n(\theta_0)+\langle\theta-\theta_0,\eta\rangle|^2 \\ &= |\pi\xi_n(\theta_0)+\langle\theta-\theta_0,\eta\rangle|^2 + |(1-\pi)\xi_n(\theta_0)|^2\end{aligned}$$

Therefore, the minimum in θ should be achieved at $\hat{\theta}_n$, satisfying $\pi\xi_n(\theta_0)+\langle\hat{\theta}_n-\theta_0,\eta\rangle = 0$. This means that

$$\xi(\hat{\theta}_n) = \xi_n(\theta_0)+\xi_n(\hat{\theta}_n)-\xi_n(\theta_0) \doteq \xi_n(\theta_0)+\langle\hat{\theta}_n-\theta_0,\eta\rangle = \xi_n(\theta_0)-\pi\xi_n(\theta_0)$$

from which follows (2.9).

(2.13) EXAMPLE. Suppose Θ is one-dimensional, so $\langle\eta,t\rangle = t\eta = T(t)$. Then, if $x \in B$, $\pi x = \lambda\eta$ where $\lambda = \langle x,\eta\rangle_B/|\eta|_B^2$. Since T^{-1} is trivial to compute in this case, it follows from (2.10) that

(2.14a) $$n^{1/2}(\hat{\theta}_n-\theta_0) = n^{1/2}\langle\xi_n(\theta_0),\eta\rangle_B/|\eta|_B^2 + o(1)$$

(2.14b) $$n^{1/2}(\hat{\theta}_n-\theta_0) \Rightarrow \langle W,\eta\rangle_B/|\eta|_B^2$$

Assume further that $B = L^2(\mu)$ where μ is a finite measure on some Euclidean space. Typically, W will then be a Gaussian process with mean 0 and covariance $R(s,t) = EW(s)W(t)$. If so, it then follows that $n^{1/2}(\hat{\theta}_n-\theta_0)$ is asymptotically normal with mean 0 and variance

$$\iint R(s,t)\eta(s)\eta(t)\mu(ds)\mu(dt) \Big/ \int \eta^2(s)\mu(ds)$$

Applications of this particular example appear in section XII.4.

(2.15) REMARKS. (a) It is possible to obtain a result like (2.8) if Θ is only a Hilbert space. In applications, the most difficult hypothesis to satisfy is (2.5). It turns out to be satisfied in certain problems involving time series (Veitch (1982)) and so the basic argument of Theorem (2.9) can be carried through. In most problems involving infinite dimensional parameter sets, (2.6) will fail; see a recent analysis of regression problems (Millar (1982)) to see a possible way to deal with the difficulty. See section X.7 for a variant of the basic result that (a) is free of the $n^{1/2}$-scaling and (b) does not require the hypothesis of non-singularity.

3. Minimum distance estimators based on the empirical cdf

Let $\{P_\theta, \theta \in \Theta\}$ be a family of probability measures on the line, indexed by Θ, an open subset of $\mathbb{R}^d$. Denote by

(3.1) $$P_\theta(t) = P_\theta\{(-\infty,t]\}\ ,\ \text{cdf of } P_\theta$$

(3.2) $$P_\theta^n\ ,\ \text{n-fold product measure of } P_\theta$$

Let $X_1,\dots,X_n$ be iid real random variables, each with distribution P_θ; then P_θ^n is the distribution of the random vector $(X_1,\dots,X_n)$ on $\mathbb{R}^n$. Let

(3.3) $$\hat{F}_n(t) = n^{-1} \sum_1^n I_{(-\infty,t]}(X_i)$$

the empirical cdf. The statistical problem here is to estimate the unknown parameter θ. Let μ be a finite measure on the line and let B be the Hilbert space $L^2(\mu)$; denote the norm by $|\cdot|_B$ as in section 3.2. Of course $\hat{F}_n(t)$, $F_\theta(t)$ are elements of B and it is well known that, under P_θ^n,

(3.4) $$n^{1/2}(\hat{F}_n - F_\theta) \Rightarrow W_\theta$$

weak convergence in B where W_θ is the stochastic process $W_\theta(t) = W^0(F_\theta(t))$, $t \in \mathbb{R}^1$, W^0 being the standard Brownian Bridge. See Billingsley (1968) for a proof of convergence in $C(-\infty,\infty)$, if P_θ is continuous; for convergence in L^2, a proof can be based on the central limit theorem for Hilbert space (Prohorov (1956), Parthasarathy (1968)).

With the foregoing choice of B, define the minimum distance estimate of θ to be $\hat{\theta}_n$, a point of Θ satisfying

(3.5) $$\inf_\theta |\hat{F}_n - F_\theta|_B = |\hat{F}_n - F_{\hat{\theta}_n}|_B$$

To establish asymptotic normality, fix θ_0 and introduce the following assumptions:

(3.6) If F_{θ_n} converges to F_θ as elements of B, then $\theta_n \rightarrow \theta$ in $\mathbb{R}^d$

(3.7) there exists $\eta = (\eta_1, \ldots, \eta_d)$, η_i linearly independent elements of B, such that $F_\theta - F_{\theta_0} = \langle \theta - \theta_0, \eta \rangle + o(|\theta - \theta_0|)$

Of course (3.7) merely asserts Fréchet differentiability of the map $\theta \rightarrow F_\theta$ at θ_0. Define

(3.8) $$\xi_n(\theta,\omega) = \hat{F}_n(\cdot) - F_\theta(\cdot)$$

a random element of B. With this choice of ξ_n it is easy to see that the $\hat{\theta}_n$ of (3.5) is the same as that of (2.1) and that (3.6), (3.7) imply, respectively, the identifiability and differentiability hypotheses of section 2. Since (3.4) implies the hypothesis of convergence, the theorem (2.9) immediately implies the following result:

(3.9) THEOREM. *Let* $\{F_\theta\}$ *satisfy (3.6), (3.7); let* π *be projection to span* $\{\eta_i\}$ *in* $L^2(\mu)$. *Then under* $P^n_{\theta_0}$

$$n^{1/2}[F_{\hat{\theta}_n} - F_{\theta_0}] \Rightarrow \pi W_{\theta_0}$$

$$n^{1/2}[\hat{\theta}_n - \theta_0] \Rightarrow T^{-1}\pi W_{\theta_0}$$

where W_{θ_0} *was defined in (3.4).*

(3.10) EXAMPLE (The location model). Let $\Theta = \mathbb{R}^1$ and fix a density f on the line, with respect to Lebesgue measure. For each θ, let P_θ be the probability with density $f_\theta(x) = f(x-\theta)$. The statistical experiment $\{P_\theta, \theta \in \Theta\}$ is called the location model. Since

$$F_\theta(t) = \int_{-\infty}^{t} f(x-\theta)dx$$

under mild conditions it is clear that, if θ_0 is fixed, F_θ is differentiable at θ_0 in the sense of (3.7) and that

$$\eta(t) = \frac{d}{d\theta}F_\theta(t)\bigg|_{\theta=\theta_0} = -f(t-\theta_0) \tag{3.11}$$

It is now possibly to apply (2.14a) and obtain an asymptotic expansion of $n^{1/2}(\hat{\theta}_n - \theta_0)$. Using $\xi_n(\theta_0) = \hat{F}_n - F_{\theta_0}$ and using the derivative implied by (3.11) together with the form of the $L^2(\mu)$ inner product, one sees easily from (2.14a) and an integration by parts that

$$n^{1/2}(\dot{\theta}_n - \theta_0) = n^{-1/2}\sum_{i=1}^{n}\psi(\theta_0;X_i) \tag{3.12}$$

where $\psi(\theta_0;\cdot)$ is a function on the line given by $\psi(\theta_0;t) = b[G(t)-c]$, where $G(t) = G(\theta_0;t) = \int^t f(s-\theta_0)\mu(ds)$, $c = \int G(\theta_0;t)f(t-\theta_0)\mu(dt)$, $b^{-1} = \int f^2(t-\theta_0)\mu(dt)$.

If an *arbitrary* estimate $\hat{\theta}_n$ of θ has an asymptotic expansion of the form (3.12) for some ψ, the theory of robust statistics calls the function ψ an "influence curve" (cf. Huber (1977)). In some theories of robustness, the "quality" of the estimator is alleged to depend on the shape of ψ; this is discussed here briefly in Chapter XII. In our development, one obtains such a ψ for each measure μ used to construct $L^2(\mu) \equiv B$; as discussed in Chapter XII, one can choose μ so that ψ behaves extremely well according to the criteria of "classical" robust analysis. The asymptotic expansion (3.12) of the minimum distance estimate $\hat{\theta}_n$ can coincide--if μ is chosen properly--with the asymptotic expansion of a number of "famous" estimators. For example, proper choices of μ give minimum distance estimators $\hat{\theta}_n$ that are asymptotically equivalent (i.e., same ψ) to sample mean, sample median, trimmed mean, Hodges-Lehmann estimator, and so forth; see Millar (1981) for description of the μ needed to achieve the various possibilities, together with a number of further examples; see also XII.4.

(3.13) REMARKS. Several extensions of the basic result (3.9) are possible:

(a) It is possible to formulate a result if the μ in $L^2(\mu) = B$ is only sigma finite. Typically, the distribution functions F_θ will then not belong to $L(\mu)$. However it may well happen that $\int F_\theta(t)[1-F_\theta(t)]\mu(dt) < \infty$. In this case, the theory goes through; the main point is to prove (3.4) which can be done via the central limit theorem for iid Hilbert-valued random variables (see Millar (1981) for a version appropriate for robustness theory).

(b) For understanding the asymptotic minimax character of the minimum distance estimators of this section, the following extension is necessary. Let F_θ, P_θ, $\theta \in \Theta$, satisfy (3.6), (3.7). Let G_n be a sequence of cdf's on the line such that

$$\sup_n n^{1/2}|G_n - F_{\theta_0}| < \infty \tag{3.14}$$

Let P^n be the product measure of the G_n and assume that

$$n^{1/2}(\hat{F}_n - G_n) \Rightarrow W_{\theta_0}$$

For any distribution G, denote by $\pi_0 G$ the element of $\{F_\theta, \theta \in \Theta\}$ closest to G in the Hilbertian distance $|\ |_B$. Let $\hat{\theta}_n$ satisfy $F_{\hat{\theta}_n} = \pi_0 \hat{F}_n$ and θ_n satisfy $\pi_0 G_n = G_{\theta_n}$. Then

$$n^{1/2}(\pi_0 \hat{F}_n - \pi_0 G_n) \Rightarrow \pi W_{\theta_0} \tag{3.15}$$

$$n^{1/2}(\hat{\theta}_n - \theta_n) \Rightarrow T^{-1}\pi W_{\theta_0} \tag{3.16}$$

Applications of this particular extension appear in sections below and also in Chapter XI.

To prove (3.15), for example, apply the basic theorem to the process $\xi_{1n}(\theta) = \hat{F}_n - F_\theta$ and to the process $\xi_{2n}(\theta) = G_n - F_\theta$; then note that within $o(n^{-1/2})$

$$\begin{aligned}
\pi_0 \hat{F}_n - \pi_0 G_n &= -(\xi_{1n}(\hat{\theta}_{1n}) + F_n) + \xi_{2n}(\hat{\theta}_{2n}) - G_n \\
&= -(1-\pi)\xi_{1n}(\theta_0) - \hat{F}_n + (1-\pi)\xi_{2n}(\theta_0) - G_n \\
&= \pi(\hat{F}_n - G_n)
\end{aligned}$$

Here $\hat{\theta}_{1n}$ is the point achieving minimum norm for $\xi_{1n}(\theta)$.

4. Weighted minimum distance estimators

Bring in the framework of section X.3, equations (3.1)-(3.3). Again the problem is to estimate θ. Let μ be a sigma finite measure on the line and for each θ let $q_\theta(x)$ be a real non-negative function. Define the weighted minimum distance estimator $\hat{\theta}_n$ of θ by

$$(4.1)\quad \inf_\theta \int |\hat{F}_n(t)-F_\theta(t)|^2 q(\theta,t)\mu(dt) = \int |\hat{F}_n(t)-F_{\hat{\theta}_n}(t)|^2 q(\hat{\theta}_n;t)\mu(dt)$$

If one defined a (semi) norm $|\ |_\theta$ by $|g|_\theta = \int g(t)^2 q(\theta;t)\mu(dt)$, then $\hat{\theta}_n$ is supposed to satisfy

$$(4.2)\qquad \inf_\theta |\hat{F}_n-F_\theta|_\theta = |\hat{F}_n-F_{\hat{\theta}_n}|_{\hat{\theta}_n}$$

Therefore, the problem is clearly a perturbation of that of section 3, the norm now depending on θ. (The formulation (4.2) is more general than that of (4.1), but formulation (4.1) suffices for our application and has a simpler theory.) If each P_θ is absolutely continuous with respect to some measure μ and has density $f(\theta;x)$, then, with $\psi(\theta) = f(\theta)$, $\hat{\theta}_n$ is the *minimum Cramer-von Mises estimator of* θ.

Estimators $\hat{\theta}_n$ defined in (4.1) typically fall into the general framework of section 2 and will therefore be asymptotically normal. To see this, define

$$(4.3)\qquad \xi_n(\theta,\omega)(\cdot) = (\hat{F}_n(\omega)-F_\theta(\cdot))q^{1/2}(\theta,\cdot)$$

Assume

$$(4.4)\qquad \theta \longrightarrow q^{1/2}(\theta,\cdot) \text{ is continuous in } L^2(\mu)$$

and that

(4.5) $$(F_\theta - F_{\theta_0})\psi^{1/2}(\theta_0) = \langle g, \theta-\theta_0\rangle + o(|\theta-\theta_0|)$$

where $g = (g_1,\ldots,g_d)$ is a d-vector of elements of $L^2(\mu)$. Under these assumptions, it is easy to see that $\xi_n(\theta)$ satisfies the hypothesis of differentiability, with derivative $-g$. If μ is finite, then simple considerations involving the central limit theorem for Hilbert-valued random variables show that, under $P^n_{\theta_0}$,

(4.6) $$\sqrt{n}\xi_n(\theta_0) \Rightarrow W^0 \circ (F(\theta_0,\cdot))q^{1/2}(\theta_0,\cdot)$$

in $L^2(\mu)$. That is, the hypothesis of convergence is satisfied. If μ is only sigma finite, evident additional assumptions will need to be brought in (see, e.g., section 3). The hypothesis of identifiability is rather more painful to check out and it seems difficult to give tight, useful, general conditions on the q functions that guarantee it. See Pollard (1980) for a couple of general conditions; fortunately, in practical examples, identifiability is usually quite clear. This being so, the basic result of section 2 then implies the asymptotic normality of $\hat\theta_n$, defined in (4.1): the limiting distribution of $n^{1/2}(\hat\theta_n-\theta_0)$ is the projection onto span $\{g_i, 1 \leq i \leq d\}$ of the $L^2(\mu)$-valued variable given in the right side of (4.6).

(4.7) EXAMPLE (The location model). Consider once again the model of (3.10). Let μ be Lebesgue measure, q a function on the line and let $q(\theta,t) = q(t-\theta)$. Then, the expansion (3.12) becomes much simpler:

(4.8) $$n^{1/2}(\theta_n-\theta_0) = n^{-1/2}\sum_1^n \psi_0(X_i-\theta_0)$$

where $\psi_0(t) = b[G(t)-c]$, $G(t) = \int^t f(s)q(s)ds$, $c = \int G(t)f(t)dt$, $b^{-1} = \int f^2(t)q(t)dt$. Actually, a more general result is possible: one can take families of measures μ_θ, where $\mu_\theta(A) = \nu(A-\theta)$ for a fixed measure ν, and use the $L^2(\mu_\theta)$ norm in (4.2). Under some regularity, a "shift" representation as in (4.8) is still possible.

5. Minimum chi-square estimators

This section is a variant of section 4, but it has enormous practical importance. Let $X_1, X_2, \ldots, X_n$ be iid random variables, with distribution concentrated on the integers $1,2,\ldots,k$. (One can take $k = +\infty$ in our development, but for simplicity let us suppose $k < \infty$.) Let Θ be an open subset of $\mathbb{R}^d$ and let $p_i(\theta)$ be the probability that the observatior falls in cell i: i.e., for each θ, $p_i(\theta)$ is the probability that $X_1 = i$. Again, the statistical problem is to estimate θ.

To do this, let n_i be the number of observations X_i falling in cell i. Then n_i is a binomial random variable with mean $np_i(\theta)$. Define the *minimum chi-square estimate of* θ to be a point $\hat{\theta}_n$ in Θ that satisfies

$$(5.1) \qquad \inf_{\theta} \sum \frac{(n_i - np_i(\theta))^2}{np_i(\theta)} = \sum (n_i - np_i(\hat{\theta}_n))^2 / np_i(\theta)$$

This classical estimator (cf. Neyman (1948)) can be cast into the general framework of section 2. Indeed, let $|\ |_2$ denote the norm of the Hilbert space ℓ^2: the space of sequences $a = (a_1, a_2, \ldots, a_k)$ where $|a|_2^2 = \sum a_i^2$ (if $k = +\infty$, one must take a weighted ℓ^2 space where $|a|_2^2 = \sum \lambda_i a_i^2$, $\sum \lambda_i < \infty$, but we do not dwell on this.) Define

(5.2) $\xi_n(\theta)$ to be the vector whose i^{th} component is $[n_i - np_i(\theta)]/(n(p_i(\theta)))^{1/2}$

Then $\hat{\theta}_n$ defined in (5.1) is also defined by

$$(5.3) \qquad \inf_{\theta} |\xi_n(\theta)|_2 = |\xi_n(\hat{\theta}_n)|_2$$

To prove asymptotic normality of $n^{1/2}(\hat{\theta}_n - \theta_0)$, it is now only a matter of verifying the three hypotheses of section 2. If one assumes

$$(5.4) \qquad \theta \to \{p_i^{-1/2}(\theta)\} \text{ is continuous at } \theta_0 \text{ in } \ell^2$$

and also

$$(5.5) \qquad \{[p_i(\theta) - p_i(\theta_0)]/p_i^{1/2}(\theta_0)\} = \{\eta_i\} + o(|\theta - \theta_0|)$$

then the hypothesis of differentiability is satisfied. If $k < \infty$, then (5.4) asserts only that each $p_i(\theta)$ be positive and that $\theta \to p_i(\theta)$

be continuous; (5.5) asserts differentiability (in the ordinary sense of $\mathbb{R}^d$) of $\theta \to p_i(\theta)$, for each i. The hypothesis of convergence of $\xi_n(\theta_0)$ is easily satisfied under (5.4) and indeed is a variant of convergence of the empirical cdf to the Brownian Bridge. Assuming identifiability, the results of section 2 then imply the asymptotic normality of the minimum chi-square estimate.

6. Other minimum distance estimators

6A Estimates based on $\hat{F}_n(k;t)$

Let Θ again be an open subset of $\mathbb{R}^d$ and P_θ, $\theta \in \Theta$, a family of probabilities on $\mathbb{R}^q$. Let $X_1,\ldots,X_n$ be iid random variables having a distribution P_θ, for some $\theta \in \Theta$. Again the problem is to estimate θ. Let T be a Euclidean space (it is not really essential that it be Euclidean). Let $k(t,x)$ be a measurable function on $T \times \mathbb{R}^q$. Define for any probability measure G on $\mathbb{R}^k$

$$G(k;t) = \int k(t,x)G(dx) \tag{6.1}$$

If $\hat{F}_n$ is the usual empirical measure of $(X_1,\ldots,X_n)$, then one can define an estimate $\hat{\theta}_n$ of minimum distance type, based on $\hat{F}_n(k;t)$ by requiring $\hat{\theta}_n$ to satisfy

$$\inf_\theta |\hat{F}_n(k;\cdot)-P_\theta(k;\cdot)| = |\hat{F}_n(k,\cdot)-P_{\hat{\theta}_n}(k;\cdot)| \tag{6.2}$$

where the norm is that of some $L^2(\mu)$-space such that $P_\theta(k,\cdot) \in L^2(\mu)$ and also $\hat{F}_n(k;\cdot) \in L^2(\mu)$.

If the X_i are real and if $T = \mathbb{R}^1$ and if $k(t,x) = I_{(-\infty,t]}(x)$, then the minimum distance estimate of section 3 is recovered. If the X_i are $\mathbb{R}^d$-valued and if $T = \mathbb{R}^d$ and if $k(t,x) = \exp\{i\langle t,x\rangle\}$ where

$\langle\ ,\ \rangle$ is the inner product of $\mathbb{R}^d$, then one obtains a minimum distance estimate based on the *empirical characteristic function*.

If the X_i are real, if $T = \mathbb{R}^1$ and if $k(t,x) = x$ for all t, then one obtains a minimum distance estimate of θ based on picking the point $\int xP_\theta(dx)$ (mean P_θ) closest to $\bar{X} = n^{-1}(\sum_1^n X_i)$, the sample mean. Even more fun is possible. For example, one can choose the kernel k such that the distances between measures implied by (6.2) give a topology equivalent to the Vasherstein metrics. The theory outlined in section 2 will, under not very restrictive hypotheses, show such estimates to be asymptotically normal. These examples and others are discussed in Millar (1982b), where, in addition, the asymptotic minimaxity is proved.

6B Estimates based on the quantile function

Let $\{P_\theta, \theta \in \Theta\}$ be a family of probabilities on the line, indexed by Θ, an open subset of $\mathbb{R}^d$. Let F_0 be the cdf of P_θ and let F_θ^{-1} denote its right continuous inverse. Let $X_1, \ldots, X_n$ be iid random variables, with distribution P_θ for some $\theta \in \Theta$. Let $\hat{F}_n$ be the empirical cdf and let $\hat{F}_n^{-1}$ be its (right continuous) inverse: the stochastic process on $[0,1]$ defined by

$$(6.3) \qquad \hat{F}_n^{-1}(t) = \inf \{s: \hat{F}_n(s) \geq t\}$$

$\hat{F}_n^{-1}$ is called the *empirical quantile function*. If the measure μ is chosen appropriately, $\hat{F}_n^{-1} - F_\theta^{-1}$ will belong to $L^2(\mu)$. It is then possible to define a minimum distance estimate $\hat{\theta}_n$ based on the empirical quantile function: $\hat{\theta}_n$ must satisfy

$$(6.4) \qquad \inf_\theta |\hat{F}_n^{-1} - F_\theta^{-1}| = |F_n^{-1} - F_{\hat{\theta}_n}^{-1}|$$

Asymptotic normality of such estimates may be deduced from the theory of section 2. The key hypothesis, of course, is that

(6.5) $\theta \to F_\theta^{-1}$ be differentiable as a mapping of Θ to $L^2(\mu)$

These estimates have been shown asymptotically minimax by Millar (1982b); asymptotic normality has also been discussed by La Riccia (1981); in a somewhat less abstract framework; La Riccia gives some concrete instances of the hypothesis (6.5).

6C Spectral functions

The theory of section 2 makes no mention of "iid observations". This subsection is included merely to emphasize the point that the case of dependent observations can also be treated. Let $X_1,\ldots,X_n$ be a real stationary Gaussian sequence with mean 0 and spectral distribution function $F_\theta(t)$, $-\pi \leq t \leq \pi$, for some $\theta \in \Theta$, open subset of $\mathbb{R}^d$. Then

$$EX_kX_j = \int_{-\pi}^{\pi} e^{-ijk}F_\theta(dt)$$

Here $F_\theta(t)$ is the cumulative of the measure $F_\theta(dt)$. Again the problem is to estimate θ. Assume $F_\theta(0) = 0$, which entails no loss of generality by renormalization. Define $\hat{F}_n(t) = (2\pi n)^{-1}\int_0^t |\sum_1^n X_k e^{-iyk}|dy$.
The minimum distance estimate of θ is $\hat{\theta}_n$, defined by the condition

$$\inf_\theta |\hat{F}_n - F_\theta| = |\hat{F}_n - F_{\theta_n}| \tag{6.6}$$

where $|\cdot|$ denotes the norm of $L^2(\mu)$, μ a finite measure on $[0,\pi]$. Such an estimate will typically be asymptotically normal (and asymptotically minimax and robust). Indeed, arguments of Ibragimov (1963) can be used to verify the hypothesis of convergence; the hypothesis of differentiability will be satisfied under mild conditions on the map $\theta \to F_\theta$ in $L^2(\mu)$. In fact, this example is *exactly* parallel to that of section 2, except the hypothesis of convergence is harder, due to the dependence of the variables $\{X_i\}$. Asymptotic minimaxity is discussed in Millar (1982b), where more detailed discussion of the example may be found.

6D Hellinger metrics

Let Θ be an open, convex subset of $\mathbb{R}^d$, and let $\{P_\theta, \theta \in \Theta\}$ be a qmd family of probabilities on the line. Let Z be a subspace of $\mathbb{R}^d$ of dimension strictly less than d, and set $\Theta' = \Theta \cap Z$. Assume that n iid observations have been taken, and the family $\{P_\theta, \theta \in \Theta\}$ satisfies the hypotheses of VII.3; let $\hat{\theta}_n$ be the one-step MLE, defined in VII.3. A minimum distance estimate $\tilde{\theta}_n$ of $\theta \in \Theta'$ can be defined as follows: if $\zeta(P,Q)$ denotes Hellinger distance between measures P, Q (cf. IV.1.7) then $\tilde{\theta}_n$ is defined by

$$\inf_{\theta \in \Theta'} \zeta(P_\theta, P_{\hat{\theta}_n}) = \zeta(P_{\tilde{\theta}_n}, P_{\hat{\theta}_n})$$

(or as usual, if there are questions of existence, $\tilde{\theta}_n$ is defined as any point $\theta \in \Theta'$ that comes within n^{-1} of achieving this inf). Since the metric ζ is Hilbertian, the basic theory of this chapter applies (since the evident differentiability properties hold here) and shows that, e.g., under the n-fold product of P_{θ_0}, $\theta_0 \in \Theta'$, $\tilde{\theta}_n$ is asymptotically normal. It is also asymptotically minimax (locally), an advantage discussed and proved in Millar (1982b).

The particular set-up just described is of considerable practical importance. Many applied statisticians would be willing to believe that the distribution of the data is governed by (say) an exponential family with rather large dimensional parameter set. If the parameter set has extremely large dimension, there is not much hope of grasping quickly the basic shape of the data. Therefore, in such situations, it is essential to try to "reduce the dimensionality" of the problem: i.e., to "explain" the data "as best as one can" in terms of a much simpler parametric family. The minimum distance method introduced above precisely addresses this particular problem. I'm told this method is part of the folklore; well, maybe, and maybe not. In any case the LAM character of such procedures is new; this aspect is discussed in Millar (1982b).

7. A general result on the asymptotic form of the minimum distance functional

The basic result (2.9) can be reformulated so as to incorporate two improvements. First, a formulation can be given that eliminates the $n^{1/2}$ in hypothesis (2.3). Second, this formulation will not require non-singularity (2.6). Here is the formulation; for proof and applications, see Millar (1982b).

Let $\Theta = \mathbb{R}^d$; fix $\theta_0 \in \Theta$ (actually Θ could be a Hilbert space, but we do not dwell on this). Let $\{\xi_n(\theta,\omega): \theta \in \Theta\}$ be a sequence of stochastic processes on (Ω_n, F_n, P_n), with values in a Hilbert space H with norm $|\ |_H$. Assume (7.1), (7.2), (7.3):

(7.1) IDENTIFIABILITY. *As* $c \uparrow \infty$, *then under* P_n,

$$\inf_n \inf_{|\theta|>c} |\xi_n(\theta)|_H \longrightarrow +\infty$$

Measurability problems here can be avoided by taking a separable version of the process.

(7.2) DIFFERENTIABILITY. *There exists* $\eta = (\eta_1,\ldots,\eta_d)$, $\eta_i \in H$, *such that for every* c, *under* $\{P_n\}$,

$$\sup_{|\theta-\theta_0|<c} |\xi_n(\theta)-\xi_n(\theta_0)-\langle\eta,\theta-\theta_0\rangle|_H = o_{P_n}(1)$$

as $n \longrightarrow \infty$.

Some applications require only a weaker form of differentiability: the term $\langle\eta,\theta-\theta_0\rangle$ is replaced by $\langle\eta,T_n(\theta-\theta_0)\rangle$ where $\{T_n\}$ is a sequence of matrices such that $|T_n\theta| \leq K|\theta|$ for some K.

(7.3) BOUNDEDNESS. *Under* $\{P_n\}$, *the* H*-valued random variables* $\xi_n(\theta_0)$ *are bounded in probability: for every* $\varepsilon > 0$, *there exists* c *such that*

$$P_n\{|\xi_n(\theta_0)| > c\} \leq \varepsilon \quad \textit{for all } n$$

With these assumptions define $\hat{\theta}_n$ by

$$\inf_\theta |\xi_n(\theta)|_H = |\xi_n(\hat{\theta}_n)|_H \tag{7.4}$$

Let

(7.5) $\quad \pi$ be projection in H to span $\{\eta_i\}$

Then the following theorem holds:

(7.6) THEOREM. $\quad \xi_n(\hat{\theta}_n) = (1-\pi)\xi_n(\theta_0) + o(1)$

The results of section 2 may be recovered (with some effort) from (7.6) and subsidiary hypotheses. However, there are important applications (e.g., in regression) where the $n^{1/2}$ scaling is not present; see Millar (1982c) for some.

8. Existence of $n^{1/2}$-consistent estimates

In this section we apply the basic theory of sections 2 and 3 to show the existence of $n^{1/2}$-consistent estimators in a special problem. Suppose $\Theta \subset \mathbb{R}^d$ is open and $\{P_\theta, \theta \in \Theta\}$ is a family of probabilities on the line. Take n iid observations. We wish to exhibit a

$n^{1/2}$-consistent estimator of θ (cf. definition VII.1.1)--an item needed in the construction of the 1-step MLE in VII.3.

We shall construct such an estimator under the following assumptions. Let F_θ be the cdf of P_θ and P_θ^n the n-fold product of P_θ. Let μ be a finite measure on the line. Fix θ_0. Assume (3.6), (3.7). These assumptions are much weaker than those needed in Chapter VII.

(8.1) THEOREM. *Under assumptions (3.6), (3.7), there exists an estimator* $\tilde{\theta}_n$ *of* θ *such that, if* θ_0 *is fixed, then for every* $c > 0$,

$$\lim_{a\uparrow\infty} \overline{\lim_n} \sup_{n^{1/2}|\theta-\theta_0|\leq c} P_\theta^n\{n^{1/2}|\tilde{\theta}_n-\theta| > a\} = 0$$

PROOF. Let $\tilde{\theta}_n$ be the minimum distance estimate of θ, constructed in section 3. Let θ_n be any sequence in Θ, $|\theta_n-\theta| \leq cn^{-1/2}$, and satisfying

$$\sup_{n^{1/2}|\theta-\theta_0|\leq c} P_\theta^n\{n^{1/2}|\tilde{\theta}_n-\theta| > a\} = P_{\theta_n}^n\{n^{1/2}|\tilde{\theta}_n-\theta_n| > a\}$$

Let $G_n = F_{\theta_n}$, P^n the product of G_n. Then, under P^n, $n^{1/2}(\hat{F}_n-G_n)$ converges in distribution on $L^2(\mu)$ to $W^0\circ F_{\theta_0}$, where W^0 is the usual Brownian Bridge. Therefore, by (3.9), $n^{1/2}(\tilde{\theta}_n-\theta_n)$ converges to $T^{-1}\pi W\circ F_{\theta_0}$ (notations as in (3.9)). Hence

$$\lim_n \sup_{n^{1/2}|\theta-\theta_0|\leq c} P_\theta^n\{n^{1/2}|\tilde{\theta}_n-\theta| > a\} = P\{|T^{-1}\pi W\circ F_{\theta_0}| > a\}$$

hence the result.

(8.2) REMARK. As the case P_θ = uniform distribution on $[0,\theta]$ shows, this does not always give the best rate.

9. Proof of asymptotic normality

This section sketches the proof of Theorem (2.9). The proof will be broken up into several steps.

Step 1. For any neighborhood C of θ_0,

$$\inf_{\theta\in\Theta} |\xi_n(\theta)| = \inf_{\theta\in C} |\xi_n(\theta)| \tag{9.1}$$

with probability approaching 1 as $n \to \infty$.

Proof. By the triangle inequality $|\xi_n(\theta)| \geq |\xi_n(\theta)-\xi_n(\theta_0)| - |\xi_n(\theta_0)|$ so that

$$\inf_{\theta\notin C} |\xi_n(\theta)| \geq \inf_{\theta\notin C} |\xi_n(\theta)-\xi_n(\theta_0)| - |\xi_n(\theta_0)|$$

By the hypothesis of boundedness, $|\xi_n(\theta_0)| \to 0$ as $n \to \infty$, while identifiability forces $\inf_{\theta\notin C} |\xi_n(\theta)-\xi_n(\theta_0)|$ to remain positive in the limit. This proves (9.1).

Step 2. Let c_0 be the number defined in (2.6) (hypothesis of non-singularity). Let $d_n = 6n^{1/2}|\xi_n(\theta_0)|c_0^{-1}$, so $\{d_n\}$ is bounded in probability by (2.3). Then with probability approaching 1 as $n \to \infty$:

$$\inf_{\theta\in\Theta} |\xi_n(\theta)| = \inf_{\theta\in A_n} |\xi_n(\theta)| \tag{9.2}$$

where $A_n = \{\theta: n^{1/2}|\theta-\theta_0| \leq d_n\}$.

Proof. Recall the function $\varepsilon(\cdot;\cdot)$ defined in (2.5); let C be a neighborhood of θ_0 such that $\varepsilon(|\theta-\theta_0|;\omega) \leq \frac{1}{2}c_0$ when $\theta \in C$, with probability at least $1-\delta$, where δ is an arbitrary small number. Then for any θ

$$|\xi_n(\theta)| \geq |\langle\theta-\theta_0,\eta\rangle| - |\xi_n(\theta_0)| - |\theta-\theta_0|\varepsilon(|\theta-\theta_0|)$$

so for $\theta \in C$

$$|\xi_n(\theta)| \geq \frac{1}{2}c_0|\theta-\theta_0| - |\xi_n(\theta_0)|$$

This implies, by the choice of d_n, that

$$\inf_{\theta\in C} |\xi_n(\theta)| = \inf_{\theta\in A_n \cap C} |\xi_n(\theta)| = \inf_{\theta\in A_n} |\xi_n(\theta)|$$

since $A_n \subset C$ eventually. The result (9.2) now follows from Step 1.

Step 3. Let $\hat{\theta}_n$ be any point that achieves $\inf_{\theta} |\xi_n(\theta)| = |\xi_n(\hat{\theta}_n)|$. Then by step 2, $|\hat{\theta}_n-\theta| \leq d_n n^{-1/2}$, and

$$\xi_n(\hat{\theta}_n) = \xi_n(\theta_0) + \langle\eta, \hat{\theta}_n-\theta_0\rangle + o(n^{-1/2}) \tag{9.3}$$

Proof. This is immediate from step 2 and the differentiability of ξ_n.

Moreover, it follows from this that if $\hat{\theta}_n$, $\tilde{\theta}_n$ are two points that achieve the inf, then $|\hat{\theta}_n-\tilde{\theta}_n| = o(n^{-1/2})$ because of nonsingularity; that is, $\tilde{\theta}_n$ is "unique".

Step 4. $\xi_n(\hat{\theta}_n) = (1-\pi)\xi_n(\theta_0) + o(n^{-1/2})$

Proof. By step 3

$$|\xi_n(\hat{\theta}_n)|^2 = |\xi_n(\theta_0) + \langle\hat{\theta}_n-\theta_0, \eta\rangle|^2 + o(n^{-1}) \tag{9.4}$$

By step 2, and differentiability of $\xi_n(\theta)$

$$\begin{aligned}
|\xi_n(\hat{\theta}_n)|^2 &= \inf_{\theta\in A_n} |\xi_n(\theta)|^2 \\
&= \inf_{|\theta-\theta_0|<n^{-1/2}d_n} |\xi_n(\theta_0) + \langle\theta-\theta_0, \eta\rangle|^2 + o(n^{-1}) \\
&= \inf_{\theta} |\xi_n(\theta_0) + \langle\theta-\theta_0, \eta\rangle|^2 + o(n^{-1}) \\
&= |(1-\pi)\xi_n(\theta_0)|^2 + o(n^{-1})
\end{aligned} \tag{9.5}$$

the third equality because of the definition of d_n and the hypothesis of boundedness. It follows that

$$|\xi_n(\hat{\theta}_n)|^2 = |(1-\pi)\xi_n(\theta_0)|^2 + o(n^{-1}) \tag{9.6}$$

But

$$\begin{aligned}
&|(1-\pi)\xi_n(\theta_0) - \xi_n(\hat{\theta}_n)|^2 \\
&\quad = |(1-\pi)\xi_n(\theta_0) - (\xi_n(\theta_0) - \langle\theta_0-\hat{\theta}_n,\eta\rangle)|^2 + o(n^{-1}) \\
&\quad = |\xi_n(\theta_0) - \langle\theta_0-\hat{\theta}_n,\eta\rangle|^2 - |(1-\pi)\xi_n(\theta_0)|^2 + o(n^{-1}) \\
&\quad = |\xi_n(\hat{\theta}_n)|^2 - |(1-\pi)\xi_n(\theta_0)|^2 + o(n^{-1}) \\
&\quad = o(n^{-1})
\end{aligned}$$

where the second equality comes from the Pythagorean theorem. This proves step 4.

Step 5. $\hat{\theta}_n - \theta_0 = -T\circ\pi\circ\xi_n(\theta_0) + o(n^{-1/2})$

Proof. By step 4

$$\begin{aligned}
(1-\pi)\xi_n(\theta_0) &= \xi_n(\hat{\theta}_n) + o(n^{-1/2}) \\
&= \xi_n(\theta_0) + \langle\hat{\theta}_n-\theta_0,\eta\rangle + o(n^{-1/2})
\end{aligned}$$

so

$$-\pi\xi_n(\theta_0) = \langle\hat{\theta}_n-\theta_0,\eta\rangle + o(n^{-1/2})$$

from which the result is immediate.

This completes the proof of (2.9).

XI. Robustness and the minimum distance concept

1. The LAM property of minimum distance estimators

This section proves that the minimum distance estimators of section X.3 have a certain asymptotic minimax property. The minimum distance procedures mentioned in the other sections of Chapter X will also be asymptotically minimax, but the argument is more complex. See Millar (1982b) for a general, but somewhat complicated, result that covers all of these cases; the specific result of this section was proved in Millar (1981). For the rest of this section, the notations will be those of section X.3.

To formulate the local asymptotic minimax property (LAM), fix $\theta_0 \in \Theta$ and let $B = L^2(\mu)$ be the Hilbert space described in section X.3. For simplicity, assume μ finite. Define *neighborhoods* of $F(\theta_0;\ )$, the cdf of P_{θ_0}, as follows. Set

(1.1) $$N_c = \{q \in B: |q| \leq c\}$$

and let $G_{nq} \in B$ be defined by

(1.2) $$G_{nq} = F(\theta_0;\ ) + qn^{-1/2}$$

where it is *assumed* q is chosen so that $G_{n,q}$ is the cdf of a probability measure on the line, at least for all large n. Let

(1.3) G_q^n be the n-fold product measure of G_{nq}

Roughly speaking, the situation here is the following. The hypothetical model is that $X_1,\ldots,X_n$ are iid, following P_θ, for some θ; but because of small amounts of "data contamination", the actual distributions of the observations X_i follow some cdf G_{nq}. Under these circumstances, one still would like to find the measure P_θ which "best" fits the data. That is, the statistical problem is to find the cdf F_θ that is closest to the actual distribution of the data. So, in decision theoretic terms, the decision space will be $\{F_\theta, \theta \in \Theta\}$. If n observations $X_1,\ldots,X_n$ are taken, the parameter set--i.e., the possible distributions of the data, will be $\{G_q^n, q \in N_c\}$ (one can relax the iid assumption but we do not dwell on this). To describe an appropriate loss structure, define, for any cdf G,

(1.4) $\pi_0 G$ = any cdf among $\{F_\theta, \theta \in \Theta\}$ that satisfies

$$\inf_\theta |G-F_\theta|_B = |G-\pi_0 G|_B$$

Let g be an increasing function on the positive reals. Suppose ℓ, a real function on B, is defined by

$$\ell(x) = g(|x|_B^2) \tag{1.5}$$

For simplicity assume g is bounded and uniformly continuous; this assumption can be easily removed by approximating the general g from below by bounded uniformly continuous functions. If $(X_1,\ldots,X_n)$ has distribution G_q^n and if $x \in \{F_\theta\}$ is chosen from the decision space, the loss entailed shall be

$$\ell(n^{1/2}(x-\pi_0 G_{nq})) = g(n|x-\pi_0 G_{nq}|_B^2) \tag{1.6}$$

Certain of the elements q in the defining neighborhoods N_c will be of particular importance for the analysis. To describe these, let m be a sigma finite measure such that P_θ is absolutely continuous with respect to m, with density f. As in V.2.10, let H be the Hilbert space consisting of all h such that $h \in L^2(m)$ and h is orthogonal to $f^{1/2}$. Define $\tau h(t) = \int_0^t h(s) f^{1/2}(s) m(ds)$; assuming, as we do, that μ is finite, τ is a continuous linear map of H to B. It is easy to see that τ is Hilbert-Schmidt so that, if B_0 is the closure in B of τH, (τ, H, B_0) is an abstract Wiener space. For any fixed h such that $|\tau h| \leq c$, it is obvious that $G_{n,\tau h}$ is one of the possible "contaminating" distributions.

Assume that the family $\{P_\theta\}$ satisfies (X.3.6), (X.3.7); in particular $\{F_\theta\}$ has a non-singular derivative η at θ_0. Assume in addition the technical hypothesis

(1.7) B_η, the span in B of the components of η, is contained in B_0, the B-closure of τH

With these assumptions, the following theorem holds:

(1.8) THEOREM (LAM lower bound).

$$\lim_{c \uparrow \infty} \lim_n \inf_T \sup_{q \in N_c} \int \ell(n^{1/2}(T - \pi_0 G_{nq})) dG_q^n \geq E\ell(\pi \circ W^0 \circ F_{\theta_0})$$

where, for each n, *the infimum is over all estimators* T, *based on* $X_1, \ldots, X_n$ *with values in* $\{F_\theta\}$ *and* π, W^0 *have the definitions of X.3.*

(1.9) DEFINITION. For each n, let T_n^0 be a Borel function of $X_1, \ldots, X_n$ with values in $\{F_\theta, \theta \in \Theta\}$. Such a sequence of estimators $\{T_n^0\}$ is locally asymptotically minimax (LAM) at θ_0 if

$$\lim_{c \uparrow \infty} \lim_n \sup_{q \in N_c} \int \ell(n^{1/2}(T_n^0 - \pi_0 G_{nq})) dG_q^n = E\ell(\pi \circ W^0 \circ F_{\theta_0})$$

Assume all the hypotheses of (1.8), except (1.7).

(1.10) THEOREM. *If* $\hat{F}_n$ *is the empirical cdf, then* $\pi_0\hat{F}_n$, *the minimum distance estimate of* $\{F_\theta\}$ *based on* $\hat{F}_n$, *is LAM.*

Proofs of (1.8), (1.10) are based on the following simple consequence of X.3.15, which shows that *the functional* π_0 *is differentiable at* θ_0.

(1.11) LEMMA. *Fix* c; *let* $q_n \in N_c$ *for* $n = 1,2,\ldots$ *and set* $G_n = G_{n,q_n}$. *Then*

$$n^{1/2}[\pi_0 G_n - F_{\theta_0}] = \pi q_n + o(1)$$

PROOF. Take $\xi_n(\theta) = G_n - F_{\theta_0}$ in X.2.

PROOF OF (1.8). Fix $c > 0$. Then

$$\ell(n^{1/2}(T-\pi_0 G_{nq})) = \ell(T'-\pi q+o(1))$$

where $T' = n^{1/2}(T-F_{\theta_0})$ since $\pi_0 G_{nq} = F_{\theta_0} + n^{-1/2}\pi q + o(n^{-1/2})$. Assuming the $o(1)$ in the last display is uniform over all q of the form $\tau h \in N_c$, we find that the left side of (1.8) satisfies

$$(1.12)\qquad \inf_T \sup_{q\in N_c} \int \ell(n^{1/2}(T-\pi_0 G_{nq}))dG_q^n \geq \inf_{T'} \sup_{h:\, h\in N_c} \int \ell(T'-\pi\tau h)dG_{\tau h}^n$$

where the inf is over all B-valued procedures. (If the assumption just made does not hold, one can replace N_c in the last line above by $N_c \cap B_r$ where B_r is a subspace of B_0 of dimension r and $B_r \subset B_{r+1}$; then the argument presented below will continue to hold.) Since $|T'-\pi\tau h|^2 = |(1-\pi)T'|^2 + |\pi(T'-\tau h)|^2$, the form of ℓ shows that one need only consider procedures T' that satisfy $T' = \pi T'$ in addition. Because of (1.7), if π_1 is projection to B_0, $\pi T' = \pi\pi_1 T'$; therefore we need consider only procedures with values in B_0. Therefore, the display (1.12) can be continued:

$$(1.13) \qquad \geq \inf_{T'} \sup_{\tau h \in N_c} \int \ell(\pi(T'-\tau h))dG^n_{\tau h}$$

the infimum being over B_0-valued procedures. The function $x \to \ell(\pi x)$ is subconvex on B_0. Therefore by the asymptotic minimax theorem, the limit as $n \to \infty$ of the quantity in the display (1.13) is

$$\inf_{T} \sup_{\tau h \in N_c} \int \ell(\pi(T'-\tau h))dP_h$$

where $\{P_h, h \in H\}$ is the Gaussian shift experiment described in V.2.10. One may now let $c \uparrow \infty$ to obtain the result; a rigorous argument may be based on the minimax theorem, as in VII.2.

PROOF OF (1.10). Fix c. Let $q_n \in N_c$ be chosen to achieve (or come within n^{-1} of achieving)

$$\sup_{q \in N_c} \int \ell(n^{1/2}(\pi_0 \hat{F}_n - \pi_0 G_{nq}))dG^n_q$$

Note that, under $G^n_{q_n}$, $n^{1/2}(\hat{F}_n - G_{nq_n}) \Rightarrow W^0 \circ F_{\theta_0}$. Then by X.3

$$n^{1/2}(\pi_0 \hat{F}_n - \pi_0 G_{nq_n}) = n^{1/2}\pi(\hat{F}_n - G_{nq_n}) + o_p(1)$$
$$\Rightarrow \pi \circ W^0 \circ F_{\theta_0}$$

Since ℓ is bounded, continuous, this proves the result.

(1.14) REMARK. The proof had the following structure, heuristically. We know, first of all, that $\hat{F}_n$ is a LAM estimator of a cdf F. In the present problem, we were given a functional π_0 defined on cdf's F, and the problem was to estimate $\pi_0 F$ instead of F. It is plausible that $\pi_0 \hat{F}_n$ should be a reasonable estimate, provided π_0 is reasonably smooth. As can be seen from the form of the final results (1.8), (1.10), the key smoothness property is that π_0 be differentiable in the senses of (1.11) and of X.3. Proof of the LAM property of the other minimum distance estimates mentioned in Chapter X proceeds in a roughly similar way, one of the main difficulties being to compute the derivative of the minimum distance functional. An abstract approach to this, with many applications, is given in Millar (1982b).

2. Robustness

The classical "robustness problem" could, perhaps, be motivated by the following example. Let $X_1,\ldots,X_n$ be random variables. Perhaps physical theory suggests that they be iid, $N(\theta,1)$, $\theta \in \mathbb{R}^1$; the problem would then be to estimate θ. Unfortunately, it could happen--and usually does indeed--that the observations $X_1,\ldots,X_n$ are not precisely $N(\theta,1)$ for any choice of θ --even though the X_i should be so because of unquestioned physical principles. This phenomenon is often called "data contamination". It can arise from extremely simple reasons: (a) the data has been *rounded* off to the (say) 3rd decimal place (so now $X_1,\ldots,X_n$ has a *discrete* distribution which is *singular*--in the usual measure theoretic sense--to $N(\theta,1)$) (b) *incompetence*--recording clerks sporadically move the decimal point in the value of X_i three or four places to the right (or left), (c) *noise*--the device that records the value of the data may miss by a decimal or two. These contaminations (and more!) are familiar to most applied statisticians. Many other mechanisms for contamination can also be advanced, including those that involve dependence of the observations.

If there were no "data contamination", the hope was to "estimate θ". If $X_1,\ldots,X_n$ were originally iid $N(\theta,1)$, and if these observations were subject to contamination as suggested above, then, I suppose, one could still try to "estimate θ". A great deal of controversy has arisen as to what this should mean. Since θ is the *mean* of $N(\theta,1)$, perhaps one wants to estimate the *mean* of the data distribution G (assuming, e.g., that $X_1,\ldots,X_n$ are iid G, G not necessarily $N(\theta,1)$). On the other hand θ is the *median* of $N(\theta,1)$, so perhaps one should be estimating the median of G. If G is *not* symmetric about some point, then these two approaches will lead to radically different notions--each one of which could be backed up by a certain amount of rhetoric. But the situation is clearly unsatisfactory, and has even driven some statisticians to the desperate measure of asserting that what is "truly" being estimated is whatever the estimator in hand estimates (!) (cf. Huber (1972), Bickel-Lehmann)

If one goes back to the original framework ($X_1,\ldots,X_n$ iid $N(\theta,1)$) but subject to "contamination"--then it seems fairly reasonable to suggest that what one really wants to do is to find the distribution $P_\theta = N(\theta,1)$ which "gives the best description of the data". From this standpoint, one can even doubt the complete theoretical accuracy of the $N(\theta,1)$ model: you might be using it because of its appealing simplicity, while acknowledging that it can't possibly be exactly right--even from principles of physics.

The notion of "best fit" is, I think, an extremely helpful clarification. But there are a bewildering number of possibilities: what should be meant by "best" fit? One possibility is to let F be the true data distribution (which is unknown), and define $P_{\hat{\theta}}$, the "best fit" of $\{N(\theta,1)\}$ to the data, by defining $\hat{\theta}$ via the minimum distance recipe:

(2.1) $$\inf_{\theta} |F-P_{\theta}| = |F-P_{\hat{\theta}}|$$

where $|\ |$ is an $L^2(\mu)$ distance on cdf's. This means one wants to estimate the functional $\pi_0 F$, defined in section 1. On the other hand one could use the characteristic function ϕ_F of F and the characteristic function ϕ_{θ} of P_{θ} and define $P_{\hat{\theta}}$, the "best fit" of the model to the data, by

(2.2) $$\inf_{\theta} |\phi_F-\phi_{\theta}| = |\phi_F-\phi_{\hat{\theta}}|$$

where again an $L^2(\mu)$ distance is employed. This means that a somewhat different functional $\pi_{00}F \equiv \phi_{\hat{\theta}}$ is involved. Moreover, one could use a sup norm in either of the above suggestions, instead of the Hilbertian distances. So one has an uncountably infinite selection of distances to use, and an uncountably infinite selection of minimum distance functionals (e.g., functionals based on the cdf, the characteristic function f, or even mean, median).

Of course no rigid guidelines can be given: much depends on what statistical use you want to make of the data. If you want to estimate a few probabilities, then it seems best to use procedures which ensure the "best fit" for the probabilities you have in mind. If the probabilities concern only intervals, then one could use the set-up (2.1). Here the norm $|\cdot|$ should be taken as Hilbertian (e.g. $L^2(\mu)$) to ensure a simple theory: no really great advantage has been demonstrated for supremum norms, for example. The set-up (2.1) should be preferred over (2.2) because we agreed at the outset that we wanted only to estimate several probabilities--and it is a complicated matter to judge, in a precise quantitative way, *how* close the probabilities of interest are if only the characteristic functions are close. In short, if you want to estimate some probabilities, then it would seem best that your metric be formulated *directly*, in a perspicuous way, in terms of the quantities you thought you wanted to estimate. Incidentally, one can choose (Hilbertian!) distances so that "best fit" not only pick the (e.g.) cdf that "best fit", but also the first k-moments that "best fit" as well--see Millar (1982b) for details.

Let us agree that a reasonable approach to estimation in the presence of data contamination is to regard the problem as one involving "best fit", and examine in this light the special situation where $X_1,\ldots,X_n$ are iid--$N(\theta,1)$ allegedly but subject to contamination. The first task is to specify the contamination possible. This shall be done by specifying that at time n, the X_i are iid with some cdf F satisfying $|F-P_\theta| \le cn^{-1/2}$, where P_θ is a $N(\theta,1)$ cdf, and $|\cdot|$ denotes the $L^2(\mu)$ metric that we have chosen. The collection of all such F is called the "contamination neighborhood" (at time n). The task is then to estimate $\pi_0 F$ (notation as in section 1). We desire estimates that are "robust": this means that if F is different from $\{N(\theta,1)\}$; then the estimates should not "deteriorate" and give "unreliable" estimates. Informally, it is clear that if one uses $\bar{X}$, the sample mean, as an estimate of θ (so $P_{\bar{X}}$ as an estimate of $\pi_0 F$) then this estimate would be unstable, since just one very large "outlying" observation drastically affects the value of the estimate (it is not difficult to see that such "outliers" will indeed arise for certain F's in the contamination neighborhood described above). Technically we shall define this desirable "stability" property by insisting that we use only estimates that are LAM in the framework of section 1, i.e. we *define* an estimate to be robust if it is LAM. Of course, the theory of section 1 there shows that $\pi_0\hat{F}_n$, where $\hat{F}_n$ is the empirical cdf, is then a robust estimate.

This set-up gives a "robust" estimate of θ (via $\pi_0\hat{F}_n = P_{\hat{\theta}_n}$) for each finite measure μ used to construct the min-distance functional in (2.1). (Actually, in many problems one uses families of measures $\{\mu_\theta\}$, as described in section X.4; but we do not dwell on this technical complication here--see Millar (1981)). The question is--are these estimators any good? That is, does the abstract formulation of robustness just given lead to estimators that are reasonable. First of all, if μ has (say) a bounded density with full support, then the procedures make excellent sense globally. Second, they have the LAM property described in section 1; as the proof there indicates, this means that they enjoy a convergence to their asymptotic limit which is uniform over a rather broad class of distributions. The desirability of such uniform convergence was discussed at length in chapter VII for classical estimation; analogous considerations apply here--the neighborhoods are just allowed to be much bigger than in the classical case. Third, proper choices of the measure μ lead, in a number of situations, to estimators that are "asymptotically equivalent" to more commonly known estimators with excellent stability properties--section 4 of chapter XIII gives a few examples; more can be found in Millar (1980). That is, a certain amount of confidence in the abstract framework delineated above derives from the fact that a number of the procedures produced there are equivalent (asymptotically) to procedures (typically advanced on ad hoc grounds) that have long since proved reliable in practice. Finally, if the above setup is adopted, then estimators known not to be "robust" (in the intuitive sense) will not be robust in the technical sense either--i.e. they will not be LAM in the framework above; see Millar (1980) for some illustrations. In short, the abstract structure functions as it should: good estimates are produced, bad ones excluded.

Choices of "best fit" other than that suggested by (2.1) are of course possible. For example, Beran (198) began by asserting that "best fit" should be in terms of Hellinger distance: if G is a probability, define the functional $\pi_1 G = P_{\hat{\theta}}$ by

$$\inf_{\theta} \zeta(G,P_\theta) = \zeta(G,P_{\hat{\theta}})$$

where $\zeta(F,G)$ = Hellinger distance between probabilities F, G. (Warning: You will have to expand a bit of effort to see that this is what Beran's formulation asserts; Beran actually gives this a local formulation in terms of projections to subspaces.) The problem then is to estimate $\pi_1 G$. The "contamination neighborhoods" at time n consist, in this setup, of all probabilities Q such that $\zeta(P_\theta,Q) \leq cn^{-1/2}$. These contamination neighborhoods are much smaller than the ones in section 1: for example, no Q that is singular with respect to P_θ appears in such a setup. This means that this framework does not include "contamination" that arises from "discretization" of the data (round-off, for example)--a rather undesirable feature. Note that it is somewhat difficult to introduce directly a minimum distance procedure here (even though one is estimating a "min-distance" functional), because a suitable analogue of the empirical cdf is not available (one does not have an empirical measure $\hat{P}_n$ that is $n^{1/2}$-consistent for the Hellinger metric).

Nonetheless, Beran was able to construct an adaptively truncated version of the one-step MLE (cf. Chapter VII) to estimate the θ that "best fits the data" in the sense described earlier; moreover, Beran's estimates have theproperty that, if there is no contamination at all, then these estimates are as good as the one-step MLE ("good" in the sense of VII.3). We refer the reader to Beran's paper for the somewhat involved construction

The analysis of procedures using "neighborhoods" that "shrink" at rate $n^{-1/2}$ has a long history. It has been used for ages in the analysis of classical parametric estimators (chapter VII); it appears in the literature as early as 1943, in Wald's famous paper. Such shrinking neighborhoods have been used by many students of robustness to delineate the possible "contaminated data" distributions, with a variety of metrics --see the bibliographical notes. Despite this long and distinguished history, time worn objections continue to be voiced from time to time--especially in the context of "robustness". The complaints usually center on the objection that "it is unfair to have the 'amount of contamination' decrease with n".

Such objections have very little merit. In the first place, they can be dismissed on practical grounds: optimal procedures devised within this framework turn out to be quite excellent in practice, so even if the structure *appears* (at first sight) to be an oversimplification, it does provide a powerful, useful structure which produces estimators of high quality. Second, LAM properties are certainly necessary properties that any asymptotically respectable estimator should have--it is difficult to imagine one using estimators that fail to have at least some sort of reasonable LAM property (see the discussions in Chapter VII). Third, in no practical statistical problem does "n tend to infinity"; n may be extremely large, but it is fixed. For such an n, the theory gives a neighborhood, scaled by $\sqrt{n}$, but actually as large as you please. What is at stake is only the way one decides to embed a particular fixed experiment into a sequence of such. In robustness problems, neighborhoods chosen as described in this section are *already* adequate to exhibit the misbehavior of evidently non-robust estimates like the sample mean. Finally, in a great many problems--including those in this section--there exist $n^{1/2}$-consistent estimates of the parameter of interest: that is, one can always pin down the unknown quantity at least this closely. This being so, restricting the analysis to neighborhoods of radius $n^{-1/2}$ is quite a reasonable thing to do: it is a way of picking the best of the estimators that are $n^{1/2}$-consistent.

XII. Optimal Estimation of Real Non-Parametric Functionals

Let $X_1,\dots,X_n$ be iid observations with unknown distribution F. In Chapters VIII and IX, we examined the problem of estimating $\xi(F)$ where ξ was some functional defined on probability measures. For example, $\xi(F)$ is the cdf of F in VIII and in IX.2, $\xi(F)$ is the quantile function of F. The functionals ξ treated had the feature that ξ was differentiable in a certain sense (cf. IX.2,3) and roughly speaking, the derivative ξ' was a 1-1 linear map between two linear spaces: that is, ξ' did not reduce the "dimensionality" of the statistical problem. In this chapter we treat the opposite extreme: those functionals ξ whose derivatives ξ' are *one-dimensional*, in a sense to be described in section 1 below.

A large collection of recent statistical papers analyse functionals falling into this particular abstract framework: The first goal of this chapter is to give several important examples of such functionals; this is done in section 1. Next, the LAM framework is formulated for the problem of estimating such functionals and the asymptotic minimax bound is calculated. It will become quite evident that the calculation of the LAM bound in the present situation is *extremely simple*--in fact the least favorable distributions in the neighborhoods form a *one-dimensional* subspace, a convenient fact that leads to a simple explicit recipe for setting up the LAM framework. In particular, it follows that the many special derivations given in the literature for the LAM bound are quite unnecessary, for the form of the LAM bound in the case of one-dimensional functionals can be given in complete generality. These matters are discussed in section 2. In section 3 it is explained, in somewhat heuristic terms, why it is quite apparent--from structural considerations alone--that a number of well known estimators must *necessarily* be LAM (or else a slight smoothing of these estimators will be LAM). Finally, section 4 defines a notion of "influence curve" and discusses relationships of the present nonparametric functionals to several of the minimum distance estimators of Chapter X.

The foregoing developments are elaborated within a framework of iid observations and functionals are defined on *probability measures*. Evidently this is done for ease of exposition only: similar considerations can be applied to real functionals of the spectral distribution of a stationary process, for example. That is, our development has a purely abstract form, involving elements of several Hilbert spaces; this point of view will not be developed here.

Finally, the consideration presented here can be extended in the evident way to functionals whose derivatives map B into a d-*dimensional subspace* of some Banach space. In particular, it is of interest to review the developments concerning minimum distance procedures in this light (Chapters X, XI).

1. Functionals with 1-dimensional derivative: examples

All examples in this section will make use of the following basic notations. Let F be a probability measure on the line, having density f with respect to some sigma finite measure ds. Let H be the set of real functions h such that $\int h^2(s)f(s)ds < \infty$, $\int h(s)f(s)ds = 0$. H can be made into a Hilbert space by specifying the inner product $\langle h_1,h_2\rangle = \int h_1(s)h_2(s)f(s)ds$. Define

$$f(h;s) = f(s)[1+h(s)] \tag{1.1}$$

and let

(1.2) $F(h;ds)$ be the measure with density $f(h;\ )$

and let

$$F(h;t) = F(h;(-\infty,t]) \tag{1.3}$$

If h is chosen so that $f(h;\ ) \geq 0$ a.e., then $f(h;\ )$ will be a probability on the line. Define

$$(\tau h)(t) = \int^{t} h(s)f(s)ds = \langle k_t, h\rangle \tag{1.4}$$

where $k_t(s) = 1$ if $s \leq t$, $= 0$ otherwise. Then

$$F(h;t) = F(0;t) + \tau h(t) \tag{1.5}$$

In this section, we shall regard $F(h;t)$ as an element of some Banach space B; this is for the convenience of these examples only, since in some problems it is better to put the *measures* $F(h;ds)$ in an appropriate normed linear space. B^* will denote the dual of B. Next, fix a measure F and suppose given a functional ξ defined on $F(\theta h)$ for some *fixed* h and all sufficiently small real θ. In most examples, ξ will *not* be defined on $F(\theta h)$ for *every* h; however, for convenience we shall often assume this so. The range of ξ is assumed to be some Banach space B_2; in all examples but one, B_2 is the real line. Let $|\ |_2$ be the norm of B_2.

(1.6) DEFINITION. ξ is differentiable, with *one dimensional derivative*, if there exists a continuous linear mapping ξ' defined on B, with values in a *one* dimensional subspace of B_2 such that

(1.7) $$\xi(F(\theta h)) - \xi(F) = \theta\xi'(\tau h) + o(|\theta|)$$

It is understood that (1.7) must hold only for those h such that $\xi(F\theta(h))$ is defined for all small θ.

Examples will be given shortly. Since ξ' is a bounded linear map with 1-dimensional range,

(1.8) $$\xi'(x) = c(x)\chi \ , \quad x \in B$$

where χ is a fixed element of B_2, and $x \longrightarrow c(x)$ is a bounded linear functional on B. Therefore,

(1.9) $$\xi'(x) = \langle m_0, x\rangle\chi \ , \quad x \in B$$

where m_0 is a fixed element of B^*. We can, and shall, arrange henceforth that $|\chi|_B = 1$.

Examination of (1.7) shows that, however, it is really only necessary to perceive what the form of $h \longrightarrow \xi'(\tau h)$ is, rather than $x \longrightarrow \xi'(x)$. In this situation $h \longrightarrow \xi'(\tau h)$ is a bounded linear functional on H so

(1.10) $$\xi'(\tau h) = b(h)\chi$$

where b is a bounded linear functional on H, so that

(1.11) $$\xi'(\tau h) = \langle h_0, h\rangle\chi$$

for some $h_0 \in H$ by Riesz representation. Comparison with (1.9) shows

(1.12) $$h_0 = \tau^* m_0$$

This fact gives some perspective on the usual "centering" of "influence curves" common throughout the literature--to be described presently.

For studying the form of the LAM lower bound, it is important to perceive the form of h_0. There are other routes (than (1.12)) to this goal. Perhaps the most important of these involves the so-called influence curve.

(1.13) DEFINITION. Fix a probability F. Define $IC(x;F)$ by

$$IC(x;F) = \lim_{\varepsilon \to 0} \frac{\xi(F+\varepsilon(\delta_x - F)) - \xi(F)}{\varepsilon}$$

where δ_x is the unit mass at the point $\{x\}$. The *Gateaux derivative* in direction G is defined to satisfy

$$\lim_{\varepsilon \downarrow 0} \frac{\xi(F+ \ (G-F)) - \xi(F)}{\varepsilon} = \int \phi(x)(G-F)(dx)$$

It is standard to choose ϕ so that

$$\int \phi(x)F(dx) = 0 \tag{1.14}$$

which is accomplished by a simple centering of ϕ. The Gateaux derivative *exists at* F, if ϕ can be chosen independently of G.

Assuming the Gateaux derivative exists, one sees immediately that

$$IC(x;F) = \phi(x). \tag{1.15}$$

In this case

$$\xi' \circ \tau h = \int IC(x;F)h(x)f(x)dx = \langle IC,h \rangle \tag{1.16}$$

so that, in the notation of (1.11)

$$h_0(x) = IC(x;F) \tag{1.17}$$

Notice that condition (1.11) ensures that

$$(1.18)\qquad IC(x;F) \in H$$

(i.e. the *orthogonality condition is met*), provided only that

$$|IC|_H < \infty ,$$

a property that is not automatic, unfortunately. From our point of view the "real" reason for the customary centering of influence curves is merely the abstract necessity that we deal with elements of H -*by definition satisfying* $\int hf = 0$.

We now present important examples of such functionals ξ and compute their derivatives. The first three examples play a key role in classical robustness theory (a subject given only scant attention in these notes); the fourth example (bootstrap) is a common technique among certain applied statisticians.

(1.19) M-FUNCTIONALS. For this example, ξ will be real valued (although extensions are possible, and even useful). Let

$$(1.20)\qquad \psi(x,t) \text{ be a real function on } R^{1}\times R^{1}$$

For a probability F, $\xi(F)$ is *defined* to be the solution of the equation

$$(1.21)\qquad \int \psi(x;t)F(dx) = 0$$

There are obvious problems of existence and uniqueness of this functional. If $\psi(x;t)$ is strictly increasing in t, then the uniqueness problem is soluble; otherwise, some restriction will have to be made on F (and then also on the Fh to be considered below). Assuming such difficulties have been dealt with, $\xi(F)$ satisfies

$$(1.22) \qquad \int \psi(x;\xi(F))F(dx) = 0$$

To see that such difficulties of existence and uniqueness are truly present, consider the two most common examples:

(a) $\psi(x,t) = x-t$, which leads, *if* $\int x dF(x) < \infty$, to $\xi(F)$ = expected value of F

(b) $\psi(x,t) = \text{sign}(x-t)$, which leads *if* F *has a unique median* to $\xi(F)$ = median F.

To see what the form of ξ' should be in such a situation, there is a *standard computational recipe*. Choose h. You *must assume* that $\xi(F(\theta h))$ is defined for all small θ. Then take the derivative (with respect to θ) in the identity

$$\int \psi(x,\xi(F\theta h))dF(\theta h) = 0$$

to find, if $\psi'(x,\theta) = \frac{\partial}{\partial\theta}\psi(x,\theta)$:

$$\xi' \circ \tau h \int \psi'(x,\xi(F))F(dx) + \int \psi(x,\xi(F))d\tau h = 0$$

so that

$$(1.23) \qquad \xi'(\tau h) = -\int \psi(x,\xi(F))d\tau h(x) \Big/ \int \psi'(x,\xi(F))dF(x) \, .$$

The derivation is obviously heuristic; it is not rigorous at all, for example, if $\xi(F)$ = median F and it won't work for *many* h, if $\xi(F)$ is the mean of F. Fortunately, this is not very serious. The important thing is to see--heuristically or otherwise--what the form of ξ' is on B. For the purpose of establishing the LAM lower bound, it is necessary to carry out this calculation (or a variant of it) rigorously *only for one specific* h, whose form will be specified in the next section, after the appropriate preliminary heuristics have been understood. The equation (1.23), of course, asserts that

$$\xi'(\tau h) = \langle h_0, h\rangle \tag{1.24}$$

where

$$h_0 = \lambda_0 \psi(x,\xi(F)) ,$$
$$\lambda_0 = -\left[\int \psi'(x,\xi(F))dF(x)\right]^{-1} .$$

(1.24) L-FUNCTIONALS. Let $a(dt)$ be a finite (signed) measure on $[0,1]$. The functional $\xi(F)$, F a probability on the line, is defined (heuristically) by

$$\xi(F) = \int F^{-1}(t)a(dt) \tag{1.25}$$

where $F^{-1}(t)$ is the quantile function defined in Chapter IX. For convenience, assume $a(dt)$ is a probability. The choice $a(dt)$ = unit mass at 1/2 forces $\xi(F)$ to be the median of F (assuming a unique median). If $\alpha < 1/2$, the choice

$$a(dt) = (1-2\alpha)^{-1}dt ,$$

dt = Lebesgue measure on $[\alpha,1-\alpha]$, leads to

$$\xi(F) = (1-2\alpha)^{-1}\int_{\alpha}^{1-\alpha} F^{-1}(t)dt ,$$

the α-*trimmed mean of* F.

To find the form of the derivative of ξ in (1.23), set $F_\theta = F + \theta\tau h$ and note that if $\xi(G) = G^{-1}(t)$ for a *fixed* t, G = cdf, then

$$F_\theta(\xi(F_\theta))(t) = t ,$$

so

$$F_0(F_\theta^{-1}(t)) + \theta\tau h(F_\theta^{-1}(t)) = t$$

Differentiate the above with respect to θ at $\theta = 0$: if f is density of $F_0 \equiv F$,

$$f(F_0^{-1}(t))(\frac{d}{d\theta}F_\theta^{-1})(t)\Big|_{\theta=0} + \tau h(F_0^{-1}(t)) = 0$$

so if $\xi' = (\frac{d}{d\theta}F_\theta^{-1})(t)\Big|_{\theta=0}$ we have

$$\xi' = \frac{-\tau h(F_0^{-1}(t))}{f(F_0^{-1}(t))},$$

the form announced in Chapter IX.2. This is the form of the derivative when $a(ds)$ = point mass at $\{t\}$. For general a, the above calculation leads to

(1.26) $$\xi'\circ\tau h = \int\{\tau h(F_0^{-1}(t))/f(F_0^{-1}(t))\}a(dt)\ .$$

This derivative may be put in the form $\xi'\circ\tau h = \langle h_0,h\rangle$, upon taking

(1.26a) $$\begin{aligned} h_0(x) &= IC(x;F_0) \\ &= \int\frac{t}{f(F^{-1}(t))}a(dt) - \int_{F(t)}^{1}\frac{1}{f(F^{-1}(t))}a(dt) \end{aligned}$$

(1.27) R-FUNCTIONALS. Here, for a cdf F on the line $\xi(F)$ is defined by

$$\int_0^1 J\{\tfrac{1}{2}[s+1-F(2\xi(F)-F^{-1}(s))]\}ds = 0$$

where J is some "score function", i. e., an integrable function on $[0,1]$. Using the method of the previous example, it may be shown that $\xi'\circ\tau h$ has the form

$$\int V(F;x)d\tau h(x)$$

where $$V(F;x) = \frac{U(x) - \int U(x)f(x)dx}{\int U'(x)f(x)dx}$$

$$f = \text{density } F$$

$$U'(x) = J'\{\tfrac{1}{2}[F(x) + 1 - F(2\xi(F)-x)]\}f(2\xi(F)-x)$$

See Huber (1977). Evidently, this example also has the form (1.11). For future reference, let us note here two "famous name" R-functionals:

(1.28) HODGES LEHMANN FUNCTIONAL is obtained on taking $J(t) = t - \frac{1}{2}$, which leads, if F is symmetric, to

$$IC(x;F) = \frac{F(x) - \frac{1}{2}}{\int f(x)^2 dx}$$

(1.29) NORMAL SCORES FUNCTIONAL is obtained on taking $J(t) = \Phi^{-1}(t)$, Φ = cdf of $N(0,1)$, which leads, if F is symmetric, to

$$IC(x;F) = \Phi^{-1}(F(x)) / \int \{f^2(x)/\phi[\Phi^{-1}(F(x))]\}dx$$

(1.30) BOOTSTRAP. This particular functional is enjoying considerable vogue among a number of applied statisticians. To describe this functional, certain preliminary motivations are necessary. Suppose $X_1,\ldots,X_n$ are iid F, with F unknown. Suppose $\hat{T}_n$ is a real valued statistic, available at time n, depending on $X_1,\ldots,X_n$, and satisfying the condition that $n^{1/2}[\hat{T}_n - T_n(F)]$ be asymptotically normal for a sequence of centering constants $T_n(F)$. Let $H_n(x,F)$ be the *exact* differentiable function of $n^{1/2}[\hat{T}_n - T_n(F)]$. The statistical problem is to estimate $H_n(x,F)$; the *bootstrap* estimate is $H_n(x,\hat{F}_n)$ where $\hat{F}_n$ is the empirical cdf. This problem constitutes a slight variant of the preceding ones, since the functional to be estimated varies with n; we shall see that this *does not matter* since the functional is differentiable, with 1-dimensional derivative--and the *derivative does not depend on* n. The examples also differs from the preceding ones in that the functional $F \to H_n(\cdot;F)$ is *not* real valued, although its *derivative* is still only *one* dimensional.

To describe the differentiable nature of the functionals $F \to H_n(\cdot,F)$, let B be the Hilbert space $L^2(\mu)$, where μ is a finite measure on the line. We regard $H_n(\cdot;F)$ as an element of B. Under certain regularity conditions, Beran (1982) has noticed that, if $|\cdot|$ is the norm of B,

$$(1.31)\quad |n^{1/2}\{H_n(\cdot;F(n^{-1/2}h;\ ))-H_n(\cdot;F)\} - \langle h_0,h\rangle\chi| = o(1) \quad \text{for each } h \in H.$$

Here h_0 is a specific element of H that has the form $h_0 = \tau^* m_0$ for $m_0 \in B^* = B$, and χ is a fixed element of B that depends on F. See Beran (1982), eg (2.8), for details; the identity (1.31) is even uniform in h, in a certain sense.

(1.32) CLASSICAL PARAMETRIC FUNCTIONAL. Let Θ be an open subinterval of the line, and $\{P_\theta, \theta \in \Theta\}$ a qmd family (cf. Chapter VII). Take n iid observations and try to estimate θ.

This particular problem can be cast into the framework developed above. It is worthwhile to do this because (i) the notion that the classic problem of estimating θ is *really* the estimation of a (rather decent) *functional* of the *measures* $\{P_\theta, \theta \in \Theta\}$ is a useful, clarifying notion that has not received the attention it deserves; (ii) it is illuminating to view this classical problem as a special case of a general, abstract problem having many ramifications; (iii) the peculiarities of this special problem serve as a corrective to the euphoric point of view sketched in section 3.

To carry out the aforementioned development, consider a functional ξ, defined on $\{P_\theta, \theta \in \Theta\}$ by

$$(1.33) \qquad \xi(P_\theta) = \theta$$

Under standard identifiability assumptions, this functional is well defined. If $\{P_\theta\}$ is qmd, and if f_θ is the density of P_θ, then for fixed θ_0,

$$(1.34) \qquad f_{\theta+\theta_0} = f_{\theta_0}[1 + \theta h_0] + o(|\theta|)$$

where $h_0 = 2\eta/f_{\theta_0}^{1/2}$, and η = qmd at θ_0. Therefore, asymptotically, the structure is exactly that of (1.1), et seq.

This functional ξ is *differentiable along* $\{\tau h_0\}$ provided that ξ' is defined by $\xi' \circ \theta\tau h_0 = \theta$. For this functional ξ, the development of section 2 leads to the same LAM lower bound as in Chapter VII. On the other hand, this functional does not permit "natural" extensions to other measures as required by the heuristics of section 3; for example, if $\hat{F}_n$ is the empirical measure, it is not clear how to define $\xi(\hat{F}_n)$ in a natural way.

2. LAM lower bound

This section examines the LAM structure of the problem of estimating $\xi(F)$ when ξ is a functional with 1-dimensional derivative.

Let us begin somewhat heuristically. Fix F_0, cdf with density f and construct the "neighboring" measures $F(h;\cdot)$ as in section 1. Assume, for the purposes of these heuristics, that ξ is defined on all $F(h;)$ (this will hardly ever be the case, but as we shall see, it does not matter). We seek a lower bound for

$$(2.1) \qquad \lim_{c\uparrow\infty} \lim_{n} \inf_{T_n} \sup_{h:|h|\leq c} \int \ell(\sqrt{n}(T_n - \xi(F(hn^{-1})))) F^n(h,dx)$$

where $F^n(h,dx)$ is the n-fold product of $F(hn^{-1/2},dt)$

ℓ is a loss function on B_2 of the form $\ell(x) = g(|x|^2)$,

g an increasing function on the line

$\inf_{T_n}$ is computed over all estimates of $\xi(F)$

This is the evident LAM framework for the present estimation problem, and, of course, it is structurally analogous to that of previous chapters for other kinds of functionals ξ. The approach used in Chapters VII, VIII, IX and XI can be employed here as well to get the lower bound: since ξ is differentiable, $\xi(F(hn^{-1/2})) \doteq \xi(F) + n^{-1/2}\xi\circ\tau h$, so

$$(2.3) \qquad \ell(n^{1/2}(T_n - \xi(F(hn^{-1/2})))) \doteq \ell(T_n' - \xi'\circ\tau h)$$

where $T_n' = \sqrt{n}(T_n - \xi(F_0))$. Since $\xi'\circ\tau h = c(\tau h)\chi$ by (1.9), it is clear that it is necessary only to consider estimates T_n' that lie in the 1-dimensional subspace of B_2 spanned by χ, so $T_n' = T_n''\chi$, T_n'' *real.* If we take $|\chi|_{B_2} = 1$, as we may, then (2.3) may be contained, using (1.9) :

$$(2.4) \qquad \ell(T_n' - \xi'\circ\tau h) = g(|T_n'' - \langle h_0,h\rangle|)$$

Since $h \in H$, there is an *expansion* $h = \sum_1^\infty a_i e_i$ where $e_1 = h_0$ and $\{e_2,\ldots\}$ is any orthogonal system that spans H. Therefore $\langle h_0,h\rangle = |h_0|^2 a_1$ and so, given h, *the loss (2.4) depends on* h *only through the coefficient* a_1 in its expansion. Of course, T_n'' can then be rewritten $T_n'' = \hat{\theta}_n |h_0|^2$, $\hat{\theta}_n$ real random variable. This suggests that in (2.1) the collection of $h = |h| \leq c$, over which a supremum is being computed, could be replaced without any real loss by a supremum over a 1-dimensional subset. More precisely, let $\ell_0(x) = g(|h_0|^2|x|)$, x real.

Then according to the above argument, (2.1) exceeds

$$(2.5) \qquad \lim_{c\uparrow\infty} \lim_{n} \inf_{T_n} \sup_{\theta:|\theta|\le c} \int \ell_0(T_n-\theta)dF^n(\theta h_0)$$

where the inf is over all real valued statistics T_n, and $F^n(\theta h_0)$ is the n-fold product of $F(n^{-1/2}\theta h_0)$. If $P_\theta^n = F^n(\theta h_0)$, then it is easy to see that the experiments $\{P_\theta^n\}$ converge to $\{P_\theta\}$ where P_θ is the Gaussian measure on the line with mean $\theta|h_0|$, variance 1. By a now familiar argument, the quantity in (2.5) is

$$(2.6) \qquad \inf_T \sup_\theta \int \ell_0(T-\theta)dP_\theta = Eg(|X|)$$

where X is $N(0,|h_0|^2)$. (The last equality is deduced from VI.4 as follows. Let Q_θ be $N(\theta,1)$, $a = |h_0|$. Then, since $P_\theta = Q_{a\theta}$,

$$\begin{aligned}\inf_T \sup_\theta \int \ell_0(T-\theta)dP_\theta &= \inf_{T'} \sup_\theta \int \ell_0(\tfrac{1}{a}(T'-a\theta))dQ_{a\theta}\\ &= \inf_{T'} \sup_\theta \int \ell_0(\tfrac{1}{a}(T'-\theta))dQ_\theta\\ &= \int \ell_0(\tfrac{x}{a})Q_0(dx)\end{aligned}$$

and the result now follows from the form of ℓ_0.)

The next proposition summarizes this development. Part (a) asserts that, at time n, the local minmax risk is really determined on a 1-dimensional subset of the Hilbertian neighborhood of F. Part (b) gives the local asymptotic minmax lower bound; section 3 shows that it typically is attained by a properly chosen sequence of estimators.

(2.7) PROPOSITION. *If* h_0 *if given by (1.9) and* $\ell(x) = g(|x|)$, g *uniformly continuous, then*

(a) $$\inf_{T_n} \sup_{|h|\le c} \int \ell(\sqrt{n}(T_n - \xi(Fh)) dF^n(h)$$

$$= \inf_{T_n} \sup_{|\theta| \le \frac{c}{|h_0|}} \int \ell(\sqrt{n}(T_n - \xi F(\theta h_0))) dF^n(\theta h_0) + o(1)$$

(b) $$\liminf_{c\uparrow\infty} \inf_{T_n} \sup_{|h|\le c} \int \ell(\sqrt{n}(T_n - \xi(Fh))) dF^n(h) \ge Eg(|X|)$$ *where* g *is* $N(0,|h_0|^2)$

Here it is understood that the supremum is computed only over h such that $\xi(F(\theta h))$ is defined, differentiable, and that such h include h_0.

(2.8) ILLUSTRATION (M-functionals). In this case, referring to section 1, h_0 is given by

$$h_0 = \lambda_0 \psi(x,\xi(F)) \ , \quad \lambda^{-1} = \int \psi'(x,\xi(F)) f(x) dx$$

so that

$$|h_0|^2 = \frac{\int \psi(x,\xi(F)^2 f(x) dx}{\int \psi'(x,\xi(F)) f(x) dx} = |IC|^2$$

This is the variance in the LAM asymptotic normal distribution of Proposition 2.7. Of course this doesn't work if $\psi \notin L^2(fdx)$. As a special case, the *mean* is determined by

$$\psi(x,\xi(F)) = x - \mu \ , \quad \mu = \text{mean } F$$

and so $\psi' = 1$ and hence

$$|h_0|^2 = \int (x-\mu)^2 f(x) dx = \sigma_F^2 \ , \quad \text{the variance of } F$$

(2.9) ILLUSTRATION (L-functionals). Let $\xi(F)$ be the *median* of F. Then in this case (cf (1.24))

$$h_0(x) = -(2f(F^{-1}(\tfrac{1}{2})))^{-1} \quad \text{of} \quad x < F^{-1}(\tfrac{1}{2})$$
$$= (2f(F^{-1}(\tfrac{1}{2})))^{-1} \quad \text{of} \quad x > F^{-1}(\tfrac{1}{2})$$

where f is the density of F. Then, of course,

$$|h_0|^2 = 1/4f^2(\mu) \ , \quad \mu = F^{-1}(\tfrac{1}{2})$$

gives the variance in the LAM result (2.7).

3. The LAM property of $\xi(\hat{F}_n)$

In this section we discuss heuristically why it is that $\xi(\hat{F}_n)$, $\hat{F}_n$ = empirical, is often LAM. Assuming that $\xi'(\tau h) = \langle \tau^* m_0, h\rangle = \langle h_0, h\rangle$, we must see why $\sqrt{n}[\xi(\hat{F}_n) - \xi F(h/\sqrt{n})]$ converges under the product measure $F^n(h)$ to $N(0, |h_0|^2)$.

Since

$$(3.1) \qquad \xi' \circ \tau h = \langle h_0, h\rangle = \int h_0(s)h(s)f(s)ds = \int h_0(s)d\tau h(s)$$

it seems likely that ξ' should be defined on certain signed measures μ (including $\hat{F}_n - F$, at least) with a definition that looks like

$$(3.2) \qquad \xi'(\mu) = \int h_0(s)\mu(ds)$$

If h_0 is bounded, there is no difficulty with this extension. Presumably then

$$(3.3) \qquad \sqrt{n}[\xi(\hat{F}_n) - \xi(F(h/\sqrt{n}))] \doteq \xi'\sqrt{n}(\hat{F}_n - F(h/\sqrt{n})) + o(1)$$

and since $\sqrt{n}(\hat{F}_n - F(h/\sqrt{n}))$ converges (under $F^n(h)$) to the same limit as

$\sqrt{n}(\hat{F}_n - F_0)$ under $F^n(0)$, we should need to look only at the limit of

(3.4) $$\xi'(\sqrt{n}(\hat{F}_n - F_0)) = \sqrt{n}\int h_0(s)d(\hat{F}_n - F_0)$$

But $\int h_0(s)F_0(ds) = 0$ *since* $h_0 \in H$ and so (3.4) may be continued as

(3.5) $$\sqrt{n}\int h_0(s)d\hat{F}_n = \frac{1}{\sqrt{n}}\sum_1^n h_0(X_i)$$

Since $h_0 \in H$, the random variables $h_0(X_0)$ are iid mean 0, variance $|h_0|^2$, so by the central limit theorem, it is plausible that

(3.6) $$\sqrt{n}[\xi\hat{F}_n - \xi F(h/\sqrt{n})] \Rightarrow N(0,|h_0|^2)$$

"proving" that $\xi(\hat{F}_n)$ is LAM.

The argument is evidently not rigorous at all. In fact, carrying through this argument even for M-functionals in a rigorous way can be quite involved--see Reed's treatise. But it does give a place to start. One could take smoothed variants of $\xi(\hat{F}_n)$, for example.

(3.7) ILLUSTRATIONS. Despite the heuristic nature of the argument, it often gives the right answer. As a simple illustration, return to (2.8) where the problem is to estimate the mean of an unknown distribution. Here $\xi(\hat{F}_n)$ is the sample mean, which evidently satisfies $\sqrt{n}[\xi(\hat{F}_n) - \xi(F)] \Rightarrow N(0,\sigma_F^2)$, so it is LAM. As a further example, consider the estimation of the median (cf. (2.9)); here $\xi(\hat{F}_n)$ is the sample median, and it is well-known that $\sqrt{n}[\xi(\hat{F}_n) - \xi(F)]$ is asymptotically $N(0,[4f(F^{-1}(\frac{1}{2}))]^{-1})$ (cf. the elementary text of Bickel-Doksum, for example); so the median is also LAM.

Note finally, however, that the foregoing heuristics do not apply to the classical parametric functional (1.32).

4. Comparison with minimum distance functionals

The heuristics of section 3 show that the optimal estimator $\xi(\hat{F}_n)$ has an *asymptotic expansion* of the form

$$(4.1) \qquad n^{1/2}[\xi(\hat{F}_n)-\xi(F)] = n^{-1/2}\sum h_0(X_i) + o_p(1)$$

In such a situation, one could *define* h_0 to be the *influence curve* of the functional ξ (or rather, of the estimator $\xi(\hat{F}_n)$), and hence get a notion of what influence curves should mean for general estimators that have similar asymptotic expansions. Assuming an asymptotic expansion of the form (4.1), one may read off certain *robustness properties* of the estimator. For example, if h_0 is bounded, then $\xi(\hat{F}_n)$ is "insensitive to outliers"; that is, the value of the estimate is not drastically changed if (because of "data contamination") one of the X_i happens to be several orders of magnitude bigger than it should be. Thus, for example, the median has this robustness property, but the sample mean does not (since in the latter case its h_0 is unbounded). See sections 2 and 3.

Consider now the *location model*. A distribution F on the line is fixed; assume F has density f. For each real θ, let P_θ be the probability with density $f(x-\theta)$. Let $X_1,\ldots,X_n$ be iid P_θ, where θ is unknown. The problem is to estimate θ.

In Chapter X, we introduced certain minimum distance estimates of θ. Fix a sequence of measures μ_θ on the line; denote by $|\ |_\theta$ the $L^2(\mu_\theta)$ norm. The minimum distance estimate $\hat{\theta}_n$ of θ, based on $\{\mu_\theta\}$, is defined by

$$(4.2) \qquad \inf_\theta |\hat{F}_n-F_\theta|_\theta = |\hat{F}_n-F_{\hat{\theta}}|_\theta$$

where $\hat{F}_n$ is the empirical cdf, F_θ is the cdf of P_θ. See X.3.10. Assume that $\mu_\theta(A) = \mu(A-\theta)$ for some measure μ. Then according to X.3, X.4

$$(4.3) \qquad n^{1/2}(\hat{\theta}_n - \theta_0) = n^{-1/2}\sum_1^n \psi(X_i - \theta_0)$$

where ψ is given by X.4.8. That is, the estimator $\hat{\theta}_n$ has an expansion like (4.1), with "influence curve" ψ *which depends on* μ.

On the other hand, if f is *symmetric* about 0, then many of the L, M, R functionals ξ of section 2 satisfy $\xi(F_\theta) = \theta$. This is true, for example, if ξ is mean, median, trimmed mean, Hodges-Lehmann functional, normal scores functional, etc. Therefore, in such cases, $\xi(\hat{F}_n)$ will also be an estimate of θ. It is of some interest to compare these estimators with the minimum distance estimators--the basic observation is that, if μ is chosen properly, then the minimum distance estimate based on μ will have the same asymptotic expansion (i.e. same "influence curve") as a particular "nonparametric" functional. Here are some examples; in each case it is a simple computation based on the recipe for ψ given in Chapter X.4.

(4.3) EXAMPLES. (a) Take $\mu(dx) = \frac{1}{f(x)}$, if $|x| < \alpha$; $= 0$ otherwise. Then ψ corresponds to the influence curve of a *trimmed mean*:
$\psi(t) = -\alpha/p$, $t < -\alpha$; $= t/p$, $-\alpha \le t \le \alpha$; $= \alpha/p$, $t > \alpha$, where $p = \int_{-\alpha}^{\alpha} f(t)dt$.

(b) Take $\alpha = +\infty$ in example (a); then ψ is the influence curve of the *mean*.

(c) Let $\mu(dx)$ be point mass at $\{0\}$. Then ψ is the influence curve of the *median*.

(d) Take $\mu(dx)$ to be Lebesgue measure on the line. Then ψ is the influence curve of the Hodges-Lehmann functional.

And so forth. We emphasize that for these comparisons, f is assumed *symmetric*; for then all of these nonparametric functionals are estimates of θ. If f is not symmetric, these functionals are then in general different, and $\xi(\hat{F}_n)$ estimates *different* things for different ξ.

5. Extensions

The structure of sections 1-3, involving a particular (τ, H, B) and the empirical cdf, was selected mainly for ease of exposition. The basic considerations given in our explicit development apply to a number of other interesting situations.

For example, one could consider $X_1, \ldots, X_n$ a stationary Gaussian sequence with spectral function F. Bring in the Hilbert space of IX.1 and the mapping τ given there; ξ will then be a differentiable functional defined on spectral functions, and having 1-dimensional derivative. Results analogous to those of sections 2 and 3 continue to hold.

For another example, suppose $X_1, \ldots, X_n$ are iid F, F is an unknown measure on [0,1] having a density f *symmetric* about 1/2 (cf. VIII.4.1). Let H be the Hilbert space of symmetric functions h given in VIII.4 (after display VIII.4.2); define τ as in that section. Let ξ be a functional defined on symmetric F, having one-dimensional derivative. Then analogues of the results of sections 2 and 3 continue to hold: (a) the least favorable family of measures is on a one-dimensional subspace of H, (b) the estimator $\xi(\hat{F}_n^s)$ is typically LAM, where $\hat{F}_n^s$ is the symmetrized empirical cdf. The details will be left to the reader.

Other problems where the distribution of the data F is *known* to belong to some class C can also be dealt with. For example, following the development of VIII and of sections 2 and 3 of the present chapter, one easily proves a LAM result for $\xi(\hat{F}_n)$ where ξ is a functional with one-dimensional derivative, and F is *known* to belong to the class of distributions with decreasing density on $[0,\infty)$ (or to the class IFR, DFR, etc.).

Finally, it should be evident that the considerations of this chapter extend, with some minor changes, to functionals ξ which are "differentiable" and whose derivatives belong to R^d, $d < \infty$ fixed. The representation of the LAM lower bounds, of course, can be given in terms of a d-dimensional normal family. The minimum distance functionals of Chapters X and XI give examples. Functionals ξ which are "differentiable", having derivative ξ' that is infinite dimensional, were met in Chapter VIII; see the section on the quantile functional especially. Evidently, there is an abstract result covering all of these cases, but it is a bit cumbersome to state; it is perhaps better to understand the basic approach than to ensconce it in a ponderous formulation.

XIII. Further applications of the asymptotic minimax theory

As further illustration of the scope of the basic theory, this chapter describes (1) δ_n-consistency and (2) regression.

1. $n^{1/2}$-consistency in qmd families, revisited

We begin discussion of this particular application by showing how the $n^{1/2}$-consistency result for qmd families (VII.1) can be derived as a consequence of the asymptotic minimax theorem. For convenience let us restate that consistency result. Let $\{P_\theta, \theta \in \Theta\}$ be a qmd family. Fix $\theta_0 \in \Theta$ where Θ is an open subset of R^d. Assume the matrix $\Gamma(\theta_0)$ defined in Chapter III is nonsingular. Let P_θ^n denote the product measure of P_{θ_n}, where $\theta_n = \theta_0 + \theta n^{-1/2}$. For the formulation below, let $b(x,dy)$ denote any (generalized) procedure with decision space Θ, i.e., the statistical problem is to estimate θ. Then according to VII.1.2

(1.1) It is impossible that there exists a sequence of estimators T_n such that

$$\lim_n \sup_{|\theta| \le c} P_\theta^n\{n^{1/2}|T_n - \theta_n| > \varepsilon\} = 0$$

for any $\varepsilon > 0$ and any $c > 0$.

To prove (1.1) it will be more than enough to show

$$\text{(1.2)} \qquad \liminf_n \inf_b \sup_{|\theta| \le c} \int \ell(n^{1/2}(x-\theta_n)) b(y,dx) P_\theta^n(dy) > 0$$

where ℓ is the real function on R^d, defined by

$$\begin{aligned} \ell(x) &= 1 \quad \text{if} \quad |x| > \varepsilon \\ &= 0 \quad \text{if} \quad |x| \le \varepsilon . \end{aligned}$$

Of course ℓ is subconvex. By a familiar argument obtained by writing $\theta_n = \theta_0 + \theta n^{-1/2}$ and re-labeling procedures, (1.2) is equal to

$$(1.3) \qquad \liminf_{n} \inf_{b} \sup_{|\theta|\le c} \int \ell(x-\theta) b(y,dx) P_\theta^n(dy) \ge \inf_{b} \sup_{|\theta|\le c} \int \ell(x-\theta) b(y,dx) P_\theta(dy)$$

by the asymptotic minimax theorem; here $\{P_\theta\}$ is the Gaussian shift family with mean vector θ, covariance $\frac{1}{4}\Gamma^{-1}(\theta_0)$. Of course $\{P_\theta^n\}$ converges to $\{P_\theta\}$ (cf. IV, VI) so application of the asymptotic minimax theorem is justified. It is easy to see that the last expression in (1.3) is strictly positive for every $c > 0$. (It is obviously positive for all large c, since as $c \uparrow \infty$, it approaches $\int \ell(x) P_0(dx)$, by VI.2; it is then positive for all c by a simple scaling argument.)

2. δ_n-consistency

The preceding section demonstrated that, in a simple problem, the determination of the best 'rate' at which one could estimate a parameter can be reformulated as an asymptotic minimax property. In this section we give a general formulation of the problem and illustrate its use in the particular problem of estimating the mode of a density.

Let $\{P_\theta^n, \theta \in \Theta\}$ be a sequence of experiments, where Θ is a metric space, with metric d. The statistical problem is to estimate θ.

(2.1) DEFINITION. Let $\{\delta_n\}$ be a non-decreasing sequence of positive integers and let $\{T_n\}$ be a sequence of estimators of θ. Fix $\theta_0 \in \Theta$ and let $N_n(c) = \{\theta \in \Theta: \delta_n d(\theta, \theta_0) \le c\}$. $\{T_n\}$ is δ_n-*consistent* if there exists $\varepsilon > 0$ such that

$$1 > \overline{\lim_n} \sup_{\theta \in N_n(c)} P_\theta^n\{\delta_n d(T_n,\theta) > \varepsilon\}$$

$$\geq \underline{\lim_n} \sup_{\theta \in N_n(c)} P_\theta^n\{\delta_n d(T_n,\theta) > \varepsilon\} > 0$$

(2.2) REMARKS. There are many variants of this definition. For example, in some applications, the neighborhoods of θ_0 are fixed and do not shrink to θ_0 at rate δ_n^{-1}. More rare is the variant where $N_n(c)$ is replaced by Θ and θ_0 is not singled out in advance. It should be emphasized that, whether or not there exist δ_n-consistent estimators (for a particular choice of δ_n like $n^{1/2}$, say) depends on the metric d used. If there are no such estimators in your particular problem, you might question whether you are using a reasonable metric on Θ.

Several examples of δ_n consistent estimators have appeared so far in these notes. If Θ is an open subset of R^d and if $\{P_\theta\}$ is a qmd family then the 1-step MLE is a $\delta_n = n^{1/2}$-consistent estimator of θ. If P_θ is the uniform $[0,\theta]$-family, then $T_n \equiv \max\{X_i\}$, where the X_i are iid P_θ, is a $\delta_n = n$-consistent estimator of θ. Finally, if Θ consists of all continuous cdf's on the line, and if the metric d on Θ is $d(\theta,\theta') = \sup_t |\theta(t)-\theta'(t)|$, then under the usual hypothesis of iid observations, $\hat{F}_n$, the empirical cdf, is a $\delta_n = n^{1/2}$-consistent estimator of θ.

A basic statistical problem is, for a given asymptotic statistical estimation problem, to find the "right" sequence δ_n: that is, to find the sequence δ_n such that it is "impossible to find a sequence of estimators T_n converging to θ further than δ^n. We shall define this rigorously in a moment. Having found the "right" sequence δ_n, one then seeks explicit estimators T_n that are δ_n-consistent. As in the simple problem of section XII.1, the notion of a sequence $\{\delta_n\}$

being the "right rate" for a statistical problem is most conveniently formulated as an asymptotic minimax property.

(2.3) DEFINITION. Let Θ be a metric space, $\{P_\theta^n, \theta\in\Theta\}$ a sequence of experiments. A sequence $\{\delta_n\}$ is called an *optimal rate* for $\{P_\theta^n\}$ if (a) and (b) below hold:

(a) for each θ_0, there exists $\varepsilon > 0$ such that

$$\varliminf_n \inf_{T_n} \sup_{\theta\in N_n(c)} P_\theta^n\{\delta_n d(T_n,\theta) > \varepsilon\} > 0$$

where the infimum is over all possible estimators of θ and $N_n(c)$ was defined in (2.1);

(b) there is a sequence $\{T_n\}$ such that $\delta_n(T_n,\theta)$ is P_θ^n-bounded in probability (i.e. the $\overline{\lim}$ condition in (2.1) holds).

Evidently, $\{T_n\}$ in (2.3b) will be δ_n-consistent if (2.3a) holds. The condition (2.3a) of course is a (local) asymptotic minimax property and presumably in many problems our general theory should help analyze it. On the other hand, the methods for finding estimators that satisfy (b) vary radically with the problem at hand, and there does not seem to be much general theory available except in the case where $\delta_n = n^{1/2}$ (cf. Chapter IX). Accordingly in this section, we shall dwell only on property (2.3a).

We illustrate now these general notions in the particular problem of estimating the mode of a probability density on the line, following the development of Hasminskii.

Let F denote the collection of all unimodal densities f that are twice differentiable in a neighborhood of the mode; if $\xi(f)$ is the mode of f, assume $f''(\xi(f)) < 0$. Fix a particular $f_0 \in F$ and define a neighborhood $N(\varepsilon;f_0)$ of f_0 by

$$N(\varepsilon;f_0) = \{f \in F\colon f(x) = f_0(x) \text{ if } |x-\mu| > \varepsilon,$$

$$\sup_{|x-\xi(f_0)|\le\varepsilon} \{|f(x)-f_0(x)| + |f'(x)-f'(x)|\} < \varepsilon\}$$

The following proposition asserts that one can estimate $\xi(f)$ no faster than rate $\delta_n = n^{1/5}$. (There indeed exist estimators that achieve this rate--see Venter, AMS 38, 1446-55 (1967)).

(2.4) PROPOSITION. *Under the conditions just given, if* P_f^n *is the* n-*fold product measure of the density* f*:*

$$\underline{\lim}\ \inf_{T_n}\ \sup_{f\in U(\varepsilon,f_0)} P_f^n\{|\xi(f)-T_n|n^{1/5} > \lambda\} > 0$$

for each $\lambda > 0$.

The infimum is taken over all estimates T_n of the mode. Here is an outline of the proof. Without loss of generality we may take the mode $\xi(f)$ of f to be 0. Let $-a = f_0''(0)$. Define $g(x) = x$ if $|x| < a^{-1}$ and to be otherwise arbitrary, except that $|g''(x)| \le a/2$, $g(x) = g(-x)$ and that the support of g be compact. Set

$$f(n;\theta;x) = f_0(x) + \theta n^{-2/5} g(xn^{1/5}) \tag{2.5}$$

One may then show $f(n;\theta;\cdot)$ is a unimodal density belonging to $U(\varepsilon,f_0)$, for all large n. Moreover,

$$\xi(f(n;\theta)) = \theta a^{-1} n^{-1/5} + o(n^{-1/5}) \tag{2.6}$$

which may be checked by a slightly tedious calculation. If

$\ell(x) = I\{|x| > \lambda\}$, then the expression in (2.4) is bounded below by

$$(2.7) \qquad \underline{\lim} \inf_{T_n} \sup_{\theta} \int \ell(T_n - a^{-1}\theta) dP_\theta^n$$

where P_θ^n is the product of $f(n;\theta;\cdot)$. Finally, it is simple to check that the experiments $\{P_\theta^n, \theta \in R'\}$ converge to a normal shift experiment $\{P_\theta\}$ on the line. The asymptotic minimax theorem then immediately implies that the expression (2.7) is bounded below by

$$\inf_{T} \sup_{\theta} \int \ell(T - a\theta) dP_\theta$$

which is strictly positive by the evaluation of the minimax risk in Gaussian shifts given in Chapter VI.

3. Regression

This section shows how minimum distance methods can be applied to very general regression problems; the asymptotic minimax theorem can then be employed to demonstrate certain local optimality properties of such methods. The problem treated here is very general and extremely technical; accordingly we shall content ourselves with only a brief outline of the development. This section should be read only after mastering Chapters VIII, IX and X.

(a) *The basic problem.* Let Θ be a subinterval of the line and let $\{F(\theta;dx), \theta \in \Theta\}$ be a fixed family of probability measures on the line. If n is a positive integer, let $a_n = (a_{n1},\ldots,a_{nn})$ be a vector in R^n, $a_{ni} \in \Theta$. Let $X_{n1},\ldots,X_{nn}$ be independent random variables, the distribution of X_{ni} being $F(a_{ni};dx)$. This set-up will be called a regression model, and the vector a_n a regression function. The regression function a_n may be entirely unknown, or else may be known to lie in some lower dimensional subset of R^n. The statistical problem is to estimate a_n.

(3.1) EXAMPLE (Shift models). Fix F, a distribution on the line, and define $F(\theta;dx) = F(dx-\theta)$. Then each $X_{ni} = Z_i + a_{ni}$, where $Z_1,\ldots,Z_n$ are iid F. If a_{ni} is known to have the form $a_{ni} = a(i/n) + b$ and if F is $N(0,1)$, then this is the familiar straight line regression problem. On the other hand, the method of this section will provide an optimal estimate of a_n even if nothing at all is assumed about its shape.

Basically the proposed estimator has this form--a more rigorous description is given later. Define

$$\hat{F}_n(s,t) = n^{-1} \sum_{i \le ns} I_{(-\infty,t]}(X_{ni}), \quad 0 \le s \le 1, \quad t \subset R' \tag{3.1a}$$

analogue of the empirical cdf for regression problems. Let m be a measure on R' and let $|\ |_m$ be the L^2 norm on $[0,1] \times R'$, relative to the measure $ds\, m(dt)$. Let $\bar{F}(a_n;s,t) = n^{-1} \sum_{i \le ns} F(a_{ni};t)$ where $F(\theta;t)$ is the cdf of $F(\theta;dx)$. The proposed estimate $\hat{a}_n$ of a_n is the vector $\hat{a}_n$ for which $\bar{F}(\hat{a}_n;\ )$ is closest to $\hat{F}_n$ in the distance $|\ |_m$. We now give a somewhat more precise description of this estimator and its optimality properties.

(b) *Reformulations*. First of all, it is necessary to get a more systematic representation of the regression functions a_n. To get a start on this, define for any $g \in L^2[0,1]$

$$c(n;i;g) = n\int_{(i-1)/n}^{i/n} g(s)ds \tag{3.2}$$

Let Γ be a fixed subset of $L^2[0,1]$; it is assumed henceforth that all the allowable a_n have the form (3.2), for some $g \in \Gamma$. This involves *no restriction* on the generality of the problem, since we are free to choose Γ. Let F_n be the sigma field on $[0,1]$ generated by the subintervals $(i/n, i+1/n]$; define

$$(3.3) \qquad T(n;g) = E\{g|F_n\}, \quad g \in \Gamma$$

Then the numbers $c(n;i;g)$, $0 < i < n$, are given by the successive values of $T(n;g)$. Accordingly, we now regard the problem of estimating the regression functions a_n, as equivalent to estimating the functions $T(n;g)$, $g \in \Gamma$; i.e. we must pick the 'right' $g \in \Gamma$. In particular, the $\{T(n,g); g \in \Gamma\}$ give the effective parameter set at time n: $X_{n1}, \ldots, X_{nn}$ are independent, X_{ni} has distribution $F(a_{ni};dt)$ where a_{ni} is the value of the ith step of some $T(n,g)$, $g \in \Gamma$.

To carry this out, define for $g \in \Gamma$ a function Fg on $[0,1] \times R$ by

$$(3.4) \qquad Fg(s,t) = \int_0^s F(g(u);t)du$$

Since this discussion is informal, we assume the appropriate measure theoretic flourishes have been announced so that this definition makes sense. Assume *identifiability*:

$$(3.5) \qquad \text{if } F(\theta;t) = F(\theta';t) \text{ for all } t, \text{ then } \theta' = \theta$$

Then if $Fg = Fh$ on $[0,1] \times R$, then $g = h$ a.e. That is, knowing Fg is tantamount to knowing g. Therefore, if $g_n = T(n;g)$, we shall try to estimate $\{Fg_n, g \in \Gamma\}$ first; this will then produce an estimate of g_n.

To define the estimator, let D_n be an increasing family of subsets of Γ, $D_n \uparrow \Gamma$. Choice of the family D_n is complicated (see Millar (1982) for details); however, if Γ is finite dimensional, then the choice $D_n = \Gamma$ is possible. Define $\hat{F}_n$ by (3.1) and $\pi_n \hat{F}_n$, $\hat{g}_n \in \{T(n,g), g \in \Gamma\}$

$$\text{(3.6)} \qquad \inf_{g \in D_n} |\hat{F}_n - F(T(n,g))|_m = |\hat{F}_n - \pi_n \hat{F}_n|_m = |\hat{F}_n - F\hat{g}_n|_m$$

so $\pi_0 \hat{F}_n = F\hat{g}_n$. (As in Chapter X, if the inf is not achieved, it does not matter--take any points that come within n^{-1} of achieving it; asymptotically it will not matter.)

(c) *Asymptotic normality*. Asymptotic normality of $\pi_n \hat{F}_n$ may be derived by methods similar to those of Chapter X, under a long list of fairly technical hypotheses. A key difficulty arises because of the failure of X.2.2 (non-singularity); it fails because of the infinite dimensional character of the present problem and the subset D_n mentioned in subsection (b) must be chosen to (among other things) evade this difficulty. Fix now $g_0 \in \Gamma$; the desired result is the asymptotic normality of $n^{1/2}[\pi_n \hat{F}_n - F(T(n,g_0))]$ when the distributions of the X_{ni} are given by $T(n,g_0)$, as described in section (b).

To begin to carry out arguments similar to those of ch X, one lets $\xi_n(\theta) = [\hat{F}_n - F(T(n,\theta))]n^{1/2}$, $\theta \in \Gamma$. The *hypothesis of convergence* (X.2.4) can be relatively simply verified, since by the CLT in Hilbert space

$$n^{1/2}[\hat{F}_n - FT(n,g_0)] \text{ converges in } L^2([0,1]\times R', dsdm)$$

to a two parameter Gaussian process $\{W(g_0;s,t), (s,t) \in [0,1]\times R'\}$ with covariance

$$K(s_1,t_1;s_2,t_2) = Fg_0(s_1 \wedge s_2;t_1 \wedge t_2) - \int_0^{s_1 \wedge s_2} F(g_0(u);t_1)F(g_0(u);t_2)du$$

To see the form of the *differentiability* assumption (X.2.5), proceed heuristically to see

$$\begin{aligned} F(g+h)(s,t) - Fg(s,t) &= \int_0^s F((g+h)(u);t) - F(g(u);t)du \\ &= \int_0^s F_1(g(u);t)h(u)du \\ &\equiv A(g;h)(s,t) \end{aligned}$$

where $F_1(\theta;t) = \frac{\partial}{\partial\theta}F(\theta;t)$. Therefore, it should be that $F(T(n,g)) \doteq FT(n,g_0) + A(T(n,g_0);T(n;g-g_0))$ which is a variant of the differentiability assumption in X.2.5. Since $T(n,g_0)$ converges in $L^2[0,1]$ to g_0, the argument of ch. X, suggests that asymptotically

$$\pi_n\hat{F}_n - FT(n,g_0) \doteq \pi[\hat{F}_n - FT(n,g_0)]$$

where π is projection to $\{A(g_0;h): h \in \text{span}\,\Gamma\}$ (Γ is regarded as a subset of $L^2[0,1]$). This, together with the convergence result mentioned above, suggests that

(3.7) $$\sqrt{n}[\pi_n\hat{F}_n - FT(n,g_0)] \Rightarrow \pi W(g_0;\cdot)$$

which is the desired asymptotic normality result.

An asymptotic normality for $\hat{g}_n$ (defined in (3.6)) can be deduced from the previous result. Define a map $V: \{A(g_0;h): h \in L^2[0,1]\}$ to $\{h: h \in L^2[0,1]\}$ by $V: A(g_0;h) \to h$. This will be well defined if things are set up properly; in fact there will be a norm $|\ |_0$ on $\{h: h \in L^2[0,1]\} \equiv L$ making L a separable, normed space (but *not* Banach) such that V is a continuous linear to the $|\ |_0$-completion of

L. (In general $|\;|_0$ depends on g_0). The result (3.7) implies

$$n^{1/2}[\pi_n\hat{F}_n - FT(n;g_0)] \doteq A(g_0;n^{1/2}T(n;\hat{g}_n - g_0))$$

is asymptotically normal. Since V is continuous, linear,

$$VA(g_0;n^{1/2}T(n;\hat{g}_n - g_0)) = n^{1/2}[g_n - T(n;g_0)]$$

must be asymptotically normal too. Much work is required to carry through this outline rigorously.

(d) *Asymptotic minimax property*. This subsection discusses the LAM character of $\pi_n\hat{F}_n$ as an estimate of $FT(n,g)$. Establishing this property also involves a fair amount of work. Here is an outline. The first step is to find an abstract Wiener space (τ,H,B), $B \subset L^2([0,1]\times R',dsdm)$, such that if $\{P_h, h\in H\}$ is the standard Gaussian shift experiment then P_0 is the distribution of the process $W(g_0;\cdot)$ given in section (c). To describe such a set-up, suppose the measures F_θ are dominated by a measure μ. Then, roughly speaking, H will be the closure in $L^2(dsd\mu)$ of all finite linear combinations of elements of L^2 having the form

$$q(u,v) = e(u)h(g_0(u);v) \tag{3.8}$$

where e is defined on $[0,1]$, h is defined on $\Theta\times R'$ and in addition satisfies the condition that

$$\int h(g_0(u);v)f^{1/2}(g_0(u);v)\mu(dv) = 0 \text{ a.e. } u \tag{3.9}$$

Here $f(\theta;x)$ is the density of $F(\theta;dx)$ with respect to μ. One may then define $\tau\colon H\rightarrow L^2(dsdm)$ by

(3.10) $$(\tau q)(s,t) = \int_0^s\int^t f^{1/2}(g_0(u);v)q(u,v)dud\mu$$

for $q \in H$. It then turns out (see Millar (1982)) that this (τ,H,B) has the desired property.

Next, it is necessary to invent experiments $\{P_h^n, h\in H\}$ that converge to $\{P_h, h\in H\}$, the Gaussian shift experiment of (τ,H,B). For this purpose, one defines P_h^n to be the product of densities $f(n;i;h;x_i)$, $1 \le i \le n$, where

$$f^{1/2}(n;i;h) = (1-|q_{nih}|_\mu^2)^{1/2} f^{1/2}(c(n;i;g_0);x_i) + q_{nih}(x_i)$$

$$h(u,v) = \sum a_{ij} e_i(u) h_j(g_0(u);v) \in H$$

h_i, $1 \le i \le n$, satisfies the condition (3.9)

e_i, $1 \le i \le n$, are defined on $[0,1]$

$$q_{n,i,h}(x_i) = \frac{1}{2}\sum_{j,k} a_{kj} b(n,i,k) h_j(c(n;i;g_0);x_i)$$

$$b(n;i;k) = n^{1/2}\int_{(i-1)/n}^{i/n} e_k(u)du$$

and $c(n;i;g_0)$ was defined in subsection (b) above. It may be shown that, for h in a dense subspace H_0 of H, the experiments $\{P_h^n, h\in H_0\}$ converge to $\{P_n, h\in H_0\}$.

Finally, let $F(n,h,i;t)$ be the cdf of $f(n,i,h)$ and set $\bar{F}(n;h;s,t) = n^{-1} \sum_{i<ns} F(n,h,i;t)$. Then $\bar{F}(n,h) = FT(n,g_0) + \tau h n^{-1/2}$, approximately. If we make the guess $FT(n,g)$ when the 'true' distribution of the data is given by $\prod_i f(n,i,h)$, then the *loss* will be

(3.11) $$\ell(n^{1/2}|\pi_n\bar{F}(n,h)-FT(n,g)|)$$

where ℓ is an increasing function on $[0,\infty)$ (see Chapter X for interpretation of such a loss structure in a simpler case). Considerations reminiscent of those in Chapter XI suggest that (3.11) should be replaceable, asymptotically, by

$$(3.12) \qquad \ell(\pi(\Delta-\tau h))$$

where $\Delta = [FT(n,g_0)-FT(n,g)]\sqrt{n}$. With (3.12) granted, the LAM lower bound is deduced exactly as in XI.1: under a number of technical assumptions

$$\lim_n \inf_{T_n} \sup_{h\in H} \int \ell[n^{1/2}|\pi_n\bar{F}(n,h)-T_n|]dP_h^n \geq E\ell(|\pi W(g_0)|)$$

where the inf is over all estimates of $\{FT(n,g), g\in\Gamma\}$. Finally, argument more or less as suggested in subsection (c) above shows $\pi_n\hat{F}_n$ to be LAM.

XIV. Bibliographical notes

Chapter II. As remarked in the text, the basic decision theoretic framework is due to Wald (1950) and Le Cam (1955). The notion of a generalized procedure was introduced by Le Cam (1955). Theorems 1.8 and 1.9 are special cases of results of Le Cam (1964) and, in the present form, are fairly elementary (I have given them as exercises in second year theoretical statistics courses). Canonical measures were introduced by Blackwell (1953); connected to Choquet conical measures by Le Cam (1972). The important notion of convergence in section 2 is due to Le Cam (1964, 1972). The notion of contiguity, invoked in (2.3), is due to Le Cam (1960); expositions may be found in Hajek-Sidak and Roussas. Section 3--the Δ-metric and its characterizations--is due to Le Cam (1964, 1972); (3.1c) is well-known from Markov process theory--see, e.g., Neveu. Theorem 3.4 is elementary.

Chapter III. The asymptotic minimax theorem is due, in its present form, to Hajek (1972) and Le Cam (1972). It has a long history--see Hajek's paper and references there. Proof given in the text appeared in Millar (1979) and is based on a minimal amount of theory and calculation; a generalization appears in Millar (1982c). Other proofs can be based on the difficult Theorem 3.8. The original convolution theorem appeared in Hajek (1970); a simpler proof, due to Bickel, was given in Roussas. The proof in the text is based on ideas of Le Cam (1972), who was the first to notice that the proof should be based on fixed point theorems. The notion of "distinguished" statistic is due to Le Cam (1972).

Chapter IV. The fundamental notion of qmd has been discussed in considerable depth by Le Cam (1969, 1974); see also references in his latter paper. The clever device in proving the asymptotic expansion of section 2 (namely to look at $(\sqrt{\ }) - 1$ instead of log) is Le Cam's (1969); it has been used by many subsequent authors.

Chapter V. This chapter summarizes standard results, but in a form easily applicable to statistical application. For sections 1 and 3, see Gross, Kuo, Skorokhod, Badrikian, Badrikian-Chevet. I guess (2.6) is well known but I can't find an explicit statement in the published literature; in any case it is very simple, and quite important for a "probabilistic" understanding. Method (2.13) has been used several times in applications (cf. papers of Millar). Converse to (2.21), which is *not needed* in statistical applications, can be proved using the powerful Prohorov theorem that gives NASC for a cylinder probability on a Hilbert space to be countably additive (Prohorov, Kuo, Badrikian, for example).

Chapter VI. The general form of the minimax risk on R^d (2.4) is now a standard fact in common texts (cf. Bickel-Doksum, Ferguson) usually with much less general loss function though. The form of the minimax risk for the abstract Wiener case appears in Millar (1979) (with slightly obscure proof and some misprints). It was also known to Le Cam, who formulated it in a purely Hilbertian situation (no τ and no B). The convolution theorem for the abstract Wiener case comes from Millar (1982a).

Chapter VII. The simple result of section VII.1 employs ideas of Le Cam (1973). Hodges' example appeared in Le Cam (1953). The faulty notion of efficiency in VII.2 is still, unfortunately, the standard one in common texts (cf. Bickel-Doksum). The LAM lower bound is, in the form here, due to Hajek (1972). The one-step MLE was introduced by Le Cam

(1956, 1969) who studied its asymptotic properties; see also Roussas, Beran (1981) for further aspects on this particular estimator.

Chapter VIII. The first proof of the asymptotic minimax theorem of section 1 is due to Dvoretsky-Kiefer-Wolfowitz (1956); the proof in the text follows Millar (1979). The basic problem of sections 2 and 4 was discussed in Millar (1979) where many other classes C are introduced; see also Kiefer-Wolfowitz (1976). The efficiency result of section 3 is due to Beran (1977) with a proof similar to that in Roussas; proof based on the infinite dimensional convolution theorem of Chapter VI is due to Millar (1982a). Section 4 is based on Millar (1979).

Chapter IX. The LAM result of section 1 is due to Levit-Samarov; that of section 2 is due to Millar (1982b) and that of section 3 is based on Wellner (1982). The similar abstract structure of these results was emphasized in Millar (1982b). The convolution theorems for sections 1, 2 and 3 appear in Millar (1982a); Wellner (1982) proves a convolution theorem for the problem of section 3, based on Beran's method (cf. Chapter VIII).

Chapter X. The basic result on asymptotic normality appears in Millar (1982b) and in slightly less general form in Millar (1981). Proof is based on arguments in those papers; see Pollard, Bolthausen, Wolfowitz for closely related developments. The examples of sections 3 and 4 are discussed at length in Millar (1981); the example of section 5 is well-known. Examples 6A, 6C and 6D are new; example 6B was discussed in somewhat less generality by La Riccia. Asymptotic minimaxity of the examples in section 6 was proved by Millar (1982b). Empirical studies of some of the minimum distance methods of sections 3 and 4 have been undertaken recently by Parr-Schucany, Parr-deWet.

Chapter XI. The LAM property of section 1 was proved by Millar (1981); for the LAM character of the other procedures of Chapter X, see Millar (1982b). General discussions of robustness can be found in Huber (1977) and Bickel (1979). The "infinitesimal" approach to robustness (involving neighborhoods shrinking at rate $n^{-1/2}$) has a long history, appearing, for example, in papers of Beran, Bickel, Huber-Carol, Jaeckel, Millar, Rieder, etc. etc. See Millar (1981), Bickel (1979) for precise references.

Chapter XII. The importance of various "differentiability" properties of functionals has been recognized since the work of von Mises in the late 1940's. Typically, the "differentiability" of the functional in question was employed to establish its asymptotic normality; see the treatise of Reeds for a modern description of some of the possibilities. The derivatives of L, M, R functionals have long been familiar to students of robustness; see the expositions of, e.g., Huber (1977), Serfling. For analysis of the LAM character of certain M-estimates, see Levit. It has long been part of the folklore that real non-parametric functionals should achieve their 'worst' behavior (locally) on certain one-dimensional parametric families; this dates at least to some rather obscure remarks of Stein in 1956 and of course has been noticed in particular cases. The development of section 2 (cf. 2.7a) gives one formulation that is quite general. See Koshevnik-Levit for related developments. Section 4 comes from Millar (1981).

Chapter XIII. The result of section 1 is part of the statistical folklore. The right rate for estimation of the mode (section 2) is due to Hasminskii. Section 3 is taken from Millar (1982c).

XV. References

Badrikian, A.: Seminaire sur les Fonctions Aleatoires Lineaires et les Mesures Cylindriques. Springer Lecture Notes 139 (1970).

______, and Chevet, S.: Mesures cylindriques, Espaces de Wiener et Fonctions Aleatoires Gaussienes. Springer Lecture Notes 379 (1974).

Beran, R. J.: Estimating a distribution function. Ann. Stat. 5, 400-404 (1977).

______: Efficient robust estimation for parametric models. Z. Wahr. (1981).

______: Estimated sampling distributions: the bootstrap and its competitors. Preprint, 1981.

Bickel, P.: Quelques aspects de la statistique robuste. Ecole d'Ete St. Flour. To appear, 1979.

______, and Doksum, K.: Mathematical Statistics: Basic Ideas and Selected Topics. Holden-Day, San Francisco, 1977.

______, and Lehmann, E.: Descriptive statistics for nonparametric models I, II. Ann. Stat. 3, 1038-1069 (1975).

Bolthausen, E.: Convergence in distribution of minimum distance estimators. Metrika 24, 215-227 (1977).

Billingsley, P.: Convergence of Probability Measures. Wiley, New York 1968.

Blackwell, D.: Equivalent comparisons of experiments. Ann. Math. Stat. 24, 265-272 (1953).

Davies, R. B.: Asymptotic inference in time series. Adv. Appl. Prob. 5, 469-497 (1973).

Dunford, N., and Schwartz, J.: Linear Operators, vol. I. Interscience, New York, 1964.

Dvoretsky, A., Kiefer, J., and Wolfowitz, J.: Asymptotic minimax character of the sample distribution function and the classical multinomial estimator. Ann. Math. Stat. 27, 642-669 (1956).

Ferguson, T. S.: Mathematical Statistics. Academic Press, New York, 1967.

Gross, L.: Abstract Wiener spaces. Proc. Fifth Berkeley Symp. 2, 31-42 (1965).

Hajek, J.: Local asymptotic minimax and admissibility in estimation. Proc. Sixth Berkeley Symp. I 175-194 (1972).

______: A characterization of limiting distributions of regular estimates. Z. Wahr. 14, 323-330 (1970).

______, and Sidak, Z.: Theory of Rank Tests. Academic Press, New York, 1967.

Hasminskii, R. Z.: Lower bound for the risk of nonparametric estimates of the mode. Contrib. to Stat. (Hajek memorial volume) 91-97, Prague, 1979.

Huber, P. J.: Robust Statistical Procedures. SIAM Series in Appl. Math. (1977).

______: Robust statistics: a review. Ann. Math. Stat. 43, 1041-1067 (1972).

Ibragimov, A.: Estimation of the spectral function of a stationary Gaussian process. Theory of Prob. 8, 391-430 (1963).

Johanssen, S.: Introduction to the Theory of Regular Exponential Families. University of Copenhagen, 1979.

Kato,

Kiefer, J., and Wolfowitz, J.: On the deviation of the empirical distribution function of a vector chance variable. T. A. M. S. (1958).

______: Asymptotically minimax estimation of concave and convex distribution functions. Z. Wahr. 34, 73-85 (1976).

Koshevnik, Ya. A, and Levit, B. Ya.: On a non parametric analogue of the information matrix. Theor. Prob. Appl. 21, 738-753 (1976).

Kuo, Hui-Hsiung: Gaussian Measures in Banach Space. Springer, 1975.

La Riccia, U. N.: Asymptotic properties of weighted L^2 quantile distance estimators. Preprint, 1981.

Le Cam, L.: On some asymptotic properties of maximum likelihood estimates and related Bayes estimates. Univ. of Calif. Publ. Stat. I, 277-330 (1953).

______: An extension of Wald's theory of statistical decision functions. A.M.S. 26, 69-81 (1955).

______: On the asymptotic theory of estimation and testing hypotheses. Proc. 3rd Berkeley Symp. I, 129-156 (1956).

______: Sufficiency and asymptotic sufficiency. A.M.S. 35, 1419-1455 (1964).

______: Theorie asymptotique de la decision statistique. Presses de l'Universite de Montreal (1969).

______: Limits of experiments. Proc. 6th Berkeley Symp. I (1972).

______: Convergence of estimates under dimensionality restrictions. Ann. Stat. 1, 38-53 (1973).

______: On the assumptions used to prove asymptotic normality of maximum likelihood estimates. A.M.S. 41, 802-828 (1970).

Levit, B. Ya.: Efficiency of a class of non parametric estimates. Theor. Prob. Appl. 20, 723-740 (1975).

______: Infinite dimensional information lower bounds. Theor. Prob. Appl. 23, 371-377 (1978).

______, and Samarov, A. M.: Estimation of spectral functions. Problems of Information Transmission, 120-124 (1978).

Loeve, M.: Probability. Van Nostrand.

Millar, P. W.: Asymptotic minimax theorems for the sample distribution function. Z. Wahr. 48, 233-252 (1979).

______: Robust estimation via minimum distance methods. Z. Wahr. 55, 73-89 (1971).

______: Non parametric applications of an infinite dimensional convolution theorem. Preprint, 1982.

______: An abstract minimax theorem, with applications to minimum distance estimation. Preprint, 1982.

______: Optimal estimation of a general regression function. To appear, Ann. Stat., 1982.

______: Robust tests of statistical hypotheses. Preprint, 1982.

Neveu, J.: Mathematical Foundations of the Calculus of Probability. Holden-Day, San Francisco, 1965.

Parr, W. C.: Minimum distance estimation: a bibliography. Preprint, 1980.

______, and Schucany, W. R.: Minimum distance and robust estimation. J.A.S.A. (1981).

______, and de Wet, T.: On minimum weighted Cramér-von Mises statistic estimation. Preprint, 1980.

Parthasarathy, K. R.: Probability Measures on Metric Spaces. Academic Press, 1967.

Pollard, D.: The minimum distance method of testing. Metrika 27, 43-70 (1980).

Prohorov, Yu. V.: Convergence of random processes and limit theorems in probability theory. Theor. Prob. Appl. (1956).

Reeds, J.: On the definition of von Mises functionals. To appear, Univ. of Chicago Press.

Roussas, G. G.: Contiguous Probability Measures: Some Applications in Statistics. Cambridge Univ. Press, 1972.

Schaeffer, H. H.: Topological Vector Spaces. Springer, 1970.

Serfling, R. J.: Approximation Theorems of Mathematical Statistics. John Wiley & Sons, New York, 1980.

Shorack, G.: Convergence of quantile and spacings processes with applications. A.M.S. 43, 1400-1411 (1972).

Skorokhod, A. V.: Integration in Hilbert Space. K. Wickwire, transl. Springer, New York, 1974.

Staudte, R. G.: Robust estimation. Queen's Papers in Pure and Applied Math. #53 (1980).

Veitch, J.: Thesis, Univ. of Calif., 1982.

Wald, A.: Statistical Decision Functions. Wiley, New York, 1950.

Wellner, J. A.: Estimating a distribution function with random censorship. Preprint, 1981.

Wolfowitz, J.: The minimum distance method. A.M.S. 28, 75-88 (1957).

SOME APPLICATIONS OF STOCHASTIC CALCULUS TO PARTIAL DIFFERENTIAL EQUATIONS

PAR Daniel W. STROOCK

Research sponsored in part by N.S.F. Grant MCS 80-07300

0. Introduction:

In these lectures I have attempted to provide an introduction to certain aspects of the application of stochastic processes to the study of partial differential equations (P.D.E.'s). It has been my intention to present this subject in such a way that it will be accessible to the non-expert. However, in spite of this intention, I have had to assume quite a good deal. In particular, an understanding of these notes will demand that the reader is comfortable with Wiener measure and has some familiarity with Itô's theory of stochastic integration (stochastic integral equations, Itô's formula, and the like). On the other hand, I have tried to make as few demands as possible on the reader's knowledge of P.D.E. theory. Indeed, what I hope is that these lectures will be convincing evidence that many of the P.D.E. theorems most useful to probabilists can be derived from probabilistic considerations.

Section 1) is an introduction to the problems treated in these notes. Included in this section is a heuristic explanation of the connection between the Cauchy initial value problem for parabolic P.D.E.'s and stochastic integral equations. Although I hope that this introduction will be useful to the non-expert, I am thoroughly aware that it is no substitute for the kind of firm background that can be obtained from a book like that by H.P. Mckean [Mckean].

In section 2), I give a quick resumé of the basic theory of stochastic integration with respect to Brownian motion. The only proofs provided are of theorems about differentiating solutions of stochastic integral equations with respect to the starting place. Our treatment here follows the ideas of H. Kunita [Kunita]. I use these results to give a probabilistic proof that solutions to the Cauchy initial value problem exist when the coefficients and initial data are smooth.

With section 3), we turn to the topic with which the rest of these notes is concerned, namely: the study of diffusion transition functions as a function of their "forward" variable. The main purpose of section 3) is to give a criterion for absolute continuity (with respect to Lebesgue measure) of measures on R^N. Although this criterion (cf. Lemma (3.2)) may be viewed as a part of Sobolev theory, its proof requires very little work and its statement is somewhat novel.

Having established in section 3) the criterion with which I want to work, I try in section 4) to explain, by means of an example, how one might attempt to use this criterion. It is at this point that the basic ideas under discussion start coming from P. Malliavin [Malliavin]. In fact, even the example which I use is really due to him.

Sections 5) through 7) are devoted to the development of the program suggested by the example in section 4). The treatment here follows [S.,J.F.A.] very closely. For the purpose of gaining an understanding of the subject, the reader might be well-advised to leave most of the details for a later time. As I said before, the basic idea is already contained in section 4); and it is there that understanding may be most easily achieved.

Section 8) is a monster. What I do in section 8) is use the methodology developed in the preceding sections to derive the renowned theorem of Hörmander [Hörmander] about hypoelliptic parabolic equations (cf. Theorem (8.43)). After attempting to wade through this section, no reader could be blamed for asking "why"? I am afraid that my only answer is that it had to be done. This is by no means the first time that Hörmander's theorem has been studied with these techniques. Malliavin himself did so in [Malliavin]. His ideas were refined and expanded by Watenabe [Ikeda and Watenabe]. More recently, Bismut has made a valiant effort in this direction. In fact, [Bismut] contains a treatment of Hörmander's full theorem. The approach adopted in section 8) is based on ideas developed by S. Kusuoka and me. A paper by us in which these ideas are exploited more fully is now in preparation. I am grateful to Kusuoka for allowing me to present this as yet unpublished work here for the first time.

Because so much space is devoted in section 8) to recovering results better proved by other methods, I decided to devote the final section, number 9), to an example for which I know no other technique. I hope that this example will serve as a stimulus for further investigation. The direction suggested by this example strikes me as a potentially fruitful line of research.

Finally, it is my pleasure to thank Professor Hennequin and the Ecole d'Été for giving me the opportunity to present these lectures. I never have a better audience than when I lecture in France. My only regret is that I am unable to do so in French; but maybe I can be forgiven since the ideas on which I was lecturing <u>are</u> French.

1. Second Order Parabolic P.D.E.'s and S.D.E.'s

Suppose that $L = 1/2 \sum_{i,j=1}^{N} a^{ij}(x)\frac{\partial^2}{\partial x_i \partial x_j} + \sum_{i=1}^{N} b^i(x)\frac{\partial}{\partial x_i}$ is a second order elliptic differential operator having smooth coefficients (i.e. for each $x \in R^N$, the matrix $a(x)$ is symmetric and non-negative definite). Given $\phi \in C_b(R^N)$ ($\equiv$ the space of bounded continuous functions on R^N), consider the Cauchy initial value problem of finding a $u \in C^{1,2}([0,\infty)\times R^N)$ satisfying

$$(1.1) \qquad \frac{\partial u}{\partial t} = Lu \quad , \quad t > 0 \quad , \quad \text{and} \quad u(0,\cdot) = \phi(\cdot) \quad .$$

Assuming that $a(\cdot)$ and $b(\cdot)$ are bounded, analysts have shown (cf. Chapter 3 of [S. & V.] for example) that for each $\phi \in C_b^2(R^N)$ ($\equiv$ the space of twice continuously differentiable functions on R^N satisfying $D^\alpha\phi \in C_b(R^N)$ for $\alpha \in (\mathbb{N})^N$ with $|\alpha| \equiv \sum_1^N \alpha_i \leq 2$) there is a unique $u_\phi \in C^{1,2}([0,\infty) \times R^N) \cap C_b([0,\infty) \times R^N)$ satisfying (1.1) . Moreover, if $\phi \geq 0$, then $u_\phi \geq 0$; and if $\phi \equiv 1$, then $u_\phi \equiv 1$. Finally, if $\eta \in C_0^\infty(R^N)$ ($\equiv$ the space of infinitely differentiable functions on R^N having compact support) satisfies: $0 \leq \eta \leq 1$, $\eta \equiv 1$ on $B(0,1)$ $\equiv \{x \in R^N : |x| < 1\}$ and $\eta \equiv 0$ off $B(0,2)$, then $u_{\eta_k} \uparrow 1$ as $k \uparrow \infty$ where $\eta_k(x) = \eta(x/k)$. From these facts (all of which, with the exception of the existence assertion, follow from the weak maximum principle) one can easily deduce the existence of a transition probability function $P(t,x,\cdot)$ on R^N such that

$$(1.2) \qquad u_\phi(t,x) = \int_{R^N} \phi(y)P(t,x,dy) \quad .$$

As a consequence of the preceding considerations, it is clear that (1.1) can be interpreted as the "backward equation" for the Markov process

$X(\cdot)$ having transition probability function $P(t,x,\cdot)$. In order to get some insight into the structure of $X(\cdot)$, note that for all $\phi \in C_b^2(R^N)$:

$$E[\phi(X(t+u)) - \phi(X(t)) | X(s), s \leq t]$$

$$= \int \phi(y)P(h,X(t),dy) - \phi(X(t))$$

$$= hL\phi(X(t)) + o(h), \quad h \downarrow 0 .$$

That is:

$$(1.3) \quad E[\phi(X(t+h)) - \phi(X(t)) | X(s), s \leq t] = hL\phi(X(t)) + o(h), \quad h \downarrow 0,$$

for all $\phi \in C_b^2(R^N)$.

Following P. Lévy, K. Itô set about constructing for each $x \in R^N$ a process $X(\cdot .x)$ satisfying (1.3) plus the initial condition $X(0,x) = x$. Itô's construction goes as follows. Let $\Theta = \{\theta \in C([0,\infty),R^N) : \theta(0) = 0\}$, give Θ the topology of uniform convergence on compacts, denote by $\mathcal{B}$ the Borel field over Θ, and set $\mathcal{B}_t = \sigma(\theta(s) : 0 \leq s \leq t)$ (i.e. the σ-algebra over Θ generated by the maps $\theta \to \theta(s)$ for $0 \leq s \leq t$). Finally, use $\mathcal{W}$ to denote Wiener measure on $(\Theta,\mathcal{B})$. It is then a simple mater to check that, for any $x \in R^N$, $x + \theta(\cdot)$ under $\mathcal{W}$ satisfies (1.3) when $L = 1/2\Delta$ (i.e. $a(\cdot) \equiv I$ and $b(\cdot) \equiv 0$). More generally, if $X_n(\cdot)$, $n \geq 1$, is defined by

$$X_n(0) = x, \quad n \geq 1$$

$$X_n(t) = X_n\left(\frac{[nt]}{n}\right) + a^{1/2}\left(X_n\left(\frac{[nt]}{n}\right)\right)\left(\theta(t) - \theta\left(\frac{[nt]}{n}\right)\right)$$

$$+ b\left(X_n\left(\frac{[nt]}{n}\right)\right)\left(t - \frac{[nt]}{n}\right), \quad n \geq 1 \text{ and } t > 0,$$

(here $[s]$ denotes the largest integer $\leq s$) then

$$(1.4)\qquad E^{\mathbb{W}}[\phi(X_n(t+h)) - \phi(X_n(t))\,|\,\mathcal{B}_t] = hL\phi\left(X_n\left(\tfrac{[nt]}{n}\right)\right) + o(h)\ , \quad h \downarrow 0\ ,$$

for all $\phi \in C_b^2(R^N)$. Moreover, it is easy to show that

$$\lim_{m\to\infty} \sup_{n\geq m} E^{\mathbb{W}}\left[\sup_{0\leq t\leq T} |X_n(t) - X_m(t)|^2\right] = 0 \quad , \quad T > 0$$

Thus we can define $X(\cdot,x) = \lim_{n\to\infty} X_n(\cdot)$ and it is clear from (1.4) that:

$$(1.5)\qquad E^{\mathbb{W}}[\phi(X(t+h,x)) - \phi(X(t,x))\,|\,\mathcal{B}_t] = hL\phi(X(t)) + o(h)\quad , \quad h \downarrow 0\ .$$

Since $X(0,x) \equiv x$ and $\mathcal{B}_t \supseteq \sigma(X(s) : 0 \leq s \leq t)$, we see that this procedure has lead us to a process $X(\cdot)$ which satisfies (1.3) and the initial condition $X(0) \equiv x$.

Returning now to the original problem, it is not hard to connect the process $X(\cdot,x)$ with the equation (1.1) . Indeed, from (1.5) , one can easily show that for any $\phi \in C_b^2(R^N)$ and $T > 0$: $(u(T-t \wedge T, X(t,x)),$ $\mathcal{B}_t, \mathbb{W})$ is a martingale. In particular,

$$u_\phi(T,x) = E^{\mathbb{W}}[\phi(X(T,x))]\ .$$

Combining this with (1.2) , we arrive at:

$$(1.6)\qquad P(T,x,\cdot) = \mathbb{W} \circ X(T,x)^{-1}$$

The equation (1.6) is one of the prime motivations for studying <u>stochastic differential equations</u> (<u>S.D.E..'s</u>). What (1.6) does is provide us with a somewhat concrete representation for the otherwise rather abstract quantity $P(t,x,\cdot)$. With this representation as the starting point, one may ask several questions.

(<u>Q.1</u>) The existence of u_ϕ plays no role in the construction of

$X(\cdot,x)$. Thus it is reasonable to ask if one cannot get an independent derivation of the existence of u_ϕ directly from the existence of $X(\cdot,x)$. That is, can one show directly that $(t,x) \to E\,[\phi(t,x)]$ is a solution to (1.1) ?

(Q2) One knows quite a good deal about $\mathbb{W}$ and the construction of $X(\cdot,x)$ is reasonably explicit. Can one use the known properties of $\mathbb{W}$ together with the construction of $X(\cdot,x)$ to derive properties of $r(t,x,\cdot)$?

In the next section we will delve more deeply into the construction of $X(\cdot,x)$ and the theory of S.D.E.'s . As a consequence, we will show that if $a(\cdot)$ admits a smooth square root, then the answer to Qi) is yes. The remainder of these lectures will be devoted to answering Qii) . As we will see, the answer is not complete but the possibilities for further investigation are vast.

2. Elements of the Theory of Stochastic Integrals and S.D.E.'s:

Let $\Theta = \{\theta \in C([0,\infty),R^d) : \theta(0) = 0\}$ and define $\mathcal{B}$, $\{\mathcal{B}_t : t \geq 0\}$, and $\mathbb{W}$ accordingly (i.e. as in section 1). A function α on $[0,\infty) \times \Theta$ is said to be progressively measurable (prog. meas.) if, for each $T > 0$, $(t,\theta) \to \alpha(t \wedge T,\theta)$ is $\mathcal{B}_{[0,T]} \times \mathcal{B}_T$-measurable ($\mathcal{B}_{[0,T]}$ is the Borel field over $[0,T]$). If $\alpha : [0,\infty) \times \Theta \to R^d$ is prog. meas. and $E^{\mathbb{W}}[\int_0^T |\alpha(t)|^2 dt] < \infty$, $T > 0$, the reader is expected to know how the Itô stochastic integral $\xi(T) = \int_0^T \alpha(t)\cdot d\theta(t)$ is defined (cf., for example, [McKean] or Chapter IV of [S. & V.]). In particular, it is assumed that the reader is familiar with the following properties of $\xi(\cdot)$:

i) there is a right-continuous, prog. meas. version of $\xi(\cdot)$ which is continuous (a.a., $\mathbb{W}$);

ii) $(\xi(t), \mathcal{B}_t, \mathbb{W})$ and $(\xi^2(t) - \int_0^t |\alpha(s)|^2 ds, \mathcal{B}_t, \mathbb{W})$ are martingales;

iii) for $p \in [2,\infty)$ there is a $C_p < \infty$ such that

$$E^{\mathbb{W}}[\sup_{0 \le t \le T} |\xi(t)|^p] \le C_p E^{\mathbb{W}}[(\int_0^T |\alpha(t)|^2 dt)^{p/2}] \quad . \tag{2.1}$$

From now on, we will always assume that the version for our stochastic integrals is the one described in i) . For those readers who are not familiar with iii) , we point out that (2.1) is one fourth of Burkholder's inequalities and suggest (4.6.12) on p. 116 of [S. & V.] as a source for a simple derivation due to A. Garsia.

Undoubtedly the most useful single fact about stochastic integrals is <u>Itô's formula ((2.2) below)</u>. Before stating Itô's formula, we introduce the definition of $\int_0^T \sigma(t)d\theta(t)$ where $\sigma : [0,\infty) \times \Theta \to R^N \otimes R^d$ is a prog. meas. function satisfying $E^{\mathbb{W}}[\int_0^T \|\sigma(t)\|^2_{H.S.} dt]$, $T > 0$. ($\|A\|^2_{H.S.} = \sum_{i,j} |A_{ij}|^2$ is the Hilbert-Schmidt norm of the matrix A .) Indeed, $\int_0^T \sigma(t)d\theta(t)$ is simply the right-continuous, prog. meas., $\mathbb{W}$-almost surely continuous R^N-valued function satisfying:

$$w\cdot\int_0^T \sigma(t)d\theta(t) = \int_0^T \sigma(t)^* w\cdot d\theta(t) \ , \quad T > 0 \ ,$$

for all $w \in R^N$. The version of Itô's formula which we will need states that if

$$\xi(T) = \xi_0 + \int_0^T \sigma(t)d\theta(t) + \int_0^T \beta(t)dt \quad , \quad T \ge 0 \ ,$$

where $\xi_0 \in R^N$ and $\sigma : [0,\infty) \times \Omega \to R^N \otimes R^N$ and $\beta : [0,\infty) \times \Omega \to R^N$ are prog. meas. functions satisfying $E^{\mathcal{W}}[\int_0^T (\|\sigma(t)\|^2_{H.S.} + |b(t)|)dt] < \infty$, then for any $F \in C^{1,2}([0,\infty) \times R^N)$ having bounded first order spacial derivatives:

$$(2.2) \qquad F(T,\xi(T)) = F(\xi_0) + \int_0^T \sigma(t)\mathrm{grad}_x F(t,\xi(t))\cdot d\theta(t) + \int_0^T \left(\frac{\partial F}{\partial t} + L_t F\right)(t,\xi(t))dt \ , \quad T \geq 0 \ ,$$

where $L_t = 1/2 \sum_{i,j=1}^N (\sigma(t)\sigma(t)^*)^{ij} \frac{\partial^2}{\partial x_i \partial x_j} + \sum_{i=1}^N b^i(t)\frac{\partial}{\partial x_i}$. Since it is nearly impossible to overstate the importance of Itô's formula, the reader to whom it is not known is urged to consult any one of the many books in which it is derived (e.g. see (4.6.10) on p. 115 of [S. & V.] for the beautiful derivation due to H. Kunita and S. Watenabe). Furthermore, the reader should be aware that (2.2) continues to hold for $F \in C^{1,2}([0,\infty) \times R^N)$ whenever the quantities on the right hand side are defined. Thus, (2.2) is true whenever $E^{\mathcal{W}}[\int_0^T \|\sigma(t)\mathrm{grad}_x F(t,x(t))\|^2_{H.S.} dt] + E^{\mathcal{W}}[\int_0^T \left|\left(\frac{\partial F}{\partial t} + L_t F\right)(t,\xi(t))\right| dt] < \infty$, $T > 0$. Such extensions of (2.2) are easy and will be used without further comment.

Itô's formula enables us to give a very succinct treatment of the ideas introduced in section 1) . To be precise, let L be the operator on the R.H.S. of (1.1) and let $\sigma : R^N \to R^N \otimes R^d$ be a "nice" function satisfying $a(\cdot) = \sigma(\cdot)\sigma(\cdot)^*$. (Exactly how "nice" it is possible to choose $\sigma(\cdot)$ in general is not entirely clear. If $a(\cdot) \geq \varepsilon I$, then with $d = N$ we may take $\sigma(\cdot) = a(\cdot)^{1/2}$ and know that $\sigma(\cdot)$ will be as smooth as $a(\cdot)$. Even if $a(\cdot)$ is allowed to degenerate, one can still show that $a(\cdot)^{1/2}$ is Lipschitz continuous so long as $a(\cdot) \in C_b^2(R^N)$. in general it is not possible to choose $\sigma(\cdot)$ any smoother than Lipschitz continuous even when $a(\cdot)$ is real-analytic. See section 2 of Chapter 5 in [S. & V.] for a discussion of this point.) Next, consider the stochastic integral equation:

$$(2.3)\qquad X(T,x) = x + \int_0^T \sigma(X(t,x))d\theta(t) + \int_0^T b(X(t,x))dt \quad , \quad T > 0 \ .$$

When $d = N$ and $\sigma(\cdot) = a(\cdot)^{1/2}$, it is, of course, precisely Euler's approximation scheme for solving (2.3) that we were using when we introduced the processes $X_n(\cdot)$ in section 1) ; and it is clear that in general (so long as $\sigma(\cdot)$ and $b(\cdot)$ are Lipschitz) the same scheme can be used to construct a solution to (2.3) . Moreover, an application of Itô's formula proves that $(u_\phi(T - T\wedge T , X(t\wedge T,x)),\mathcal{B}_t,\mathbb{W})$ is a martinagle for any $\phi \in C_b^2(R^N)$. Thus, just as in section 1) , we can conclude that (1.6) holds when $X(\cdot,x)$ is the solution to (2.3) .

For the purpose of answering question (Q.1) from section 1) and also for later applications, we will now make a slightly more thorough study of equations like (2.3) .

In the next two lemmas we will be using γ to denote a function on $[0,\infty) \times R^D \times \Theta$ into R^M having the following properties:

a) $(t,X) \to \gamma(t,X)$ is -almost surely continuous;

b) $\gamma(\cdot,X)$ is prog. meas. for each $X \in R^D$;

c) for each $T > 0$ and $p \in [2,\infty)$ there is a $C_p(T) < \infty$

(2.4) such that $E^{\mathbb{W}}[\sup_{0\leq t\leq T}|\gamma(t,0)|^p] \leq C_p(T)$;

d) for each $T > 0$, $R > 0$, and $p \in [2,\infty)$ there is a $C_p(T,R) < \infty$ such that $E^{\mathbb{W}}[\sup_{0\leq t\leq T}|\gamma(t,X)-\gamma(t,X)|] \leq C (T,R)|X-X|$

Next, $\sum : R^M \times R^N \to R^N \otimes R^d$ and $B : R^M \times R^N \to R^N$ will denote smooth functions satisfying:

a) the functions $\gamma \in R^M \to \sum(\gamma,0) \in R^N \otimes R^d$,

(2.5) $\gamma \in R^N \to B(\gamma,0) \in R^N$, and all derivatives of these two functions are slowly increasing (i.e. bounded by a polynomial)

b) $\max_{1\leq j\leq N} \sup_{(\gamma,\eta)\in R^M\times R^N} \|\frac{\partial\Sigma}{\partial\eta_j}(\gamma,\eta)\|_{H.S.} \left|\frac{\partial B}{\partial\eta_j}(\gamma,\eta)\right| < \infty$.

(2.6) Lemma: Let $\gamma : [0,\infty) \times R^D \times \Theta \to R^N$, $\sum : R^M \times R^N \to R^N \otimes R^d$ and $B : R^M \times R^N \to R^N$ be given functions. Assume that γ satisfies (2.4) and that $\sum$ and B satisfy (2.5). Let $F : R^D \to R^N$ be a smooth function having uniformly bounded first order derivatives. Then, for each $X \in R^D$ there is a prog. meas. function $\eta(\cdot,X) : [0,\infty) \times \Theta \to R^N$ satisfying

$$\eta(T,X) = F(X) + \int_0^T \sum(\gamma(t,X),\eta(t,X))d\theta(t) + \int_0^T B(\gamma(t,X),\eta(T,X))dt \quad , \quad T \geq 0 \quad . \tag{2.7}$$

Moreover, $\eta(\cdot,X)$ is unique up to a $\mathfrak{W}$-null set. Finally, there is a version of η having the properties listed in (2.4) .

Proof: In spite of the formidable list of hypotheses made, this lemma is quite easy to prove. Indeed, we use Picard's iteration scheme to generate functions $\eta_\nu(\cdot,X)$, $\nu \geq 0$, so that $\eta_0(\cdot,X) \equiv F(X)$ and $\eta_{\nu+1}(\cdot,X)$ is defined by the R.H.S. of (2.7) after all occurences of $\eta(\cdot,X)$ have been replaced by $\eta_\nu(\cdot,X)$. Using (2.1) , one easily shows that for each $p \in [2,\infty)$ and $T > 0$: $\lim_{\mu\to\infty} \sup_{\nu\geq\mu} E\left[\sup_{0\leq t\leq T} |\eta_\nu(t,X)-\eta_\mu(t,X)|^p\right] = 0$. Thus, $\eta(\cdot,X) = \lim_{\nu\to\infty} \eta_\nu(\cdot,X)$ solves (2.7) . Moreover, again applying (2.1) , one easily derives for each $p \in [2,\infty)$, $T > 0$, and $R > 0$ the existence of a $\overline{C}_p(T,R) < \infty$ such that:

$$(2.8) \qquad E^{\mathcal{W}}[\sup_{0\le t\le T}|\eta(t,X)-\eta(t,X')|^p] \le \overline{C}_p(T,R)|X-X'|^p \quad , \quad X,X' \in B(0,R) \ .$$

By standard techniques, one can easily pass from (2.8) to the existence of a $\mathcal{W}$-almost surely continuous version of $(t,X) \to \eta(t,X)$. The important tool to have is the Kolmogorov continuity criterion for multi-parameter processes (cf. (2.4.1) in [S. & V.]). The interested reader may find more details in Chapter I of [S] . ☒

(2.9) Lemma: Let Υ , Σ , and B be as in (2.6) and let η be the $\mathcal{W}$-almost surely continuous version of the solution to (2.7) . Assume that $\frac{\partial \Upsilon}{\partial X_\ell}$ exists $\mathcal{W}$-almost surely for each $1 \le \ell \le D$ and that for each $1 \le \ell \le D$ the function $\frac{\partial \Upsilon}{\partial X_\ell}$ has a version satisfing (2.4) . Also assume that the second order as well as the first order derivatives of F are bounded. Then: $\eta \in C^{0,1}([0,\infty) \times R^D)$ (a.s. $\mathcal{W}$) and, for each $1 \le \ell \le D$, $\frac{\partial \eta}{\partial X_\ell}$ admits a version which satisfies (2.4) . In fact, if

$$\hat{\Upsilon}(\cdot,X) = \begin{pmatrix}\Upsilon(\cdot,X)\\ \eta(\cdot,X)\end{pmatrix} \text{ and } \hat{\eta}(\cdot,X) = \begin{pmatrix}\frac{\partial\eta}{\partial x_1}(\cdot,X)\\ \vdots\\ \frac{\partial\eta}{\partial X_D}(\cdot,X)\end{pmatrix}, \text{ then there exist}$$

$\hat{\Sigma} : R^{M+N} \times R^{N\cdot D} \to R^{N\cdot D} \otimes R^d$ and $B^{M+N} : R^{N\cdot D} \to R^{N\cdot D}$ satisfying (2.5) such that

$$(2.10) \qquad \hat{\eta}(T,X) = \begin{pmatrix}\frac{\partial F}{\partial X_1}(X)\\ \vdots\\ \frac{\partial F}{\partial X_D}(X)\end{pmatrix} + \int_0^T \hat{\Sigma}(\hat{\Upsilon}(t,X),\hat{\eta}(t,X))d\theta(t)$$

$$+ \int_0^T \hat{B}(\hat{\gamma}(t,X),\hat{\eta}(t,X))dt \quad , \quad T \geq 0$$

Proof: Choose and fix $X \in R^D$ and $1 \leq \ell \leq D$. Using e_ℓ to denote the unit vector in R^D along direction ℓ , define:

$$\Delta_h \eta(t,X) = \eta(t,X+he_\ell) - \eta(t,X) \quad ,$$

$$\Delta_h \gamma(t,X) = \gamma(t,X+he_\ell) - \gamma(t,X) \quad ,$$

and

$$\Delta_h F(X) = F(X+he_\ell) - F(X)$$

for each $h \in R^1 \setminus \{0\}$. Next, define

$$\hat{\gamma}(t;X,h) = \begin{pmatrix} \hat{\gamma}^{(0)}(t;X,h) \\ \hat{\gamma}^{(1)}(t;X,h) \\ \hat{\gamma}^{(2)}(t;X,h) \\ \hat{\gamma}^{(3)}(t;X,h) \\ \hat{\gamma}^{(4)}(t;X,h) \end{pmatrix} \equiv \begin{pmatrix} \gamma(t,X) \\ \Delta_h \gamma(t,X) \\ 1/h\Delta_h \gamma(t,X) \\ \eta(t,X) \\ \Delta_h \eta(t,x) \end{pmatrix}$$

for $(t,X,h) \in [0,\infty) \times R^D \times R^1$, where $1/h\Delta_h \gamma(t,X) \equiv \frac{\partial \gamma}{\partial X_\ell}(t,X)$ and $\Delta_h \eta(t,X) \equiv 0$ for $h = 0$. Finally, define

$$\hat{\Sigma}(\hat{\gamma},\xi) = \sum_{i=1}^{M} \hat{\gamma}_i^{(2)} \int_0^1 \frac{\partial \Sigma}{\partial \gamma_i}(\hat{\gamma}^{(0)} + \omega\hat{\gamma}^{(1)},\hat{\gamma}^{(3)} + \omega\hat{\gamma}^{(4)})d\omega$$

$$+ \sum_{j=1}^{N} \xi_j \int_0^1 \frac{\partial \Sigma}{\partial \eta_j}(\hat{\gamma}^{(0)} + \omega\hat{\gamma}^{(1)},\hat{\gamma}^{(3)} + \omega\hat{\gamma}^{(4)})d\omega$$

and

$$\hat{B}(\hat{\gamma},\xi) = \sum_{i=1}^{N} \hat{\gamma}_i^{(2)} \int_0^1 \frac{\partial B}{\partial \gamma_i}(\hat{\gamma}^{(0)} + \omega\hat{\gamma}^{(1)}, \hat{\gamma}^{(3)} + \omega\hat{\gamma}^{(4)})d\omega$$

$$+ \sum_{j=1}^{N} \xi_j \int_0^1 \frac{\partial G}{\partial \eta_j}(\hat{\gamma}^{(0)} + \omega\hat{\gamma}^{(1)}, \hat{\gamma}^{(3)} + \omega\hat{\gamma}^{(4)})d\omega \quad .$$

Now consider the equation:

$$(2.11) \qquad \xi(T;X,h) = 1/h\ \Delta_h F(X) + \int_0^T \hat{\sum}(\hat{\gamma}(t;X,h),\xi(t;X,h))d\theta(t)$$

$$+ \int_0^T \hat{B}(\hat{\gamma}(t;X,h),\xi(t,X,h))dt \quad , \quad T \geq 0 \quad ,$$

where $1/h\ \Delta_h F(X) = \frac{\partial F}{\partial X_\ell}(X)$ for $h = 0$. Because (2.11) has precisely the same form as (2.7), the results of Lemma (2.6) apply to $\xi(\cdot;X,h)$. In particular, for each (X,h), (2.10) admits a solution ξ and this solution as $\mathfrak{W}$-almost surely unique. Moreover, there is a version of ξ such that $(t,X,h) \to \xi(t;X,h)$ is $\mathfrak{W}$-almost surely continuous. But if $h \neq 0$, then it is clear that $1/h\ \Delta_h \eta(\cdot,X)$ solves (2.11). Hence, by uniqueness, $\xi(t;X,h) = 1/h\ \Delta_h \eta(\cdot,X)$, $h \neq 0$, $\mathfrak{W}$-almost surely. This means that $(t,X,h) \in [0,\infty) \times R^D \times (R^1\backslash\{0\}) \to 1/h\Delta_h \eta(t,X)$ $\mathfrak{W}$-almost surely admits a continuous extension to $[0,\infty) \times R^D \times R^1$. In particular, $\frac{\partial \eta}{\partial X_\ell}(t,X)$ exists $\mathfrak{W}$-almost surely and is equal to $\xi(t;X,0)$. Clearly this observation completes the proof.

□

(<u>2.12</u>) <u>Theorem:</u> Suppose that $\sigma : R^N \to R^N \otimes R^d$ and $b : R^N \to R^N$ are smooth functions having bounded first order derivatives and slowly increasing derivatives of all orders. Then for each $x \in R^N$ there is a prog. meas. $X(\cdot,x)$ satisfying

$$(2.13)\qquad X(T,x) = x + \int_0^T \sigma(X(t,x))d\theta(t) + \int_0^T b(X(t,x))dt \quad , \quad T \geq 0 \ .$$

Moreover, $X(\cdot,x)$ is unique up to a $\mathcal{W}$-null set and admits a version such that $(t,x) \to X(t,x)$ is continuous $\mathcal{W}$-almost surely; and this version has the properties that $X \in C^{0,\infty}([0,\infty) \times R^N)$ (a.s. $\mathcal{W}$) and that for each $p \in [2,\infty)$, $T > 0$, $R > 0$, and $k \geq 0$:

$$(2.14)\qquad \sup_{|x|\leq R} E^{\mathcal{W}}[\max_{|\alpha|\leq k} \sup_{0\leq t\leq T} |D_x^{\alpha} X(t,x)|^p] < \infty \ .$$

Finally, if $\tau : \Theta \to [0,\infty]$ is a stopping time, then the conditional distribution of $X(\cdot + \tau,x)$ given $\mathcal{B}_\tau$ on $\{t < \infty\}$ coincides, $\mathcal{W}$-almost surely, with the distribution of $X(\cdot,y)$, where $y = X(\tau,x)$.

Proof: The existence and uniqueness assertions as well as the existence of a continuous version are immediate consequences of Lemma (2.6) . Moreover, the differentiability proeprties of X and the estimate (2.14) follow inductively from repeated applications of Lemma (2.9) . Finally, the last part of the theorem is a consequence of the following two facts: $\theta(\cdot + \tau) - \theta(\tau)$ has the same conditional distribution on $\{\tau < \infty\}$ given $\mathcal{B}_\tau$ as $\theta(\cdot)$ and $X(\cdot + \tau,x)$ is the same functional of $\theta(\cdot + \tau) - \theta(\tau)$ as $X(\cdot,y)$ is of $\theta(\cdot)$. (This latter fact is a result of uniqueness.) Details of this argument may be found in the proof of (5.15) in [S. & V.] .

□

As we pointed out during our discussion of Itô's formula, if $f \in C^{1,2}([0,\infty) \times R^d)$ and $\frac{\partial f}{\partial t} + Lf$ is bounded, then $(f(t,X(t,x)) - \int_0^t (\frac{\partial f}{\partial s} + Lf)(s,X(s),x))ds, \mathcal{B}_t, \mathcal{W})$ is a martingale, where

$L = 1/2 \sum_{i,j=1}^{N} (\sigma\sigma^*)(\cdot)\frac{\partial^2}{\partial x_i \partial x_j} + \sum_{i=1} b^i(\cdot)\frac{\partial}{\partial x_i}$; and from this observation one can easily derive

$$(2.15) \qquad u_\phi(T,x) = E^{\mathcal{W}}[\phi(X(t,x))] \quad ,$$

where u_ϕ is the solution to (1.1) . We now want to show directly that if u is defined as the R.H.S. of (2.15) , then u satisfies (1.1) . To this end, first note that if ϕ $C_b^k(R^N)$ and $|\alpha| \leq k$, then, by Theorem (2.12) :

$$(2.15) \qquad D_x^\alpha E\,[\phi(X(t,x))] = \sum_{\beta\leq\alpha} E^{\mathcal{W}}[D^\beta\phi(X(t,x))\Xi_\beta^{(\alpha)}(t,x)] \quad ,$$

where $\Xi_\beta^{(\alpha)} : [0,\infty) \times R^N \times \Theta \to R^1$ is a prog. meas. function which is $\mathcal{W}$-almost surely continuous with respect to (t,x) and satisfies:

$$(2.17) \qquad \sup_{|x|\leq R} E^{\mathcal{W}}[\sup_{0\leq t\leq T} |\Xi_\beta^{(\alpha)}(t,x)|^p] < \infty$$

for all $p \in [2,\infty)$, $R > 0$, and $T > 0$. In particular, $(t,x) \to E^{\mathcal{W}}[\phi(X(t,x))]$ is in $C^{0,k}([0,\infty) \times R^N)$.

Now set $u(t,x) = E^{\mathcal{W}}[\phi(X(t,x))]$ where $\phi \in C_b^2(R^N)$. Then, as we have just seen, $u \in C^{0,2}([0,\infty) \times R^N)$. Moreover, if $\sigma(\cdot)$ and $b(\cdot)$ are bounded, then a closer look at the derivation of (2.14) reveals that $u \in C_b^{0,2}([0,T] \times R^N)$ for all $T > 0$. In particular, when $\sigma(\cdot)$ and $b(\cdot)$ are bounded, $E^{\mathcal{W}}[u(t,X,=(h,x))] - u(t,x) = E^{\mathcal{W}}[\int_0^h Lu(t,X(s,x))ds]$ and so $1/h(u(t,X(h,x))-u(t,x)) \to Lu(t,x)$ boundedly. At the same time, by the last part of Theorem (2.12) :

$$u(t+h,x) - u(t,x) = E^{\mathcal{W}}[u(t,X(h,x))] - u(t,x) \,.$$

Hence, if $\sigma(\cdot)$ and $b(\cdot)$ are bounded, we have shown that:

$$1/h(u(t+h,x)-u(t,x)) \to Lu(t,x) \quad .$$

In other words, when $\sigma(\cdot)$ and $b(\cdot)$ are bounded, when we have given a direct proof (i.e. without invoking the theory of P.D.E.'s) that $(t,x) \to E[\phi(X(t,x))]$ solves (1.1) .

In order to get away from the condition that $\sigma(\cdot)$ and $b(\cdot)$ are bounded, recall where we used this assumption in the preceding discussion. Namely, we wanted to know that $u(t,\cdot) \in C_b^2(R\ )$ in order to apply Itô's formula and get: $E^{\mathfrak{W}}[u(t,X(h,x))] - u(t,x) = E^{\mathfrak{W}}[\int_0^h Lu(t,X(s,x))ds]$. However, as we pointed out earlier, it suffices to know that $\sup_{0\leq s\leq h} E^{\mathfrak{W}}[|Lu(t,X(s,x))|] < \infty$ in order to reach this conclusion. Thus, one way around our problem is to be more careful about the derivation of (2.14) and prove that under the conditions of (2.12) , $E^{\mathfrak{W}}[\max_{|x|\leq x}\sup_{0\leq t\leq T}|D_x^\alpha X(t,x)|^P]$ is at worst of polynomial growth in x . Then we would know that for $\phi \in C_b^2(R^N)$, $\sup_{0\leq t\leq T}|D_x^\alpha u(t,x)|$ is slowly increasing in x for each $T > 0$ and $|\alpha| \leq 2$. In particular, it would be obvious that $\sup_{0\leq s\leq h} E^{\mathfrak{W}}[|Lu(t,X(s,x))|] < \infty$ for all $h > 0$; and so the argument given in the preceding paragraph could be applied to the general case.

A second route around the difficulty discussed above is to observe that if $f \in C^2(R^N)$, $x \in R^N$, $R > 0$, and $\tau = \inf\{t \geq 0 : |X(t,x) - x| \geq 1\}$, then

$$(2.18) \qquad E^{\mathfrak{W}}[f(X(T\wedge\tau,x))] - f(x) = E^{\mathfrak{W}}[\int_0^{T\wedge\tau} LfX(t,x)dt]$$

for all $T > 0$. (Indeed, (2.18) is a simple consequence of Itô's formula applied to any $\bar{f} \in \{g \in C_b^2(R^N) : g = f$ on $B(x,R)\}$ plus Doob's stopping theorem.) In particular, since $\mathbb{P}(\tau = 0) = 0$, $1/h(E^{\mathbb{P}}[f(X(T\ \ \tau),x))] - f(x)) = E^{\mathbb{P}}[\frac{1}{h}\int_0^{h\wedge\tau} Lf(X(t,x))dt] \to Lf(X)$ as $t \downarrow 0$. Thus, if we knew that $\mathbb{P}(\tau \leq h)/h \to 0$ as $h \downarrow 0$, then we would have that $1/h(E^{\mathbb{P}}[f(X(h,x))] - f(x)] \to Lf(x)$ as $h \downarrow 0$ for $f \in C^2(R^N) \cap C_b(R^N)$. Applying this line of reasoning to $u(t,\cdot)$, we would then have a second proof that $\frac{\partial u}{\partial t} = Lu$. The following lemma contains the missing ingredient needed to make the above argument complete.

(2.19) Lemma: Let $\sigma(\cdot)$ and $b(\cdot)$ be as in (2.12) and let $X(\cdot,x)$ be the solution to (2.13) . Set $M_\sigma(x,R) = \sup_{y\in B(x,R)} \|\sigma(y)\|_{H.S.}$ and $M_b(x,R) = \sup_{y\in B(x,R)} |b(y)|$. Then for $T > 0$ and $R > 0$ sastifying $R > N^{1/2}TM_b(x,R)$:

$$(2.20)\quad \mathbb{P}(\sup_{0\leq t\leq T} |X(t,x) - x| \geq R) \leq 2N\exp\left(-\frac{(R/N^{1/2} - TM_b(x,R))^2}{2TM_\sigma(x,R)^2}\right) .$$

In particular, for every $R > 0$, $\mathbb{P}(\sup_{0\leq t\leq T} |X(t,x) - x| \geq R)$ goes to zero as $T \downarrow 0$ at least as fast as $\exp(-R^2/2TNM_\sigma(x,R)^2)$.

Proof: Fix $R > 0$ and set $\tau = \inf\{t \geq 0 : |X(t,x)-x| \geq R\}$. Localizing Itô's formula as above, we see that for $v \in R^N$, $(Y_v(t),\mathcal{B}_t,\mathbb{P})$ is a martingale, where

$$Y_v(t) \equiv \exp[v\cdot(X(t\wedge\tau,x) - x - \int_0^{t\wedge\tau} b(X(s,x))ds)$$
$$- 1/2\int_0^{t\wedge\tau} |\sigma(X(s,x))^* v|^2 ds] .$$

Thus, by Doob's inequality, if $\omega \in S^{N-1}$ and $\lambda > 0$:

$$\mathbb{P}(\sup_{0\leq t\leq T} \omega\cdot(X(t,x)-x) \geq R/N^{1/2})$$

$$= \mathbb{W}\left(\sup_{0\le t\le T} \omega\cdot(X(t\wedge\tau,x)-x) \ge R/N^{1/2}\right)$$

$$\le \left(\sup_{0\le t\le T} Y_{\lambda\omega}(t\wedge\tau) \ge \exp[\lambda(R/N^{1/2} - TM_b(x,R)) - \frac{\lambda^2}{2} TM_\sigma(x,R)\ P)\right.$$

$$\le \exp[-\lambda(R/N^{1/2} - TM_b(x,R)) + \frac{\lambda^2}{2} TM_\sigma(x,R)^2] \quad .$$

In particular, if $R > N^{1/2}TM_b(x,R)$, we may take $\lambda = (R/N^{1/2} - TM_b(x,R))/TM_\sigma(x,R)^2$ and thereby obtain:

$$\mathbb{W}\left(\sup_{0\le t\le T} \omega\cdot(X(t,x) - x) \ge R/N^{1/2}\right)$$

$$\le \exp\left(- \frac{(R/N^{1/2} - TM_b(x,R))^2}{2TM_\sigma(X,R)^2}\right) .$$

Choosing $\omega_1,\ldots,\omega_N \in S^{N-1}$ to be an orthogonal basis, we see that

$$\mathbb{W}\left(\sup_{0\le t\le T}\left|X(t,x)-x\right| \ge R\right) \le \sum^{N} \mathbb{W}\left(\sup_{0\le t\le T}\left|\omega_j\cdot(X(t,x)-x)\right| \ge R/N^{1/2}\right) \le$$

$$2N\exp\left(- \frac{(R/N^{1/2} - TM_b(x,R))^2}{2TM_\sigma(x,R)^2}\right) .$$

□

Summarizing these considerations, we now state the following theorem, which answers (Q.1) of section 1) , since our proof relies on probability reasoning alone.

(2.21) Theorem: Let $\sigma(\cdot)$ and $b(\cdot)$ be as in (2.12) and set $L = 1/2 \sum_{i,j=1}^{N} (\sigma(\cdot)\sigma(\cdot)^*)^{ij} \frac{\partial^2}{\partial x_i \partial x_j} + \sum_{i=1}^{N} b^i(\cdot)\frac{\partial}{\partial x_i}$. Given $\phi \in C_b^k(R^N)$, define $u(t,x) = E^{\mathbb{W}}[f(X(t,x))]$, where $X(\cdot,x)$ is the solution to (2.13) . Then u is the one and only element of $C^{1,2}([0,\infty) \times R^N) \cap C_b([0,\infty) \times R^N)$ which satisfies (1.1) .

(2.22) Remark: Although (2.21) represents a reasonably satisfactory answer to (Q.1) , it must be admitted that it does not give as good a result as the analysts can prove. The deficiency in (2.21) comes from the fact that it rests on assumptions about $\sigma(\cdot)$ instead of $\sigma(\cdot)\sigma(\cdot)^*$. As we noted in section 1), $\sigma(\cdot)\sigma(\cdot)^*$ in general will be smoother than $\sigma(\cdot)$. Thus the existence result proved in Chapter 3 of [S. & V.] is stronger than the one just given.

3. A Criterion for Absolute Continuity of a Measure on R^N:

Let $\sigma: R^N \to R^N \otimes R^d$ and $b: R^N \to R^N$ be as in (2.12) and let $X(\cdot,x)$ be the solution to (2.13) . As we saw in section 1) , and reconfirmed in section 2) , the fundamental solution $P(t,x,\cdot)$ to (1.1) with $L = 1/2 \sum_{i,j=1}^{N} (\sigma(\cdot)\sigma(\cdot)^*)^{ij} \frac{\partial^2}{\partial x_i \partial x_j} + \sum_{i=1}^{N} b^i(\cdot)\frac{\partial}{\partial x_i}$ satisfies:

$$(3.1) \qquad P(T,x,\cdot) = \mathbb{W} \circ X(T,x)^{-1} .$$

In (Q.2) we asked if it is not possible to exploit (3.1) to study the probability measure $P(T,x,\cdot)$. In particular, it ought to be possible to use (3.1) to see that $P(T,x,\cdot)$ admits a nice density when L satisfies sufficient non-degeneracy conditions. Since we are now looking for properties of $P(T,x,\cdot)$ which depend on non-degeneracy, rather than simply smoothness, of the coefficients, it should come as no surprise that more sophisticated techniques are required.

To begin with, we must develop a criterion for checking when a measure on R^N is absolutely continuous. The criterion which seems to be best suited for our purposes is the following.

(3.2) Lemma: Let μ be a probability measure on R^N. Suppose that for each $1 \leq k \leq N$ there is a $\psi_k \in L^1(\mu)$ such that:

$$\int \frac{\partial \phi}{\partial x_k} d\mu = -\int \phi \psi_k d\mu \quad , \quad \phi \in C_0^\infty(R^N) \quad . \tag{3.3}$$

Then $\mu(dx) = f(x)dx$ for some $f \in L^1(R^N)^+$. Moreover, if, for some $q \in (N,\infty)$, $\psi_k \in L^q(\mu)$, $1 \leq k \leq N$, then $f \in \hat{C}(R^N)$ ($\equiv \{g \in C(R^N) : \lim_{x\to\infty} g(x) = 0\}$), and there exists a $C_q < \infty$ (depending only on N as well as q) such that $\|f\|_{C_b(R^N)} \leq C_q\left(\sum_1^N \|\psi_k\|_{L^q(\mu)}\right)^N$ and $\|Df\|_{L^q(R^N)} \leq C_q\left(\sum_1^N \|\psi_k\|_{L^q(\mu)}\right)^{N+1-N/q}$. In particular, f is Hölder continuous of order $1 - N/q$ and $\|f\|_{Lip(1-N/q)}$ can be estimated in terms of $N,q,$ and $\sum_1^N \|\psi_k\|_{L^q(\mu)}$ alone.

Proof: Think of μ as a tempered distribution (in the sense of L. Schwartz). Then (3.3) becomes: $D_k\mu = \psi_k\mu$. Thus, if $G_\lambda(x) = \int_0^\infty \frac{e^{-\lambda t}}{(4\pi t)^{N/2}} e^{-|x|^2/4\pi t} dt$ for $\lambda > 0$ and $x \neq 0$, then

$$\mu = \lambda G_\lambda * \mu + \sum_{k=1}^N \frac{\partial G_\lambda}{\partial x_k} * (\psi_k \mu) \quad .$$

Since G_λ and $\frac{\partial G_\lambda}{\partial x_k}$, $1 \leq k \leq N$, are in $L^1(R^N)$, this proves that, as a distribution, $\mu \in L^1(R^N)$. That is, as a measure, $\mu(dx) = f(x)dx$ for some $f \in L^1(R^N)^+$.

Now think of f as a tempered distribution. Then (3.3) becomes $D_k f = \psi_k f$. Moreover, $v = f^{1/q}$ is well defined as a tempered distribution. We now want to compute $D_k v$. Given $\varepsilon > 0$, set $v_\varepsilon = (f + \varepsilon)^{1/q}$. Then it is easy to check that, as a tempered distribution, $D_k v_\varepsilon = 1/q(f + \varepsilon)^{1/q-1} D_k f = 1/q(f + \varepsilon)^{1/q-1} \psi_k f$. Clearly v_ε tends to v as tempered distributions, and therefore $D_k v_\varepsilon \to D_k v$ as tempered distributions. Also, $D_k v_\varepsilon = 1/q(f + \varepsilon)^{1/q-1} \psi_\varepsilon f \to 1/q \psi_k v$ a.e. and $|D_k v_\varepsilon| \leq 1/q |\psi_k| f^{1/q}$. Thus, $D_k v_\varepsilon \to 1/q \psi_k v$ in $L^q(R^N)$ and therefore as tempered distributions. Thus, $D_k v = 1/q\ \psi_k v$, $1 \leq k \leq N$. We now have:

$$v = \lambda G_\lambda * v + 1/q \sum_1^N \frac{\partial G_\lambda}{\partial x_k} * (\psi_k v) .$$

Since $q > N$, G_λ and $\frac{\partial G_\lambda}{\partial x_k}$, $1 \leq k \leq N$, are elements of $L^{q'}(R^N)$ $(1/q + 1/q' = 1)$. In fact:

$$\|\lambda G_\lambda\|_{L^{q'}(R^N)} = A_q \lambda^{N/2q} \quad \text{and} \quad \left\|\frac{\partial G_\lambda}{\partial x_k}\right\|_{L^{q'}(R^N)} = B_q \lambda^{N/2q-1/2} .$$

Therefore, because $v \in L^q(R^N)$ and $\psi_k v \in L^q(R^N)$ with $\|v\|_{L^q(R^N)} = 1$ and $\|\psi v\|_{L^q(R^N)} = 1/q \|\psi_k\|_{L^q(\mu)}$, we see that $v \in C(R^N)$ and that

$$\|v\|_{C_b(R^N)} \leq A_q \lambda^{N/2q} + 1/q B_q \lambda^{N/2q - 1/2} \sum_1^N \|\psi_k\|_{L^q(\mu)}$$

$$= \lambda^{N/2q}\left(A_q + 1/q B_q \lambda^{-1/2}\left(\sum_1^N \|\psi_k\|_{L^q(\mu)}\right)\right) .$$

Clearly this proves that $f = v^q \in C(R^N)$; and taking $\lambda^{1/2} = \sum_1^N \|\psi_k\|_{L^q(\mu)}$ we obtain:

$$\|f\|_{C_b(R^N)} \leq (A_q + 1/qB_q)^q (\sum_1^N \|\psi_k\|_{L^q(\mu)})^N .$$

Finally, $\|D_k f\|_{L^q(R^N)} = \|\psi_k f\|_{L^q(R^N)} \leq \|f\|^{1-1/q}_{C_b(R^N)} \|\psi_k\|_{L^q(\mu)}$ and so the estimate on $\|D_k f\|_{L^q(R^N)}$ follows immediately from the one on $\|f\|_{C_b(R^N)}$. The Hölder continuity assertion and estimate on $\|f\|_{Lip(1-q/N)}$ are now easy consequences of elementary Sobolev embedding theory (cf. [Adams])

Q. E. D.

Lemma (3.2) gives a hint about how we might proceed to study $P(T,x,\cdot)$. Indeed, suppose that we could show that for $1 \leq k \leq N$ there exist $\Psi_k(T,x) \in L^1(\mathfrak{W})$ such that

$$(3.4) \qquad E^{\mathfrak{W}}\left[\frac{\partial\phi}{\partial x_k}(X(T,x))\right] = -E^{\mathfrak{W}}[\phi(X(T,x))\Psi_k(T,x)]$$

for all $\phi \in C_0^\infty(R^N)$ and $1 \leq k \leq N$. We could then find a measurable $\psi_k(T,x;\cdot)$ on R^N such that $\psi_k(T,x;X(T,x)) = E^{\mathfrak{W}}[\Psi_k(T,x)|X(T,x)^{-1}(\mathcal{B}_{R^N})]$ (a.s., $\mathfrak{W}$) and we would then have that $\psi_k(T,x;\cdot) \in L^1(P(T,x,\cdot))$ and that

$$(3.5) \qquad \int \frac{\partial\phi}{\partial x_k}(y)P(T,x,dy) = -\int \phi(y)\psi_k(T,x;y)P(T,x,dy)$$

for all $\phi \in C_0(R^N)$. Moreover, for any $q \geq 1$:

$$\|\psi_k(T,x;\cdot)\|_{L^q(P(T,x,\cdot))} = E^{\mathfrak{W}}[|E^{\mathfrak{W}}[\Psi_k(T,x;\cdot)|X(T,x)^{-1}(\mathcal{B}_{R^N})]|^q]^{1/q}$$

$$\leq E^{\mathfrak{W}}[|\Psi_k(T,x,\cdot)|^q]^{1/q} = \|\Psi_k(T,x;\cdot)\|_{L^q(\mathfrak{W})} .$$

Thus, if, for some $q \in (N,\infty)$, $\max_{1\leq k\leq N} \|\Psi_k(T,x;\cdot)\|_{L^q(\mathfrak{W})} < \infty$, then, by Lemma (3.2), we would know that there exists a $p(T,x,\cdot) \in \hat{C}(R^N)^+$ such that $P(T,x,dy) = p(T,x,y)dy$ and that $p(T,x,\cdot) \in Lip(1-N/q)$.

Thus, we must learn how to derive expressions of the sort given in (3.4).

4. Gaussian Calculus in Finite Dimensions:

The purpose of this section is to provide motivation for the procedure with which we will eventually derive expressions like (3.4).

It should be clear that an equality like (3.4) must come from "integration by parts". In order to see how to "integrate by parts" on Wiener space, it may be helpful to first carry out a finite dimensional analogue. To be specific, consider an R^1 (alias Θ) the Gaussian measure $\gamma(dx) = g(x)dx$ (alias $\mathbb{W}$), with $g(x) = e^{-|x|^2/2}/(2\pi)^{1/2}$, and a smooth mapping Φ (alias $X(T,x)$) of R^1 into R^1. We want to produce a $\Psi : R^1 \to R^1$ such that

$$\int \phi' \circ \Phi d\gamma = -\int (\phi \circ \Phi)\Psi d\gamma \tag{4.1}$$

for all $\phi \in C_0^\infty(R^1)$. Writing $\phi' \circ \Phi$ as $(\phi\circ\Phi)'/\Phi'$, we use ordinary integration by parts and obtain:

$$\int \phi' \circ \Phi d\gamma = -\int (\phi\circ\Phi)(g/\Phi')'dx$$

$$= -\int (\phi \; \Phi)(-x/\Phi' - \Phi''/(\Phi')^2)d\gamma \quad .$$

That is, we may take $\Psi = (-x/\Phi' - \Phi''/(\Phi')^2)$. However, in its present form, the expression for Ψ relies too heavily on the structure of R^1 and does not seem to have an obvious counterpart in Wiener space. Thus we have to find another expression for Ψ in which only quantities intrinsic to the Gaussian nature of $(R^1, \mathcal{B}_{R^1}, \gamma)$ appear.

Associated with γ there is a "natural" second order elliptic operator; namely, the Ornstein-Uhlenbeck operator $\mathcal{L} = 1/2(\frac{\partial^2}{\partial x^2} - x\frac{\partial}{\partial x})$ defined on $C^\infty_\uparrow(R^1)$ (functions which, together with their derivatives, are slowly increasing). The most obvious sense in which $\mathcal{L}$ is "natural" for γ is that $\mathcal{L}$ is formally symmetric on $L^2(\gamma)$ (i.e. $\int \Phi\mathcal{L}\Psi d\gamma = \int \Phi\mathcal{L}\Psi d\gamma$). Associated with any second order elliptic operator $\mathcal{L}$ there is an important bilinear operator $\langle\cdot,\cdot\rangle_{\mathcal{L}}$ given by $\langle\Phi,\Psi\rangle_{\mathcal{L}} = \mathcal{L}(\Phi\cdot\Psi) - \Phi\mathcal{L}\Psi - \Psi\mathcal{L}\Phi$ In the present case, $\langle\Phi,\Psi\rangle_{\mathcal{L}}$ is obviously simply $\Phi'\Psi'$; but in general $\langle\cdot,\cdot\rangle_{\mathcal{L}}$ should be thought of as "the amount by which $\mathcal{L}$ fails to satisfy Leibnitz's rule" and therefore as a measure of the "ellipticity" of $\mathcal{L}$. In any case, $\langle\cdot,\cdot\rangle_{\mathcal{L}}$, being an expression involving $\mathcal{L}$ alone, is just as intrinsic to γ as is $\mathcal{L}$ itself.

We now show how to re-express Ψ in terms of $\mathcal{L}$ and $\langle\cdot,\cdot\rangle_{\mathcal{L}}$ alone. To this end, note that

$$\Psi = -x/\Phi' - \Phi''/(\Phi')^2 = -x\Phi'/\langle\Phi,\Phi\rangle_{\mathcal{L}} - \Phi''/(\Phi')^2$$

$$= 2\mathcal{L}\Phi/\langle\Phi,\Phi\rangle_{\mathcal{L}} - 2\Phi''/\langle\Phi,\Phi\rangle^2_{\mathcal{L}}$$

$$= 2\mathcal{L}\Phi/\langle\Phi,\Phi\rangle_{\mathcal{L}} - 2\Phi''(\Phi')^2/\langle\Phi,\Phi\rangle^2_{\mathcal{L}}$$

$$= 2\mathcal{L}\Phi/\langle\Phi,\Phi\rangle_{\mathcal{L}} - \langle\Phi,\langle\Phi,\Phi\rangle_{\mathcal{L}}\rangle\,/\langle\Phi,\Phi\rangle^2_{\mathcal{L}}$$

$$= 2\mathcal{L}\Phi/\langle\Phi,\Phi\rangle + \langle\Phi,1/\langle\Phi,\Phi\rangle_{\mathcal{L}}\rangle_{\mathcal{L}}\,.$$

Having carried out this somewhat tedious exercise, it is still necessary for us to find an "intrinsic" proof that Ψ given by this new expression does indeed work in (4.1) . Unless we do so, our chances of transferring the procedure to Wiener space are slim.

An "intrinsic" proof must rest only on those properties of γ , $\mathcal{L}$, and $\langle\cdot,\cdot\rangle_{\mathcal{L}}$ which are subject to generalization when we leave the Euclidean context. As we have mentioned, our choice of $\mathcal{L}$ was motived by the fact that it is symmetric in $L^2(\gamma)$. Thus symmetry is a property of $\mathcal{L}$ which we may exploit. Although it is not so obviously intrinsic as is the symmetry of $\mathcal{L}$, a second property which turns out to generalize is the fact that $\langle\phi\circ\Phi,\Psi\rangle_{\mathcal{L}} = \phi'\circ\Phi\langle\Phi,\Psi\rangle_{\mathcal{L}}$. This second property is intimately related to and derivable from the path continuity of the process generated by $\mathcal{L}$ (i.e. the Ornstein-Uhlenbeck process); and its intrinsic nature is a consequence of this fact. Be that as it may, we now show how to derive (4.1) with

$$\Psi = 2\,\Phi/\langle\Phi,\Phi\rangle_{\mathcal{L}} + \langle\Phi,1/\langle\Phi,\Phi\rangle_{\mathcal{L}}\rangle_{\mathcal{L}} \tag{4.2}$$

purely on the basis of the properties:

(4.3)
i) $\mathcal{L}$ is symmetric in $L^2(\gamma)$

ii) $\langle\phi\circ\Phi,\Psi\rangle_{\mathcal{L}} = \phi'\circ\Phi\langle\Phi,\Psi\rangle_{\mathcal{L}}$.

To this end, we use (4.3) ii) to write

$$\begin{aligned}\phi'\circ\Phi &= \langle\phi\circ\Phi,\Phi\rangle_{\mathcal{L}}/\langle\Phi,\Phi\rangle_{\mathcal{L}}\\ &= [\mathcal{L}(\Phi\cdot\phi\circ\Phi) - \phi\circ\Phi\cdot\mathcal{L}\Phi - \Phi\mathcal{L}(\phi\circ\Phi)]/\langle\Phi,\Phi\rangle_{\mathcal{L}} .\end{aligned}$$

We next use (4.3) i) to shift $\mathcal{L}$ in order to obtain an expression in which all occurrences of $\phi\circ\Phi$ are "unencumbered" and thereby obtain:

$$\int\phi'\circ\Phi d\gamma = \int\phi\circ\Phi\cdot[\Phi\cdot\mathcal{L}(1/\langle\Phi,\Phi\rangle_{\mathcal{L}}) - \mathcal{L}\Phi/\langle\Phi,\Phi\rangle_{\mathcal{L}} - \mathcal{L}(\Phi/\langle\Phi,\Phi\rangle_{\mathcal{L}})]d\gamma .$$

Since $\Phi\cdot\mathcal{L}(1/\langle\Phi,\Phi\rangle_{\mathcal{L}}) - \mathcal{L}(\Phi/\langle\Phi,\Phi\rangle_{\mathcal{L}}) = -\langle\Phi,1/\langle\Phi,\Phi\rangle_{\mathcal{L}}\rangle_{\mathcal{L}} - \mathcal{L}\Phi/\langle\Phi,\Phi\rangle_{\mathcal{L}}$, the proof is complete.

Before moving on, we will consider one more example in order to emphasize the intrinsically Gaussian nature of the preceding line of reasoning. Let $B \in R^D \otimes R^D$ be a positive definite symmetric matrix and define $\gamma(dx) = (2\pi)^{-D/2}(\det B)^{-1/2}\exp(-1/2\ x\cdot B^{-1}x)dx$ on R^D. The Ornstein-Uhlenbeck operator in this case is $\mathcal{L} = 1/2\left(\sum_{i,j=1}^{D} B^{ij}\frac{\partial^2}{\partial x_i \partial x_j} - \sum_{i=1}^{D} x_i \frac{\partial}{\partial x_i}\right)$ and the associated $\langle\cdot,\cdot\rangle_{\mathcal{L}}$ can be written as $\langle\Phi,\Phi\rangle_{\mathcal{L}} = \sum_{i,j=1}^{D} B^{ij}\frac{\partial\Phi}{\partial x_i}\frac{\partial\Psi}{\partial x_j}$. In particular, it is easy to check that (4.3) i) and ii) continues to hold. Now let $\Phi = (\Phi_1,\ldots,\Phi_N)$ be a mapping from R^D into R^N and set $A = ((\langle\Phi_k,\Phi_\ell\rangle_{\mathcal{L}}))_{1\le k,\ell\le N}$. From (4.3) i) it is easy to show that if $\phi \in C_0^\infty(R^N)$ then:

$$\langle\phi\circ\Phi,\Phi_\ell\rangle_{\mathcal{L}} = \sum_{\nu=1}^{N}\frac{\partial\phi}{\partial x_\nu}\circ\Phi\cdot A_{\nu\ell} \ .$$

Hence, if $A^{(\mu,\nu)}$ denotes the $(\mu,\nu)^{th}$ cofactor of A and $\Delta = \det A$, then

$$\left(\frac{\partial\phi}{\partial x_k}\ \Phi\right)\cdot\Delta = \sum_{\ell=1}^{N}\langle\phi\ \Phi,\Phi_\ell\rangle_{\mathcal{L}}A^{(k,\ell)}$$

or

$$\frac{\partial\phi}{\partial x_k}\ \Phi = \sum_{\ell=1}^{N}\langle\phi\ \Phi,\Phi\ \rangle_{\mathcal{L}}(A^{-1})_{k\ell} \ .$$

Starting from here and proceeding in precisely the same fashion as we did in the special case above, one arrives at:

(4.4) $$\int\frac{\partial\phi}{\partial x_k}\ \Phi d\gamma = -\int(\phi\ \Phi)\cdot\Psi_k d\gamma \ , \quad \phi \in C_0^\infty(R^N) \ ,$$

where

$$(4.5)\qquad \Psi_k = \sum_{\ell=1}^{N} (2\mathcal{L}(\Phi_\ell)\cdot(A^{-1})_{k\ell} + \langle\Phi_\ell,(A^{-1})_{k\ell}\rangle_{\mathcal{L}}) .$$

Of course we have been consistently negligent throughout the preceding about technical integrability questions and, even worse, about the danger of occasionally dividing by zero. Nonetheless, what we have done is correct at least in spirit and should bolster the reader's conviction that our program can be carried out rigorously even in Wiener space.

5. Symmetric Diffusion Semigroups:

We want now to carry out the program outlined in section (4) when R^N is replaced by Θ (with $d = 1$) and γ is replaced by $\mathcal{W}$. Proceeding by direct analogy, we should guess that the appropriate operator $\mathcal{L}$ is given by:

$$(5.1)\qquad \mathcal{L}\Phi = 1/2\left[\int_0^\infty\int_0^\infty s\wedge t \frac{\partial^2\Phi}{\partial\theta(s)\partial\theta(t)}\,dsdt - \int_0^\infty \theta(s)\frac{\partial\Phi}{\partial\theta(s)}\,ds\right] ,$$

where the indicated derivatives are taken in the sense of Frechét. Indeed, this choice of $\mathcal{L}$ is the precise analogue of the one in section 4), since $\{\theta(t) : t \geq 0\}$ under $\mathcal{W}$ is a Gaussian family with mean 0 and $E^{\mathcal{W}}[\theta(s)\theta(t)] = s\wedge t$. Assuming that (5.1) gives the correct definition of $\mathcal{L}$, we can also write down an expression for $\langle\cdot,\cdot\rangle_{\mathcal{L}}$, namely

$$(5.2)\qquad \langle\Phi,\Psi\rangle_{\mathcal{L}} = \int_0^\infty\int_0^\infty s\wedge t \frac{\partial\Phi}{\partial\theta(s)}\frac{\partial\Psi}{\partial\theta(t)}\,dsdt .$$

To check whether (5.1) is "correct", we must show that the resulting $\mathcal{L}$ and $\langle\cdot,\cdot\rangle_{\mathcal{L}}$ satisfy (4.3) i) and ii) . From (5.2) is is clear that

(4.3) ii) is satisfied by $\langle\cdot,\cdot\rangle_{\mathcal{L}}$. To see that (4.3) i) holds is not quite so simple; a proof requires some facility with Frechét derivatives and "integration by parts" in Wiener space. In any case, when the domain of the operator given by (5.1) is correctly interpreted, one can show that $\mathcal{L}$ and $\langle\cdot,\cdot\rangle_{\mathcal{L}}$ do indeed satisfy (4.3) and therefore provide a scheme on $(\Theta,\mathbb{W})$ analogous to the one developed for (R^N,γ) in section 4).

Thus, so far, everything appears to be the same as in the finite dimensional case. However, when it comes to applying this scheme to functions on Θ in which we are interested (e.g. to solutios of S.D.E.'s), we encounter a phenomenon which is less visible in the finite dimensional analogue. Namely, throughout the discussion in section 4) we were implicitly assuming that the functions on which we were operating had classical derivatives of the appropriate orders. Of course, a little thought about what we actually required about our functions in order to justify our manipulations would have made us raelize that everything we did could be justified had we taken the $L^2(\gamma)$-closures of all our operations. Such technicalities do not seem very important in finite dimensions because the domain of the $L^2(\gamma)$-closed operations is not dramatically larger than that of the unclosed ones. (This fact is a consequence of Sobolev embedding theory.) In contrast to the situation for finite dimensions, closing our infinite dimensional operations makes an enormous difference and enables us to operate a vastly richer class of functions.

There are various ways in which to arrive at the closure of the operations described in (5.1) and (5.2) . Perhaps the most direct procedure is to mimic Sobolev theory, as was done by Shigekawa [Shig.]. A second approach, the one suggested by Malliavin, is to start from martingale calculus in place of Frechét calculus (cf. [S,Sys. Th.] for the details of this method). A third technique is to carry out the required closures in finite dimensions and then use tensor products to get the desired infiite dimensional operations. It is this third approach that we will adopt here.

Our treatment will follow [S., J. Fnal. Anal.] very closely and the interested reader is urged to consult [S., J. Fnal. Anal.] for the many technical details which we will omit here.

The first step in our program is to cast everything in an abstract context. Let $(E,\mathcal{F},m)$ be a separable probability space. We will call $\{T_\tau : \tau > 0\}$ a <u>symmetric Markov semigroup</u> on $L^2(m)$ (abbr., a <u>s.m.s. on</u> $L^2(m)$) if:

(5.3)

i) $\{T_\tau : \tau > 0\}$ is a strongly continuous semigroup of self-adjoint contractions on $L^2(m)$;

ii) for each $\tau > 0$: $T_\tau \Phi \geq 0$ (a.s., m) if $\Phi \geq 0$ (a.s., m) , and $T_\tau 1 = 1$ (a.s., m) .

Examples of s.m.s.'s are plentiful. For example, any Markov process which is symmetric with respect to a probabilty measure determines one (cf. Fukashima [Fuk.]). The following lemma is completely elementary (cf. Lemma (1.2) in [S., J. Fnal. Anal.]).

<u>(5.4)</u> <u>Lemma</u>: Let $\{T_\tau : \tau > 0\}$ be a s.m.s. on $L^2(m)$. Then for each $q \in [1,\infty)$ and $\Phi \in L^2(m) \cap L^q(m)$, $\|T_\tau \Phi\|_{L^q(m)} \leq \|\Phi\|_{L^q(m)}$, $\tau > 0$. In particular, for each $\tau > 0$ and $q \in [1,\infty)$ there is a unique $L^q(m)$-contraction $T_\tau^{(q)}$ such that $T_\tau^{(q)}$ coincides with T_τ on $L^2(m) \cap L^q(m)$.

Moreover, $\{T_\tau^{(q)} : \tau > 0\}$ is a strongly continuous semigroup on $L^q(m)$, $T_\tau^{(q)}$ is non-negativity preserving for each $\tau > 0$, and $T_\tau^{(q)} \subseteq T_\tau^{(1)}$. Finally, if $A^{(q)}$ denotes the generator of $\{T_\tau^{(q)} : \tau > 0\}$, then $A^{(q)} \subseteq A^{(1)}$ and $\Phi \in \mathrm{Dom}(A^{(q)})$ if and only if $\Phi \in L^q(m) \cap \mathrm{Dom}(A^{(1)})$ and $A^{(1)}\Phi \in L^q(m)$ (in which case $A^{(1)}\Phi = A^{(q)}\Phi$).

As we have said, there are lots of s.m.s.'s. In fact, for our purposes there are too many. We are therefore going to restrict our attention to a much smaller class. Probabilistically speaking, the class we have in mind corresponds to symmetric Markov processes having continuous paths.

We say that $(\mathcal{L}, T_\cdot, \mathcal{D}, m)$ is a <u>symmetric</u> <u>diffusion</u> <u>semigroup</u> <u>on</u> <u>$L^2(m)$</u> (abbr., <u>s.d.s.</u> <u>on</u> <u>$L^2(m)$</u>) if

(5.5)

i) $\{T_\tau : \tau > 0\}$ is a s.m.s. on $L^2(m)$ and $\mathcal{L}$ is its generator;

ii) $\mathcal{D} \subseteq \bigcap_{q=1}^{\infty} \mathrm{Dom}(A^{(q)})$ is an algebra containing 1 such that graph $(\mathcal{L}|_{\mathcal{D}})$ is dense in graph$(\mathcal{L})$;

iii) for $\Phi \in \mathcal{D}$ and $F \in \mathcal{S}(R^1)$ (the real valued Schwartz test function space), $F \circ \Phi \in \mathrm{Dom}(\mathcal{L})$ and

$$\mathcal{L}(F \circ \Phi) = 1/2\langle\Phi,\Phi\rangle_{\mathcal{L}} F'' \circ \Phi + \mathcal{L}\Phi \cdot F' \circ \Phi \tag{5.6}$$

where

$$\langle\Phi,\Psi\rangle_{\mathcal{L}} = \mathcal{L}(\Phi\cdot\Psi) - \Phi \cdot \mathcal{L}\Psi - \Psi \cdot \mathcal{L}\Phi \quad . \tag{5.7}$$

The reader who is familiar with martingale calculus will recognize (5.6) as Itô's formula for continuous martingales. It is precisely in (5.6) that the aforementioned connection between s.d.s.'s and path continuity lies.

(5.8) Lemma: Let $(\mathfrak{L}, T_{\cdot}, \mathfrak{D}, m)$ be a s.d.s. and define $\langle\cdot,\cdot\rangle_{\mathfrak{L}}$ on $\mathfrak{D} \times \mathfrak{D}$ by (5.7). Then $\langle\Phi,\Phi\rangle_{\mathfrak{L}} \geq 0$ (a.s., m) for each $\Phi \in \mathfrak{D}$. Set

$$(5.9) \qquad \langle\Phi\rangle_{\mathfrak{L}} = |\langle\Phi,\Phi\rangle_{\mathfrak{L}}|^{1/2} ,$$

for $\Phi \in \mathfrak{D}$. Then for all $\Phi,\Psi \in \mathfrak{D}$, $|\langle\Phi,\Psi\rangle_{\mathfrak{L}}| \leq \langle\Phi\rangle_{\mathfrak{L}} \langle\Psi\rangle_{\mathfrak{L}}$ and $\langle\Phi + \Psi\rangle_{\mathfrak{L}} \leq \langle\Phi\rangle_{\mathfrak{L}} + \langle\Psi\rangle_{\mathfrak{L}}$ (a.s., m). Finally, for $\Phi,\Psi \in \mathfrak{D}$:

$$(5.10) \qquad E^m[\langle\Phi,\Psi\rangle] = 2E^m[\Phi\mathfrak{L}\Psi] .$$

Proof: Once we show that $\langle\Phi,\Phi\rangle_{\mathfrak{L}} \geq 0$ (a.s., m), all but the final assertion follow easily by the same reasoning as is always used in connection with non-negative bilinear forms. To see that $\langle\Phi,\Phi\rangle_{\mathfrak{L}} \geq 0$ (a.s., m), one only has to show that $T_\tau\Phi^2 \geq (T_\tau\Phi)^2$ (a.s., m) for $\Phi \in \mathfrak{D}$ and $\tau > 0$. Indeed, given this, one sees that for $\Psi \in L^2(m)^+$:

$$(\langle\Phi,\Phi\rangle_{\mathfrak{L}},\Psi)_{L^2(m)} = \lim_{\tau\downarrow 0} 1/\tau((T_\tau\Phi^2 - \Phi^2) - 2\Phi(T_\tau\Phi - \Phi),\Psi)_{L^2(m)}$$

$$\geq \lim_{\tau\downarrow 0} 1/\tau((T_\tau\Phi - \Phi)^2,\Psi) \geq 0 .$$

The proof that $T_\tau\Phi^2 \geq (T_\tau\Phi)^2$ (a.s., m) can be easily reduced to Jensen's inequality (cf. Lemma (1.5) in [S.,J.F.A.]).

Finally, the proof of (5.10) follows from the identity $\mathfrak{L}(\Phi\cdot\Psi) = \Phi\mathfrak{L}\Psi + \Psi\mathfrak{L}\Phi + \langle\Phi,\Psi\rangle_{\mathfrak{L}}$, the symmetry of $\mathfrak{L}$ on $L^2(m)$, and the fact that for any $\Xi \in \mathrm{Dom}(\mathfrak{L})$, $E^m[\mathfrak{L}\Xi] = E^m[\Xi\mathfrak{L}1] = 0$. □

(5.10) Lemma: Let $(\mathfrak{L}, T_{\cdot}, \mathfrak{D}, m)$ be a s.d.s. and define $\langle\cdot,\cdot\rangle_{\mathfrak{L}}$ as before on $\mathfrak{D} \times \mathfrak{D}$. Then $\langle\cdot,\cdot\rangle_{\mathfrak{L}}$ admits a unique graph($\mathfrak{L}$)-continuous extension as a bilinear operator from $\mathrm{Dom}(\mathfrak{L}) \times \mathrm{Dom}(\mathfrak{L})$ into $L^1(m)$. Moreover, this extension continues to satisfy (5.10) as well as the Schwartz and Minkowski inequalities mentioned in (5.8) . Finally, if $N \geq 1$, $\Phi = (\Phi_1,\ldots,\Phi_N) \in (\mathrm{Dom}(\mathfrak{L}))^N$, and $F \in C_b^2(R^N)$, then $F \circ \Phi \in \mathrm{Dom}(A^{(1)})$ and

$$(5.11) \qquad A^{(1)}(F\circ\Phi) = 1/2 \sum_{i,j=1}^{N} \langle\Phi_i,\Phi_j\rangle_{\mathfrak{L}} \frac{\partial^2 F}{\partial x_i \partial x_j} \circ \Phi + \sum_{i=1}^{N} \mathfrak{L}\Phi_i \frac{\partial F}{\partial x_i} \circ \Phi$$

<u>Proof</u>: In order to prove that the desired extension exists it suffices to show that if $\{\Phi_n\}_1^\infty \subseteq \mathfrak{D}$ is a graph($\mathfrak{L}$)-Cauchy convergent sequence, then $\{\langle\Phi_n,\Phi_n\rangle_{\mathfrak{L}}\}_1^\infty$ is an $L^1(m)$-Cauchy convergent sequence. Since $\left|\langle\Phi_n,\Phi_n\rangle_{\mathfrak{L}} - \langle\Phi_m,\Phi_m\rangle_{\mathfrak{L}}\right| = \left|\langle\Phi_n\rangle_{\mathfrak{L}} - \langle\Phi_m\rangle_{\mathfrak{L}}\right|\left|\langle\Phi_n\rangle_{\mathfrak{L}} + \langle\Phi_m\rangle_{\mathfrak{L}}\right|$ and $\sup_n E^m[\langle\Phi_n\rangle_{\mathfrak{L}}^2] = \sup_n 2E^m[\Phi_n\mathfrak{L}\Phi_m] \leq \sup_n \|\Phi_n\|_{L^2(m)} \|\mathfrak{L}\Phi\|_{L^2(m)} < \infty$, it suffices show that $\{\langle\Phi\rangle_{\mathfrak{L}}\}_1^\infty$ is an $L^2(m)$-Cauchy convergent sequence. But $\left|\langle\Phi_n\rangle_{\mathfrak{L}} - \langle\Phi_m\rangle_{\mathfrak{L}}\right| \leq \langle\Phi_n - \Phi_m\rangle_{\mathfrak{L}}$ and so we need only prove that $E^m[\langle\Phi_n - \Phi_m\rangle_{\mathfrak{L}}^2] \to 0$ as $m,n \to \infty$. The proof is therefore complete after one uses (5.10) and Schwartz's inequality to verify that $E^m[\langle\Phi_n - \Phi_m\rangle_{\mathfrak{L}}^2] \leq 2\|\Phi_n - \Phi_m\|_{L^2(m)} \|\mathfrak{L}\Phi_n - \mathfrak{L}\Phi_m\|_{L^2(m)}$.

Knowing that the extension exists, one sees immediately that the extension continues to have all the properties discussed in (5.8) . Thus it remains only to verify the final assertion. To this end, note that for fixed $F \in C_b^2(R^N)$ the set of $\Phi \in (\mathrm{Dom}(\mathfrak{L}))^N$ for which $F \circ \Phi \in \mathrm{Dom}(A^{(1)})$ and $A^{(1)}(F \circ \Phi)$ is given by (5.11) is closed under $(\mathrm{graph}(\mathfrak{L}))^N$ -convergence. Indeed, this observation is an immediate consequence of the continuity property of $\langle\cdot,\cdot\rangle_{\mathfrak{L}}$ just proved plus the fact that $A^{(1)}$

(being the generator of a semigroup) is a closed operator. Thus we may and will assume that $\Phi \in \mathcal{D}^N$. Next, for fixed $\Phi \in \mathcal{D}^N$, it is clear that the class of $F \in C_b^2(R^{N^\circ})$ for which $F \circ \Phi \in \mathrm{Dom}(A^{(1)})$ and (5.11) holds is linear and is closed under bounded point-wise convergence of $\{D^\alpha F : |\alpha| \leq 2\}$. (This is again due to the closedness of $A^{(1)}$.) Thus we may and will assume that $F(x) = f(\omega \cdot x)$, $x \in R^N$, for some $f \in \mathcal{S}(R^1)$ and $\omega \in R^N$. But in this case $F \circ \Phi = f(\sum_1^N \omega_j \Phi_j)$ and so (5.11) is nothing but (5.6). □

Given a s.d.s $(\mathcal{L}, T_\bullet, \mathcal{D}, m)$ and $q \in [2,\infty)$, set $\mathcal{K}_{(q)}(\mathcal{L}) = \{\Phi \in \mathrm{Dom}(A^{(q)}) : \langle\Phi\rangle_{\mathcal{L}} \in L^q(m)\}$ and define $\|\Phi\|_{\mathcal{K}(q)(\mathcal{L})} = E^m[(\Phi^2 + \langle\Phi\rangle_{\mathcal{L}}^2 + (\mathcal{L}\Phi)^2)^{q/2}]^{1/q}$. Because $A^{(q)}$ is closed, it is easy to check that $(\mathcal{K}_q(\mathcal{L}), \|\cdot\|_{\mathcal{K}_{(q)}(\mathcal{L})})$ is a Banach space. Next, set $\mathcal{K}(\mathcal{L}) = \bigcap_{2 \leq q < \infty} \mathcal{K}_{(q)}(\mathcal{L})$. Then $\mathcal{K}(\mathcal{L})$ can be easily turned into a countably normed Frechét space in which convergence corresponds to $\mathcal{K}_{(q)}(\mathcal{L})$-convergence for all $q \in [2,\infty)$. As a consequence of the next lemma, it will be clear that $\mathcal{K}(\mathcal{L})$ is an algebra.

<u>(5.12)</u> <u>Lemma</u>: Let $(\mathcal{L}, T_\bullet, \mathcal{D}, m)$ be a s.d.s. Given $N \geq 1$, $\Phi = (\Phi_1, \ldots, \Phi_N) \in (\mathcal{K}(\mathcal{L}))^N$, and $F \in C_\uparrow^2(R^N)$ (the space of $F \in C^2(R^N)$ such that $D^\alpha F$ is slowly increasing for each $|\alpha| \leq 2$), $F \circ \Phi \in \mathcal{K}(\mathcal{L})$ and $\mathcal{L}(F \circ \Phi)$ is given by the R.H.S. of (5.11). Moreover, if $\Psi \in \mathcal{K}(\mathcal{L})$, then

$$\langle F\circ\Phi, \Psi\rangle_{\mathcal{L}} = \sum_{i=1}^N \frac{\partial F}{\partial x_i} \circ \Phi \langle \Phi_i, \Psi\rangle_{\mathcal{L}} . \tag{5.13}$$

Finally, if $\Phi \in (\mathcal{K}(\mathcal{L}))^+$ and $1/\Phi \in \bigcap_1^\infty L^q(m)$, then $1/\Phi \in \mathcal{K}(\mathcal{L})$,

$\mathcal{L}(1/\Phi) = \langle\Phi,\Phi\rangle_{\mathcal{L}}/\Phi^3 - \mathcal{L}\Phi/\Phi^2$, and $\langle 1/\Phi,\Psi\rangle_{\mathcal{L}} = -1/\Phi^2 \langle\Phi,\Psi\rangle_{\mathcal{L}}$ for $\Psi \in K(\mathcal{L})$

Proof: The first asserion follows from (5.11) by an obvious approximation arguement. Given the first assertion, (5.13) is obtained from (5.11) by replacing F with $\tilde{F}(x,y) = F(x)y$ and Φ with $\tilde{\Phi} = (\Phi,\Psi)$. Finally, the last assertion is easily proved by replacing $1/\Phi$ with $1/(\Phi^2 + \varepsilon^2)^{1/2}$, applying the preceding to $1/(\Phi^2 + \varepsilon^2)^{1/2}$, and letting $\varepsilon \downarrow 0$. See Lemmas (1.11) and (1.13) in [S.,J.F.A.] for more details. □

We are at last ready to give a rigorous and general version of the itegration by parts procedure discussed in section 4).

(5.13) Theorem: Let $(\mathcal{L},T_{\cdot},\mathcal{D},m)$ be a s.d.s. Given $N \geq 1$ and $\Phi \in (K(\mathcal{L}))^N$, set $A = \langle\langle\Phi,\Phi\rangle\rangle_{\mathcal{L}} \equiv ((\langle\Phi_i,\Phi_j\rangle_{\mathcal{L}}))_{1\leq i,j\leq N}$. Then A is m-a.s.'ly symmetric and non-negative definite. Moreover, if $A \in (K(\mathcal{L}))^{N^2}$, then for $F \in C_0^1(R^N)$ and $\Psi \in K(\mathcal{L})$:

$$E^m[(\frac{\partial F}{\partial x_k} \circ \Phi)(\Delta\cdot\Psi)] = -E^m[(f \circ \Phi)\mathcal{H}_k\Psi] \quad , \quad 1 \leq k \leq N \quad , \tag{5.14}$$

where $\Delta = \det A$ and

$$\mathcal{H}_k\Psi = \sum_{\ell=1}^{N} (\langle\Phi_\ell,A^{(k,\ell)}\Psi\rangle_{\mathcal{L}} + 2(\mathcal{L}\Phi_\ell)A^{(k,\ell)}\Psi) \tag{5.15}$$

with $A^{(k,\ell)}$ denoting the $(k,\ell)^{th}$ cofactor of A. In particular, if $1/\Delta \in \bigcap_1^\infty L^q(m)$, then

$$E^m[(\frac{\partial F}{\partial x_k} \circ \Phi)\Psi] = -E^m[(f \circ \Phi)\mathcal{H}_k(\Psi/\Delta))] \quad . \tag{5.16}$$

Proof: The properties of A are obvious. Assuming $A \in (K(\mathcal{L}))^N$, one derives (5.15) in the same way as we did (4.5). That is:

$$\langle F\circ\Phi,\Phi_\ell\rangle_{\mathcal{L}} = \sum_{j=1}^{N}\left(\frac{\partial F}{\partial x_j}\circ\Phi\right)A_{j\ell} \quad ;$$

and so, by Cramer's rule:

$$\left(\frac{\partial F}{\partial x_k}\circ\Phi\right)\Delta = \sum_{\ell=1}^{N}\langle F\circ\Phi,\Phi_\ell\rangle_{\mathcal{L}}A^{(k,\ell)} \quad .$$

Hence, since $\Delta\Psi\in\mathcal{K}(\mathcal{L})$ and $A^{(k,\ell)}\Psi\in\mathcal{K}(\mathcal{L})$:

$$E^m\left[\left(\frac{\partial F}{\partial x_k}\circ\Phi\right)(\Delta\Psi)\right] =$$

$$= \sum_{\ell=1}^{N}E^m[(F\circ\Phi)(\Phi_\ell\mathcal{L}(A^{(k,\ell)}\Psi) - A^{(k,\ell)}\mathcal{L}\Phi_\ell - \mathcal{L}(\Phi_\ell A^{(k,\ell)}))]$$

$$= E^m[(F\circ\Phi)(\mathcal{V}_k\Psi)] \quad .$$

Clearly (5.16) is just (5.15) with Ψ replaced by Ψ/Δ ($\in\mathcal{K}(\mathcal{L})$ when $1/\Delta\in\bigcap_1^\infty L^q(m)$). □

(5.17) Corollary: Let everything be as in (5.15) and assume that $A\in(\mathcal{K}(\mathcal{K}(\mathcal{L}))^N$ and that $1/\Delta\in\bigcap_1^\infty L^q(m)$. Set $\mu = m\circ\Phi^{-1}$. Then μ is absolutely continuous with respect to Lebesgue measure on R^N ; and if $f = \frac{d\mu}{dx}$, then $f\in\hat{C}(R^N)\cap\bigcap_{\gamma\in(0,1)}\mathrm{Lip}(\gamma)$ and $\mathrm{grad}\, f\in\bigcap_1^\infty(L^q(R^N))^N$.

Proof: Set $\Psi_k = \mathcal{V}_k(1/\Delta)$, $1\le k\le N$. Then by (5.16) :

$$E^m\left[\frac{\partial\phi}{\partial x_k}\circ\Phi\right] = -E^m[(\phi\circ\Phi)\Psi_k] \quad , \quad 1\le k\le N \quad ,$$

for $\phi\in C_0^\infty(R^N)$. Next set $\nu_k = (\Psi_k m)\circ\Phi^{-1}$. Then $\nu_k \ll \mu$; and if $\psi_k = \frac{d\nu_k}{d\mu}$, then

$$E^m[(\phi\circ\Phi)(\psi_k\circ\Phi)] = \int_{R^N}\phi\psi_k d\mu = \int_{R^N}\phi d\nu_k = E^m[(\phi\circ\Phi)\Psi_k]$$

$$= E^m[(\phi\circ\Phi)E^m[\Psi_k|\Phi^{-1}(\mathcal{B}_{R^N})]]$$

for all $\phi \in C_b(R^N)$. Hence $\psi_k \circ \Phi = E^m[\Psi_k|\Phi^{-1}(\mathcal{B}_{R^N})]$ (a.s., m) . In particular, $\|\psi_k\|_{L^q(m)} = \|E^m[\Psi_k|\Phi^{-1}(\mathcal{B}_{R^N})]\|_{L^q(m)} \leq \|\Psi_k\|_{L^q(m)}$. Since $1/\Delta \in \mathcal{K}(\mathcal{L})$, $\Psi_k \in \bigcap_1^\infty L^q(m)$ and so $\psi_k \in \bigcap_1^\infty L^q(\mu)$. Finally:

$$\int_{R^N}\frac{\partial\phi}{\partial x_k}d\mu = E^m[\frac{\partial\phi}{\partial x_k}\circ\Phi] = -E^m[(\phi\circ\Phi)\Psi_k]$$

$$= -\int_{R^N}\phi d\nu_k = -\int_{R^N}\phi\psi_k d\mu \ , \quad 1 \leq k \leq N \ ,$$

for $\phi \in C_0^\infty(R^N)$. Thus Lemma (3.2) applies.

6. The Ornstein-Uhlenbeck Semigroup on Wiener Space:

We now want to put a s.d.s. over Wiener space. The one which we will construct is precisely the one associated with the operator given in (5.1) . However, from the outset, we will be presented with the $L^2(\mathfrak{W})$ completed versions of this operator. Our construction is very straightforward. We first show how to put a s.d.s. over (R^1,γ) where $\phi(dx) = (2\pi)^{-1/2}e^{-x^2/2}dx$. This step is accomplished by our simply being a little careful with the first example discussed in secion 4). Having completed the first step, our second step involves lifting the s.d.s over (R^1,γ) to (R^{Z^+},γ^{Z^+}) by taking tensor products. It is at this step that we begin to take advantage of the abstract formulation given in section 5) The point here is that having abstracted everything, it is easy to use soft functional analytic manipulations and check that the resulting quantities still have the desired properties. Finally, our third step consists of transferring our s.d.s. over (R^{Z^+},γ^{Z^+}) to $(\Theta,\mathfrak{W})$ by utilizing Wiener's measure preserving map of $(\Theta,\mathfrak{W})$ into (R^{Z^+},γ^{Z^+}). Again at this step, the abstract formulation developed in section 5) proves useful.

The necessary machinery from functional analysis is contained in the following lemmas.

(6.1) Lemma: For $n \geq 1$ let $(\mathcal{L}^{(n)},T_\cdot^{(n)},\mathcal{D}^{(n)},m^{(n)})$ be a s.d.s. over the separable probability space $(E^{(n)},\mathcal{F}^{(n)},m^{(n)})$. Set $E = \prod_1^\infty E^{(n)}$, $\mathcal{F} = \prod_1^\infty \mathcal{F}^{(n)}$, $m = \prod_1^\infty m^{(n)}$, and $T_\tau = \bigotimes_1^\infty T_\tau^{(n)}$, $\tau > 0$. Let $\mathcal{D}$ be the linear span of $\bigcup_{N=1}^\infty \mathcal{D}^{(1)} \otimes \cdots \otimes \mathcal{D}^{(N)}$. Then $\{T_\tau : \tau > 0\}$ is a s.m.s. on $L^2(m)$; and if $\mathcal{L}$ is the generator of $\{T_\tau : \tau > 0\}$, then $(\mathcal{L},T_\cdot,\mathcal{D},m)$ is a s.d.s. Moreover, if $N \geq 1$ and $\Phi^{(n)} \in \mathrm{Dom}(\mathcal{L}^{(n)})$, $1 \leq n \leq N$, then $\Phi^{(1)} \otimes \cdots \otimes \Phi^{(N)} \in \mathrm{Dom}(\mathcal{L})$, $\mathcal{L}(\Phi^{(1)} \otimes \cdots \otimes \Phi^{(N)}) =$

$$\sum_1^N \Phi^{(1)} \otimes \cdots \otimes \Phi^{(n-1)} \otimes \mathcal{L}\Phi^{(n)} \otimes \Phi^{(n+1)} \otimes \cdots \otimes \Phi^{(N)},$$

and

$$\langle \Phi^{(1)} \otimes \cdots \otimes \Phi^{(N)},\Phi^{(1)} \otimes \cdots \otimes \Phi^{(N)}\rangle = \sum_1^N (\Phi^{(1)})^2 \otimes \cdots \otimes (\Phi^{(n-1)})^2 \otimes \langle \Phi^{(n)},\Phi^{(n)}\rangle_{\mathcal{L}^{(n)}} \otimes (\Phi^{(n+1)})^2 \otimes \cdots \otimes (\Phi^{(N)})^2 .$$

(6.2) Lemma: Let $(E^{(i)},\mathcal{F}^{(i)},m^{(i)})$, $i = 1,2$, be separable probability spaces and suppose that $\Xi : E^{(2)} \to E^{(1)}$ is a measurable measure preserving map such that the isometry $\Lambda_\Xi : L^2(m^{(1)}) \to L^2(m^{(2)})$ given by $\Lambda_\Xi \Phi = \Phi \circ \Xi$, $\Phi \in L^2(m^{(1)})$ is onto. Given a s.d.s. $(\mathcal{L}^{(1)},T_\cdot^{(1)},\mathcal{D}^{(1)},m^{(1)})$ over $(E^{(1)},\mathcal{F}^{(1)},m^{(1)})$, define

$\mathcal{L}^{(2)} = \Lambda_\Xi \circ \mathcal{L}^{(1)} \Lambda_\Xi^{-1}$ on $\mathrm{Dom}(\mathcal{L}^{(2)}) = \Lambda_\Xi(\mathrm{Dom}\,\mathcal{L}^{(1)})$,

$T^{(2)} = \Lambda_\Xi \circ T^{(1)} \circ \Lambda_\Xi^{-1}$, $\tau > 0$, on $L^2(m^{(2)})$, and $\mathcal{D}^{(2)} = \Lambda_\Xi \mathcal{D}^{(1)}$.

Then $(\mathcal{L}^{(2)},T^{(2)},\mathcal{D}^{(2)},m^{(2)})$ is a s.d.s. over $(E^{(2)},\mathcal{F}^{(2)},m^{(2)})$.

Lemmas (6.1) and (6.2) are far harder to state than they are to prove. The interested reader should look at Theorems (2.1) and (2.3) in [S.,J.F.A.].

Set $p(\tau,x) = \frac{1}{(2\pi\tau)^{1/2}} e^{-x^2/2\tau}$ for $\tau > 0$ and $x \in R^1$ and define $T_\tau\Phi(x) = \int_{R^1} p(1-e^{-\tau}, y-e^{-\tau/2}x)\Phi(y)dy$ for $\tau > 0$, $x \in R^1$, and $\Phi \in C_b(R)$. Clearly $\{T_\tau : \tau > 0\}$ forms a Markovian semigroup of non-negative contractions on $C_b(R^1)$ such that $T_\tau 1 = 1$, $\tau > 0$, and $T_\tau\Phi \to \Phi$ point-wise as $\tau \downarrow 0$. Moreover, $\int \Phi T_\tau \Psi d\gamma = \int \Psi T_\tau \Phi d\gamma$, $\tau > 0$, for all $\Phi,\Psi \in C_b(R^1)$. In particular, $\int |T_\tau\Phi|^2 d\gamma \leq \int T_\tau \Phi^2 d\gamma = \int \Phi^2 d\gamma$. Thus for each $\tau > 0$ there is a unique extension T_τ^γ of T_τ as a self-adjoint operator on $L^2(\gamma)$. It is a simple matter to check that $\{T_\tau^\gamma : \tau > 0\}$ is a s.m.s. Denote by $\mathcal{L}^\gamma$ the generator of $\{T_\tau^\gamma : \tau > 0\}$.

<u>(6.3)</u> <u>Lemma</u>: Define

$$H_n(x) = \frac{(-1)^n}{(n!)^{1/2}} e^{x^2/2} D(e^{-x^2/2}) \quad , \quad n \geq 0 \; .$$

Then $\{H_n : n \geq 0\}$ is an ortho-normal basis in $L^2(\gamma)$, $\mathrm{Dom}(\mathcal{L}^\gamma) = \{\Phi \in L^2(\gamma) : \sum^\infty n^2(\Phi,H_n)^2_{L^2(\gamma)} < \infty\}$, and $\mathcal{L}^\gamma\Phi = -1/2\sum^\infty n(\Phi,H_n)_{L^2(\gamma)} H_n$ for $\Phi \in \mathrm{Dom}(\mathcal{L}^\gamma)$. Moroever, $C_\uparrow^\infty(R^1) \subseteq \mathrm{Dom}(\mathcal{L}^\gamma)$ and for $\Phi \in C_\uparrow^\infty(R^1)$:

$$\mathcal{L}^\gamma\Phi = 1/2(D^2 - xD)\Phi \; .$$

Proof: Using:

$$(6.4) \qquad e^{\lambda x-\lambda^2/2} = \sum_0^\infty \frac{\lambda^n}{(n!)^{1/2}} H_n(x) ,$$

one easily obtains:

$$T_\tau^\gamma H_n = e^{-n/2\tau} H_n , \quad \tau > 0 \text{ and } n \geq 0 .$$

Since it is well-known that $\{H_n : n \geq 0\}$ is an orthogonal basis in $L^2(\gamma)$ the proof of everything except the final assertion is complete.

To handle the first part, introduce $h_n(x) = e^{-x^2/4} H_n(x)$, $n \geq 0$ and $x \in R$. Then it is well-known that $\mathscr{S}(R^1)$ coincides with the set of $f \in L^2(R^1)$ such that $\{(f,h_n)_{L^2(R^1)} : n \geq 0\}$ is rapidly decreasing. Moreover, for $f \in \mathscr{S}(R^1)$, $\sum_0^N (f,h_n)_{L^2(R^1)} h_n \to f$ in $\mathscr{S}(R^1)$. Thus, $e^{x^2/4}\mathscr{S}(R^1)$ coincides with the set of $\Phi \in L^2(\gamma)$ such that $\{(\Phi,H_n)_{L^2(\gamma)} : n \geq 0\}$ is rapidly decreasing; and for $\Phi \in e^{x^2/4}\mathscr{S}(R^1)$: $x^k D^\ell(\sum_0^N (\Phi,H_n)_{L^2(\gamma)} H_n) \to x^k D^\ell \Phi$ in $L^2(\gamma)$ for all $k,\ell \geq 0$. In particular, if $\Phi \in C_\uparrow^\infty(R^1)$, then $\Phi \in e^{x^2/4}\mathscr{S}(R^1)$; and so, since $1/2(D^2-xD)H_n = -\frac{n}{2} H_n$:

$$\mathfrak{L}^\gamma(\sum_0^N (\Phi,H_n)_{L^2(\gamma)} H_n) = 1/2(D^2-xD)(\sum_0^N (\Phi,H_n)_{L^2(\gamma)} H_n)$$
$$\to 1/2(D^2-xD)\Phi$$

in $L^2(\gamma)$. Because $\mathfrak{L}^\gamma$ is closed, this proves both that $\Phi \in \mathrm{Dom}(\mathfrak{L}^\gamma)$ and that $\mathfrak{L}^\gamma\Phi = 1/2(D^2-xD)\Phi$.

Let $\mathcal{D}^{\gamma}$ denote the space of real polynomials on R^1. Then, with the help of (6.3), it is easy to check that $(\mathcal{L}^{\gamma}, T_{\cdot}^{\gamma}, \mathcal{D}^{\gamma}, \gamma)$ is a s.d.s. on $L^2(\gamma)$. □

Our second step is to apply Lemma (6.1) with $(E^{(n)}, \mathcal{F}^{(n)}, m^{(n)}) = (R^1, \mathcal{B}_{R^1}, \gamma)$ and $(\mathcal{L}^{(n)}, T_{\cdot}^{(n)}, \mathcal{D}^{(n)}, m^{(n)}) = (\mathcal{L}^{\gamma}, T_{\cdot}^{\gamma}, \mathcal{D}^{\gamma}, \gamma)$ for all $n \geq 1$. Letting $\Gamma = \gamma^{Z^+}$ and $T_{\tau}^{\Gamma} = (T_{\tau}^{\gamma})^{Z^+}$ and denoting by $\mathcal{D}^{\Gamma}$ the space of real polynomials on R^{Z^+}, we see that $(\mathcal{L}^{\Gamma}, T_{\cdot}^{\Gamma}, \mathcal{D}^{\Gamma}, \Gamma)$ is a s.d.s. on $L^2(\Gamma)$, where $\mathcal{L}^{\Gamma}$ denotes the generator of $\{T_{\tau}^{\Gamma} : \tau > 0\}$. Moreover, as a consequence of the last parts of Lemmas (6.1) and (6.3), we see that if $\Phi(x) = f(x_1, \ldots, x_N)$, $x \in R^{Z^+}$, where $f \in C_{\uparrow}^{\infty}(R^N)$, then $\Phi \in \mathrm{Dom}(\mathcal{L}^{\Gamma})$,

$$\mathcal{L}^{\Gamma}\Phi = 1/2 \sum_{1}^{N} \left(\frac{\partial^2 f}{\partial x_i^2} - x_i \frac{\partial f}{\partial x_i}\right) \tag{6.5}$$

and

$$\langle \Phi, \Phi \rangle_{\mathcal{L}^{\Gamma}} = \sum_{1}^{N} \left(\frac{\partial f}{\partial x_i}\right)^2 . \tag{6.6}$$

The spectral resolution of $\mathcal{L}^{\Gamma}$ is also easily derived from Lemma (6.3). Indeed, if $A \equiv \{\alpha \in \mathcal{N}^{Z^+} : |\alpha| < \infty\}$ and if for $\alpha \in A$ we define

$$H_{\alpha}(x) = \prod_{1}^{\infty} H_{\alpha_k}(x_k) ,$$

then $\{H_{\alpha} : \alpha \in A\}$ is an orthonormal basis in $L^2(\Gamma)$ and $\mathcal{L}^{\Gamma} H_{\alpha} = -\frac{|\alpha|}{2} H_{\alpha}$. Hence, if

$$\mathcal{H}^{(n)} = \overline{\mathrm{span}\{H_{\alpha} : |\alpha| = n\}}^{L^2(\Gamma)} \tag{6.7}$$

and $E_{\mathcal{H}^{(n)}}$ denotes orthogonal projection onto $\mathcal{H}^{(n)}$, then

$$\mathrm{Dom}(\mathcal{L}^\Gamma) = \{\Phi \in L^2(\Gamma) : \sum_0^\infty n\|E_{\mathcal{H}^{(n)}}\Phi\|^2_{L^2(\Gamma)} < \infty\}$$

Using these observations, we can now derive the following invariance properties of $(\mathcal{L}^\Gamma, T_\cdot^\Gamma, \mathcal{D}^\Gamma, \Gamma)$.

(6.8) <u>Theorem</u>: Let $U = ((u_{k,\ell}))_{k,\ell \in Z^+}$ be a real orthogonal matrix on $\ell^2(Z^+)$. Then there is a Γ-null set $B \in \mathcal{B}_{R^{Z^+}}$ such that $\sum_\ell u_{k,\ell}x_\ell$ converges for all $k \in Z^+$ and $x \notin B$. Define $U : R^{Z^+} \to R^{Z^+}$ so that $(Ux)_k = \sum_\ell u_{k,\ell}x_\ell$ if $k \in Z^+$ and $x \notin B$ and $Ux = 0$ if $x \in B$. Then Then U is a measure preserving transformation of $(R^{Z^+}, \mathcal{B}_{R^{Z^+}}, \Gamma)$ into itself. Moreover, if $\Lambda_U : L^2(\Gamma) \to L^2(\Gamma)$ is defined by $\Lambda_U\Phi = \Phi \circ U$, then Λ_U is unitary and $\Lambda_U^* = \Lambda_{U^*}$. Finally, for all $n \geq 0$, $\Lambda_U\mathcal{H}^{(n)} = \mathcal{H}^{(n)}$. In particular, $\{T_\tau^\Gamma : \tau > 0\}$, and therefore $\mathcal{L}^\Gamma$, are Λ_U-invariant (i.e. $\Lambda_U \circ T_\tau^\Gamma = T_\tau^\Gamma \circ \Lambda_U$, $\tau > 0$, and $\Lambda_U \circ \mathcal{L}^\Gamma = \mathcal{L}^\Gamma \circ \Lambda_U$)

<u>Proof</u>: Under Γ , the x_k , $k \in Z^+$, are independent Gaussian random variables with mean 0 and variance 1 . Hence the convergence statement follows immediately from $\sum_{\ell=1}^\infty u_{k,\ell}^2 = 1$. Moreover, for each $N \geq 1$, $\{\sum_{\ell=1}^N u_{k,\ell}x_\ell : k \in Z^+\}$ is a mean zero Gaussian family; and for each $k \in Z^+$, $\sum_{\ell=1}^N u_{k,\ell}x_\ell \to (Ux)_k$ in $L^2(\Gamma)$. Thus, $\{(Ux)_k : k \in Z^+\}$ is a mean zero Gaussian family and $E^\Gamma[(Ux)_k(Ux)_\ell] = \delta_{k,\ell}$. In other words, $\Gamma = \Gamma \circ U^{-1}$ (i.e. U is measure preserving).

To prove that Λ_U is unitary and that $\Lambda_U^* = \Lambda_{U^*}$, it suffices to show that $\Lambda_U \circ \Lambda_{U^*} = \text{identity}$. To this end, let $f \in C_b(R^N)$ and set $\Phi(x) = f(x_1,\ldots,x_N)$. For $L \geq N$, set $\Phi_L(x) = f(\sum_1^L u_{\ell;1}x_\ell,\ldots,\sum_1^L u_{\ell,N}x_\ell)$. Then $\Phi_L \to \Lambda_{U^*}\Phi$ in $L^2(\Gamma)$ as $L \uparrow \infty$. At the same time,

$$\Lambda_U\Phi_L(x) = f(\sum_{k=1}^{\infty} u_{1,k}\sum_{\ell=1}^{L} u_{\ell,k}x_\ell,\ldots,\sum_{k=1}^{\infty} u_{N,k}\sum_{\ell=1}^{L} u_{\ell,k}x_\ell) = f(x_1,\ldots,x_N) .$$ Thus $\Lambda_U \circ \Lambda_{U^*}\Phi = \lim_{L\uparrow\infty} \Lambda_U\Phi_L = \Phi$. The desired conclusion is now immediate.

In view of the preceding, combined with our spectral analysis of $\mathcal{L}^\gamma$, the proof will be complete as soon as we show that $\Lambda_U\mathcal{H}^{(n)} \subseteq \mathcal{H}^{(n)}$ for all $n \geq 0$. To this end, note that if $v \in R^{Z^+}$ with $v_\ell = 0$ for all $\ell \geq L$ and $|v| = 1$, then for any $n \geq 0$:

$$H_n(v\cdot x) = \sum_{|\alpha|=n} \binom{n}{\alpha}^{1/2} v^\alpha H_\alpha(x) \in \mathcal{H}^{(n)} .$$

This formula is easy to derive from the generating function given in (6.4). From this it is immediate that for any $\alpha \in A$: $\Lambda_U H_\alpha \in \bigoplus_0^{|\alpha|}\mathcal{H}^{(n)}$. At the same time, if $|\beta| < \alpha$, then $(\Lambda_U H_\alpha, H_\beta)_{L^2(\Gamma)} = (H_\alpha, \Lambda_{U^*}H_\beta)_{L^2(\Gamma)} = 0$ since $\Lambda_{U^*}H_\beta \in \bigoplus_0^{|\beta|}\mathcal{H}^{(n)}$. Thus $\Lambda_U H_\alpha \in \mathcal{H}^{(|\alpha|)}$ for all $\alpha \in A$. This is equivalent to $\Lambda_U\mathcal{H}^{(n)} \subseteq \mathcal{H}^{(n)}$ for all $n \geq 0$. □

Our third step is to transfer $(\mathcal{L}^\Gamma, T_\cdot^\Gamma, \mathcal{D}^\Gamma, \Gamma)$ to $(\Theta, \mathcal{B}, \mathbb{W})$ via a measure preserving map. The next lemma describes the map which we have in mind.

(6.9) **Lemma:** Given an orthonormal basis $\mathcal{F} = \{f_k\}_1^\infty$ in $L^2([0,\infty))$, define $\mathcal{F} : \Theta \to R^{Z^+}$ by $(\mathcal{F}(\theta))_k = \int_0^\infty f_k(t)d\theta(t)$, $k \in Z^+$. Then $\mathcal{F}$ is a measurable measure preserving map from $(\Theta,\mathcal{B},\mathcal{W})$ into $(R^{Z^+}, \mathcal{B}_{R^{Z^+}},\Gamma)$; and the isometry $\Lambda_{\mathcal{F}} : L^2(\Gamma) \to L^2(\mathcal{W})$ given by $\Lambda_{\mathcal{F}}F = F \circ \mathcal{F}$, $F \in L^2(\Gamma)$, is onto. Moreover, if $\mathcal{G} = \{g_k\}_1^\infty$ is a second orthonormal basis in $L^2([0,\infty))$ and $U = (((f_k,g_\ell)_{L^2([0,\infty))}))_{k,\ell\in Z^+}$, then $\Lambda_{\mathcal{G}} = \Lambda_{\mathcal{F}} \circ \Lambda_U$.

Proof: Note that $\{\int_0^\infty f(t)d\theta(t) : f \in L^2([0,\infty))\}$ is a mean zero Gaussian family under $\mathcal{W}$ and that $E[(\int_0^\infty f(t)d\theta(t))(\int_0^\infty g(t)d\theta(t))] = (f,g)_{L^2([0,\infty))}$. From this it is obvious that $\Gamma = \mathcal{W} \circ \mathcal{F}^{-1}$. Before proving that $\Lambda_{\mathcal{F}}$ is onto, we observe that the relation $\Lambda_{\mathcal{G}} = \Lambda_{\mathcal{F}} \circ \Lambda_U$ in the last assertion is obvious. This shows, in particular, that the closed subspace $\Lambda_{\mathcal{F}}(L^2(\Gamma))$ is the same for all choices of $\mathcal{F}$. Now suppose that $\Psi \perp \Lambda_{\mathcal{F}}(L^2(\Gamma))$ for some, and therefore all, $\mathcal{F}$'s . Given $N \geq 1$, $\{\lambda_n\}_1^N \subseteq R^1\setminus \{0\}$, and $0 = t_0 < \cdots < t_N$, define $\phi(t) = \sum_1^N \lambda_n \chi_{[t_{n-1},t_n)}(t)$, $t \geq 0$, and set $f_1 = \phi/\|\phi\|_{L^2([0,\infty))}$. Then we may choose $\mathcal{F}$ so that f_1 is its first entry. With this choice of $\mathcal{F}$ we see that

$$\exp(\sum_1^N \lambda_j(\theta(t_j) - \theta(t_{j-1}))) = \exp(\|\phi\|_{L^2([0,\infty))}\int_0^T f_1(t)d\theta(t))$$
$$= \Lambda_{\mathcal{F}}(F) \in \Lambda_{\mathcal{F}}(L^2(\Gamma))$$

where $F(x) = \exp(\|\phi\|_{L^2([0,\infty))} x_1)$, $x \in R^{Z^+}$. Hence $E^{\mathcal{W}}[\Psi \exp(\sum_1^N \lambda_j(\theta(t_j)-\theta(t_{j-1})))] = 0$ for all $N \geq 1$, $\{\lambda_j\}^N \subseteq R^1$, and $0 = t_0 < \cdots < t_N$. From this it is clear that $\Psi = 0$. Hence $\Lambda_{\mathcal{F}}(L^2(\Gamma)) = L^2(\mathcal{W})$. □

Let $\mathcal{F} = \{f_k\}_1^\infty$ be an orthonormal basis in $L^2([0,\infty))$ and define

$\mathcal{F} : \Theta \to R^{Z^+}$ accordingly. Define $T_\tau^{\mathcal{W}} = \Lambda_{\mathcal{F}} \circ T_\tau^\Gamma \circ \Lambda_{\mathcal{F}}^{-1}$, $\tau > 0$, and $\mathcal{L}^{\mathcal{W}} = \Lambda_{\mathcal{F}} \circ \mathcal{L}^\Gamma \circ \Lambda_{\mathcal{F}}^{-1}$ on $\Lambda_{\mathcal{F}}(\mathrm{Dom}(\mathcal{L}^\Gamma))$. By Lemma (6.9) and Theorem (6.8), the definition of $\{T_\tau^{\mathcal{W}} : \tau > 0\}$ and $\mathcal{L}^{\mathcal{W}}$ does not depend on the choice of $\mathcal{F}$ Moreover, if $\mathcal{D}_{\mathcal{F}}^{\mathcal{W}} = \Lambda_{\mathcal{F}}(\mathcal{D}^\Gamma)$, then, by Lemma (6.2) , $(\mathcal{L}^{\mathcal{W}}, T^{\mathcal{W}}, \mathcal{D}_{\mathcal{F}}^{\mathcal{W}}, \mathcal{W})$ is an s.d.s. on $L^2(\mathcal{W})$. We will call this s.d.s. the <u>Ornstein-Uhlenbeck semigroup on Wiener space</u>.

<u>(6.10)</u> <u>Theorem</u>: For $t \geq 0$, set $\mathcal{B}^t = \sigma(\theta(u)-\theta(t) : u \geq t)$. Given $t \geq 0$, $\Phi \in L^2(\Theta,\mathcal{B}_t,\mathcal{W})$, and $\Psi \in L^2(\Theta,\mathcal{B}^t,\mathcal{W})$: $\Phi\cdot\Psi \in L^2(\mathcal{W})$, $T_\tau^{\mathcal{W}}\Phi$ is $\mathcal{B}_t$-measurable, $T_\tau^{\mathcal{W}}\Psi$ is $\mathcal{B}^t$-measurable, and $T_\tau^{\mathcal{W}}(\Phi\cdot\Psi) = (T_\tau^{\mathcal{W}}(\Phi))\cdot(T_\tau^{\mathcal{W}}(\Psi))$. If, in addition, $\Phi,\Psi \in \mathcal{K}_q(\mathcal{L}^{\mathcal{W}})$ for some $q \in [2,\infty)$, $\Phi\cdot\Psi \in \mathcal{K}_q(\mathcal{L}^{\mathcal{W}})$, $\mathcal{L}^{\mathcal{W}}\Phi$ is $\mathcal{B}_t$-measurable, $\mathcal{L}^{\mathcal{W}}\Psi$ is $\mathcal{B}^t$-measurable, $\mathcal{L}^{\mathcal{W}}(\Phi\cdot\Psi) = \Phi\mathcal{L}^{\mathcal{W}}\Psi + \Psi\mathcal{L}^{\mathcal{W}}\Phi$, and $\langle\Phi,\Psi\rangle_{\mathcal{L}^{\mathcal{W}}} = 0$.

<u>Proof</u>: Since $\mathcal{B}_t$ and $\mathcal{B}^t$ are independent under $\mathcal{W}$, everything follows easily once it has been shown that $L^2(\Theta,\mathcal{B}_t,\mathcal{W})$ and $L^2(\Theta,\mathcal{B}^t,\mathcal{W})$ are $\{T_\tau^{\mathcal{W}} : \tau > 0\}$ invariant subspace and that $T_\tau^{\mathcal{W}}(\Phi\cdot\Psi) = (T_\tau^{\mathcal{W}}(\Phi))\cdot(T_\tau^{\mathcal{W}}(\Psi))$ for $\tau > 0$, $\Phi \in L^2(\Theta,\mathcal{B}_t,\mathcal{W})$, and $\Psi \in L^2(\Theta,\mathcal{B}^t,\mathcal{W})$. To prove these facts, choose $\mathcal{F} = \{f_n\}_1^\infty$ to be an orthonormal basis in $L^2([0,\infty))$ such that $\mathrm{supp}(f_{2n+1}) \subset [0,t)$ and $\mathrm{supp}(f_{2n+1}) \subset [t,\infty)$ for all $n \geq 0$. Set $A_0 = \{\alpha \in A : \alpha_{2k+1} = 0$ for all $k \geq 0\}$, $A_e = \{\alpha \in A : \alpha_{2k} = 0$ for all $k \geq 1\}$, $\mathcal{H}_0 = \overline{\mathrm{span}\{H_\alpha : \alpha \in A_0\}}^{L^2(\Gamma)}$, and $\mathcal{H}_e = \overline{\mathrm{span}\{H_\alpha : \alpha \in A_e\}}^{L^2(\Gamma)}$ It is clear that $\Lambda_{\mathcal{F}}$ maps $\mathcal{H}_0$ isometrically onto $L^2(\Theta,\mathcal{B}_t,\mathcal{W})$ and that $\Lambda_{\mathcal{F}}$ maps $\mathcal{H}_e$ isometrically on $L^2(\Theta,\mathcal{B}^t,\mathcal{W})$. Furthermore, $\mathcal{H}_0$ and $\mathcal{H}_e$ are invariant subspaces of $\{T_\tau^\Gamma : \tau > 0\}$ and $T_\tau^\Gamma(F\cdot G) = (T_\tau^\Gamma(F))\cdot(T_\tau^\Gamma(G))$ for $\tau > 0$, $F \in \mathcal{H}_0$, and $G \in \mathcal{H}_e$. Hence, the same properties hold for $L^2(\Theta,\mathcal{B}_t,\mathcal{W})$ and $L^2(\Theta,\mathcal{B}^t,\mathcal{W})$ with respect to $\{T_\tau^{\mathcal{W}} : \tau > 0\}$. ⌐

As we will see in the next section, Theorem (6.10) plays a critical role when it comes to computing with $\mathcal{L}^{\mathcal{W}}$. As a preliminary example, we give the following corollary.

(6.11) Corollary: For all $t \geq 0$, $\theta(t) \in \mathcal{K}(\mathcal{L}^{\mathcal{W}})$ and $\mathcal{L}^{\mathcal{W}}\theta(t) = -1/2\ \theta(t)$. Moreover, if $s,t \geq 0$, then $\langle\theta(s),\theta(t)\rangle_{\mathcal{L}^{\mathcal{W}}} = s \wedge t$. In particular, if $0 \leq s < t$, then $\langle\theta(t)-\theta(s),\theta(t)-\theta(s)\rangle_{\mathcal{L}^{\mathcal{W}}} = t - s$.

Proof: Choose an orthonormal basis $\mathcal{F} = \{f_n\}_1^\infty$ in $L^2([0,\infty))$ with $f_1 = 1/t^{1/2}\,\chi_{[0,t)}$. Since $H_1(y) = y$ and $H_2(y) = \frac{y^2-1}{2^{1/2}}$, $y \in R^1$, $\theta(t) = t^{1/2}\,\Lambda_{\mathcal{F}}H_\alpha$ and $\theta(t)^2 = 2^{1/2}t\Lambda_{\mathcal{F}}H_\beta$ where $\alpha = (1,0,\ldots,0,\ldots)$ and $\beta = (2,0,\ldots,0,\ldots)$. This proves that $\theta(t) \in \mathcal{K}(\mathcal{L}^{\mathcal{W}})$ and that $\mathcal{L}^{\mathcal{W}}(\theta(t)) = t^{1/2}\Lambda_{\mathcal{F}}\mathcal{L}^{\Gamma}H_\alpha = -1/2t^{1/2}\Lambda_{\mathcal{F}}H_\alpha = -1/2\theta(t)$ and $\mathcal{L}^{\mathcal{W}}(\theta(t)^2) = 2^{1/2}t\Lambda_{\mathcal{F}}\mathcal{L}^{\Gamma}H_\beta = -2^{1/2}t\Lambda_{\mathcal{F}}H_\beta = -\theta(t)^2 + t$. Hence $\langle\theta(t),\theta(t)\rangle_{\mathcal{L}^{\mathcal{W}}} = t$. Finally, if $0 \leq s < t$, then by Theorem (6.10) and the preceding: $\langle\theta(t),\theta(s)\rangle_{\mathcal{L}^{\mathcal{W}}} = \langle\theta(t)-\theta(s),\theta(s)\rangle_{\mathcal{L}^{\mathcal{W}}} + \langle\theta(s),\theta(s)\rangle_{\mathcal{L}^{\mathcal{W}}} = s$. ❑

Before moving on we owe it to ourselves to see in what sense $\mathcal{L}^{\mathcal{W}}$ is related to the operator $\mathcal{L}$ in (5.1). To this end, observe that the natural extension of the $\mathcal{L}$ in (5.1) to Gateau differentiable functions would yield:

$$\mathcal{L}\Phi = 1/2\Big(\sum_{i,j=1}^{N} t_i \wedge t_j \frac{\partial^2 f}{\partial x_i \partial x_j}(\theta(t_1),\ldots,\theta(t_N)) - \sum_{i=1}^{N} \theta(t_i)\frac{\partial f}{\partial x_i}(\theta(t_1),\ldots,\theta(t_N))\Big)$$

if $\Phi : \Theta \to R^1$ is given by:

$$\Phi(\theta) = f(\theta(t_1),\ldots,\theta(t_N))$$

for some $N \geq 0$, $f \in C_b^2(R^N)$, and $0 \leq t_1 < \cdots < t_N$. But this is precisely how $\mathcal{L}^{\mathcal{W}}$ acts on such a Φ, since

$$\mathcal{L}^{\mathcal{W}}\Phi = 1/2 \sum_{i,j=1}^{N} \langle\theta(t_i),\theta(t_j)\rangle_{\mathcal{L}^{\mathcal{W}}} \frac{\partial^2 f}{\partial x_i \partial x_j}(\theta(t_1),\ldots,\theta(t_N)) + \sum_{i=1}^{N} \mathcal{L}^{\mathcal{W}}(\theta(t_i))\frac{\partial^2 f}{\partial x_i}(\theta(t_1),\ldots,\theta(t_N)) \quad ;$$

and, by (6.11) : $\langle\theta(t_i),\theta(t_j)\rangle_{\mathcal{L}^{\mathcal{W}}} = t_i \wedge t_j$ and $\mathcal{L}^{\mathcal{W}}(\theta(t_i)) = -1/2\theta(t_i)$.

Having constructed a s.d.s. over one-dimesional Wiener space, it is an easy matter to put one over d-dimenional Wiener space for any $d \geq 1$. Indeed, if $\Theta_d = \{\theta \in C([0,\infty),R^d) : \theta(0) = 0\}$, then Θ_d is naturally isomorphic to $(\Theta_1)^d$ and the image of d-dimensional Wiener measure $\mathcal{W}_d$ on Θ_d under this isomorphism is $(\mathcal{W}_1)^d$. Hence Lemma (6.1) tells us how to put a s.d.s. over $(\Theta_d,\mathcal{W}_d)$. The properties of this s.d.s. over $(\Theta_d,\mathcal{W}_d)$ are easily read off from those of the one on $(\Theta_1,\mathcal{W}_1)$. We will again call the s.d.s. constructed in this way the Ornstein-Uhlenbeck semigroup.

(6.12) Warning about Notation: From now on we will rely more on context and less on sub- and superscripts. Thus we will write $(\Theta,\mathcal{W})$ to denote Wiener space, no matter what the dimension of the paths may be. Moreover, since the only s.d.s. with which we will be concerned is the one just described in the preceding paragraph, we will write simply $\mathcal{L}$, $\langle\cdot,\cdot\rangle_{\mathcal{L}}$, and $\{T_\tau : \tau > 0\}$ to denote the operations associated with the Ornstein-Uhlenbeck semigroup over our Wiener space. The calculus generated by these operations will be called the Malliavin Calculus.

7. The Malliavin's Calculus and Stochastic Integrals:

In order to apply the machinery just developed to (Q.2) in section 1) , it is necessary for us to learn how $\mathcal{L}$ and $\langle\cdot,\cdot\rangle_{\mathcal{L}}$ act on stochastic integrals. In particular, we must show that solutions to S.D.E.'s are in the domain of these operations.

Suppose that $\alpha : \Theta \to R^d$ is $\mathcal{B}_t$-measurable and that $\alpha \in (\mathrm{Dom}(\mathcal{L}))^d$. Given $h > 0$, Theorem (6.10) tells us that $\alpha\cdot(\theta(t+h)-\theta(t)) \in \mathrm{Dom}(\mathcal{L})$ and that $\mathcal{L}(\alpha\cdot(\theta(t+h)-\theta(t))) = (\mathcal{L}\alpha - 1/2\alpha)\cdot(\theta(t+h)-\theta(t))$. Thus if $\alpha : [0,\infty) \times \Theta \to R^d$ is progressively measurable and simple (i.e. $\alpha(t) = \alpha([Nt]/N)$, $t \geq 0$, for some $N \geq 1$, $\alpha(t) \in (\mathrm{Dom}(\mathcal{L}))$ for all $t \geq 0$ implies that $\int_0^T \alpha(t)\cdot d\theta(t) \in \mathrm{Dom}(\mathcal{L})$, $T \geq 0$, and

$$(7.1)\qquad \mathcal{L}\left(\int_0^T \alpha(t)\cdot d\theta(t)\right) = \int_0^T (\mathcal{L}\alpha(t) - 1/2\alpha(t))\cdot d\theta(t) \quad .$$

(Note that, by Theorem (6.10), $\mathcal{L}\,\alpha(t)$ is again a progressively measurable function and so the right hand side of (7.1) is a well-defined Itô integral.) Next suppose that $\beta : [0,\infty) \times \Theta \to R^1$ is a simple progressively measurable function satisfying $\beta(t) \in \mathrm{Dom}(\mathcal{L})$ for all $t \geq 0$. Then it is easy to see that $\int_0^T \beta(t)dt \in \mathrm{Dom}(\mathcal{L})$ and that

$$(7.2)\qquad \mathcal{L}\left(\int_0^T \beta(t)dt\right) = \int_0^T \mathcal{L}\beta(t)\,dt$$

for all $T \geq 0$.

In order to get away from simple functions, we require an approximation result analogous to the one used in order to extend the Itô integral beyond simple integrals. The required result is stated below; the interested reader can find its proof in [S.,J.F.A.] (cf. Lemma (4.2) of that paper).

(7.3) Lemma: Let $\alpha : [0,T] \times \Theta \to R^1$ be a progressively measurable function such that, for some $q \in [2,\infty)$, $\alpha(t) \in \mathcal{H}_{(q)}(\mathcal{L})$, $t \in [0,T]$;

and assume that $\int_0^T \|\alpha(t)\|^q_{\mathcal{K}_{(q)}(\mathcal{L})} dt < \infty$. Then there is a progressively measurable $\beta : [0,T] \times \Theta \to R^1$ such that $\beta(t) = \mathcal{L}(\alpha(t))$ (a.s., $\mathfrak{W}$) for a.e. $t \in [0,T]$. Moreover, if $q \in [4,\infty)$, then there is a progressively measurable $\gamma : [0,T] \times \Theta \to R^1$ such that $\gamma(t) = \langle \alpha(t), \alpha(t) \rangle_{\mathcal{L}}$ (a.s., $\mathfrak{W}$) for a.e. $t \in [0,T]$. Finally, there exist simple progressively measurable functions $\alpha : [0,T] \times \Theta \to R^1$ such that: $\alpha(t) \in \mathcal{K}_($ $t \in [0,T]$; $\int_0^T (\|\alpha_n(t) - \alpha(t)\|^q_{L^q(\mathfrak{W})} + \|\mathcal{L}(\alpha_n(t)) - \mathcal{L}(\alpha(t))\|^q_{L^q(\mathfrak{W})}) dt \to 0$ as $n \to \infty$; and, if $q \geq 4$, $\int_0^T \|\langle \alpha_n(t) \rangle_{\mathcal{L}} - \gamma(t)^{1/2}\|^q_{L^q(\mathfrak{W})} dt \to 0$ as $n \to \infty$.

(7.4) Warning: Given $\alpha(\cdot)$ of the sort discussed in (7.3), we will use $\mathcal{L}(\alpha(\cdot))$ and $\langle \alpha(\cdot), \alpha(\cdot) \rangle_{\mathcal{L}}$ to denote the progressively measurable functions $\beta(\cdot)$ and $\gamma(\cdot)$ whose existence (7.3) asserts. This convention can cause no difficulties so long as these quantities appear as integrands in "dt" or "d$\theta(t)$" integrals. Moreover, in all of our applications, $\alpha(\cdot)$, $\mathcal{L}(\alpha(\cdot))$, and $\langle \alpha(\cdot), \alpha(\cdot) \rangle_{\mathcal{L}}$ will be continuous (a.s., $\mathfrak{W}$), and therefore there will be one $\mathfrak{W}$-null set B such that $\mathcal{L}(\alpha(\cdot)) = \beta(\cdot)$ and $\langle \alpha(\cdot), \alpha(\cdot) \rangle_{\mathcal{L}} = \gamma(\cdot)$ off of B.

(7.5) Notation: Given $\Phi = (\Phi_1, \ldots, \Phi_N) \in (\mathcal{K}_{(q)}(\mathcal{L}))^N$, define

$$|||\Phi|||_{(q)} = E^{\mathfrak{W}}[(\sum_1^N (\Phi_n^2 + (\mathcal{L}\Phi_n)^2 + \langle \Phi_n \rangle_{\mathcal{L}}^2)^{q/2}]^{1/q} .$$

Note that, since $|\langle \Phi_m, \Phi_n \rangle_{\mathcal{L}}| \leq \langle \Phi_m \rangle_{\mathcal{L}} \langle \Phi_n \rangle_{\mathcal{L}}$, $\|\langle \Phi_m, \Phi_n \rangle\|_{L^q(\mathfrak{W})} \leq |||\Phi|||$. Also, recall the notation $\langle\langle \Phi, \Phi \rangle\rangle_{\mathcal{L}} \equiv ((\langle \Phi_m, \Phi_n \rangle_{\mathcal{L}}))_{1 \leq m, n \leq N}$. Given $\Phi : [0,T] \times \Theta \to R^N$ such that $\Phi(t) \in (\mathcal{K}_q(\mathcal{L}))^N$, $t \geq 0$, and $t \to (\Phi, \mathcal{L}(\Phi(t))$, $\langle\langle \Phi(t), \Phi(t) \rangle\rangle_{\mathcal{L}})$ is continuous (a.s., $\mathfrak{W}$), define

$$|||\Phi(\cdot)|||_{(q),T} = E^{W}[\sup_{0\le t\le T}(\sum_1^N((\Phi_n(t))^2 + (\mathfrak{L}\Phi_n(t))^2 + \langle\Phi_n(t)\rangle^2)^{q/2}]^{1/q}$$

for $T \geq 0$.

(7.6) Theorem: Let $\alpha : [0,\infty) \times \Theta \to R^d$ and $\beta : [0,\infty) \times \Theta \to R^1$ be progressively measurable functions. Assume that $\alpha(t) \in (\mathcal{K}(\mathfrak{L}))^d$ and $\beta(t) \in \mathcal{K}(\mathfrak{L})$, $t \geq 0$, and that $\int_0^T(|||\alpha(t)|||^4_{(4)} + |||\beta(t)|||^4_{(4)})dt < \infty$, $T > 0$. Set $\xi(T) = \int_0^T \alpha(t)\cdot d\theta(t) + \int_0^T \beta(t)dt$, $T \geq 0$. Then $\xi(T) \in \mathcal{K}_{(4)}(\mathfrak{L})$, $T \geq 0$; $T \to (\xi(T), \mathfrak{L}\xi(T), \langle\xi(T),\xi(T)\rangle_{\mathfrak{L}})$ is continuous (a.s., $\mathcal{W}$) ; and for each $q \in [2,\infty)$ there is a non-decreasing function $C_q : [0,\infty) \to [0,\infty)$ such that

(7.7) $$|||\xi(\cdot)|||_{(q),T} \leq C_q(T)\int_0^T(|||\alpha(t)|||^q_{(q)} + |||\beta(t)|||^q_{(q)})dt \quad , \quad T \geq 0 .$$

Finally, $\mathfrak{L}\xi(\cdot)$ is given by:

(7.8) $$\mathfrak{L}\xi(T) = \int_0^T(\mathfrak{L}(\alpha(t)) + 1/2\alpha(t))\cdot d\theta(t) + \int_0^T \mathfrak{L}(\beta(t))dt \quad , \quad T \geq 0 ,$$

and $\langle\xi(\cdot),\xi(\cdot)\rangle_{\mathfrak{L}}$ satisfies:

(7.9) $$\langle\xi(T),\xi(T)\rangle_{\mathfrak{L}} = 2\int_0^T\langle\xi(t),\alpha(t)\rangle_{\mathfrak{L}}\cdot d\theta(t) + \int_0^T(2\langle\xi(t),\beta(t)\rangle_{\mathfrak{L}} + \langle\alpha(t)\rangle^2_{\mathfrak{L}} + |\alpha(t)|^2)dt \quad , \quad T \geq 0 .$$

(In (7.9) , $\langle\xi(\cdot),\alpha(\cdot)\rangle_{\mathfrak{L}} = (\langle\xi(\cdot),\alpha_{\cdot}(\cdot)\rangle_{\mathfrak{L}} ,\ldots,\langle\xi(\cdot),\alpha_{\alpha}(\cdot)\rangle_{\mathfrak{L}})$ and $\langle\alpha(\cdot)\rangle^2_{\mathfrak{L}} = \mathrm{Trace}\langle\langle a(\cdot),\alpha(\cdot)\rangle\rangle_{\mathfrak{L}}$.)

Proof: We first assume that $\alpha(\cdot)$ and $\beta(\cdot)$ are simple. Using (7.1) and (7.2) , we see that $\xi(T) \in \mathrm{Dom}(\mathfrak{L})$ and that (7.8) holds, even when we insist only that $\alpha(t) \in (\mathcal{K}_{(2)}(\mathfrak{L}))^d$ and $\beta(t) \in \mathcal{K}_{(2)}(\mathfrak{L})$, $t \geq 0$. Also, from (7.8) and Burkholder's inequality, we see that for any $q \in [2,\infty)$:

$$(7.10) \qquad E^{\mathfrak{W}}[\sup_{0\le t\le T}|\mathfrak{L}(\xi(t))|^q]$$

$$\le C_q(T^{q/2-1}\int_0^T(|||\mathfrak{L}(\alpha(t))|||^q_{L^q(\mathfrak{W})} + \||\alpha(t)|\|^q_{L^q(\mathfrak{W})})dt$$

$$+ T^{q-1}\int_0^T \|\mathfrak{L}(\beta(t))\|^q_{L^q(\mathfrak{W})} dt) \quad .$$

Thus, by Lemma (7.3) with $q = 2$, for any progressively measurable $\alpha : [0,\infty) \times \Theta \to R^d$ and $\beta : [0,\infty) \times \Theta \to R^1$ satisfying $\alpha(t) \in (\mathcal{K}_{(2)}(\mathfrak{L}))^d$ and $\beta(t) \in \mathcal{K}_{(2)}(\mathfrak{L}(\ )$, $t \ge 0$, with $\int_0^T(||| \alpha(t)|||^2_{(2)} + ||| \beta(t)|||^2_{(2)})dt < \infty$, $T > 0$, we can use (7.10) and the closedness of $\mathfrak{L}$ to show that: $\xi(T) \in \mathcal{K}_{(2)}(\mathfrak{L})$, $T \ge 0$; $\mathfrak{L}\xi(\cdot)$ is given by (7.8); and (7.10) holds for each $q \in [2,\infty)$.

To complete the proof, we again return to the assumption that $\alpha(\cdot)$ and $\beta(\cdot)$ are simple. From Theorem (6.10), it is clear that $\xi(T) \in \mathcal{K}_{(4)}(\mathfrak{L})$, and that $\sup_{0\le t\le T}|||\xi(t)|||_{(4)} < \infty$ for all $T \ge 0$. In particular,

$$\int_0^T(|||\xi(t)\alpha(t)|||^2_{(2)} + |||\xi(t)\beta(t)|||^2_{(2)} + |||\,|\alpha(t)|^2|||^2_{(2)})dt < \infty$$

for all $T > 0$. Since, by Itô's formula,

$$\xi(T)^2 = 2\int_0^T\xi(t)\alpha(t)\cdot d\theta(t) + \int_0^T(2\xi(t)\beta(t) + |\alpha(t)|^2)dt \quad ,$$

it follows from the preceding paragraph that $\xi(T)^2 \in \mathcal{K}_{(2)}(\mathfrak{L})$, $T \ge 0$, and that

$$\mathfrak{L}(\xi(T)^2) = \int_0^T(2\mathfrak{L}(\xi(t)\alpha(t)) - \xi(t)\alpha(t))\cdot d\theta(t)$$

$$+ \int_0^T(2\mathfrak{L}(\xi(t)\beta(t)) + \mathfrak{L}|\alpha(t)|^2))dt \quad , \ T \ge 0 \ .$$

At the same time, by Itô's formula and (7.8) ,

$$\xi(T)\mathcal{L}(\xi(T)) = \int_0^T (\mathcal{L}(\xi(t))\alpha(t) + \xi(t)\mathcal{L}(\alpha(t)) - 1/2\xi(t)\alpha(t))\cdot d\theta(t)$$

$$+ \int_0^T (\mathcal{L}(\xi(t))\beta(t) + \xi(t)\mathcal{L}(\beta(t)) + \alpha(t)\cdot(\mathcal{L}(\alpha(t)) - 1/2\alpha(t)))dt \quad .$$

Hence, since $\langle\xi(T),\xi(T)\rangle_{\mathcal{L}} = \mathcal{L}(\xi(T)^2) - 2\xi(T)\mathcal{L}(\xi(T))$, we now obtain (7.9) . Moreover, since $|\langle\xi(t),\alpha(t)\rangle_{\mathcal{L}}| \leq \langle\xi(t)\rangle_{\mathcal{L}}^2 + \langle\alpha(t)\rangle_{\mathcal{L}}^2$ and $|\langle\xi(t),\beta(t)\rangle_{\mathcal{L}}| \leq \langle\xi(t)\rangle_{\mathcal{L}}^2 + \langle\beta(t)\rangle_{\mathcal{L}}^2$, an application of Burkholder's inequality to (7.9) yields an integral inequality to which Gromwall's inequality applies and yields that part of (7.7) not covered by (7.10) .

We have not completed the proof when $\alpha(\cdot)$ and $\beta(\cdot)$ are simple. The general case now follows easily from Lemma (7.3) .

Before these considerations can be put to work for us, we still have to see whether they apply to solutions of S.D.E.'s. To gain some insight into what our problem is, let $d = 1$ and suppose that $X(\cdot)$ is the solution to

$$X(T) = \int_0^T \sigma(X(t))d\theta(t) + \int_0^T b(X(t))dt \quad , \quad T \geq 0 \quad ,$$

where $\sigma : R^1 \to R^1$ and $b : R^1 \to R^1$ are smooth functions having bounded derivatives. If we know that $X(T) \in K_{(4)}(\mathcal{L})$, and that $\int_0^T |||X(t)|||_{(4)}^4 dt < \infty$, $T \geq 0$, then we could apply Theorem (7.6) and thereby show that:

$$(7.11) \qquad \langle X(t),X(T)\rangle_{\mathcal{L}} = \int_0^T 2\sigma'(X(t))\langle X(t),X(t)\rangle_{\mathcal{L}}\, d\theta(t)$$

$$+ \int_0^T (2b'(X(t))\langle X(t),X(T)\rangle_{\mathcal{L}} + \sigma'(X(t))^2\langle X(t),X(t)\rangle_{\mathcal{L}} + \sigma(X(t))^2)dt \quad .$$

(In the derivation of (7.11) we have made repeated use of the equation $\langle f\circ\Phi,\Psi\rangle_{\mathfrak{L}} = f'\circ\Phi\langle\Phi,\Psi\rangle_{\mathfrak{L}}$.) Of course, since $X(\cdot)$ is already known, (7.11) itself uniquely determines $\langle X(\cdot),X(\cdot)\rangle_{\mathfrak{L}}$ and can be used to get estimates on $\|\sup_{0\le t\le T}\langle X(t),X(t)\rangle_{\mathfrak{L}}\|_{L^q(\mathfrak{W})}$ in terms of $\|\sigma'\|_{C_b(R^1)}$ and $\|b'\|_{C_b(R^1)}$. Again assuming that we know that $\int_0^T |||X(t)|||^q_{(q)}dt < \infty$ for $q\in[4,\infty)$ and $T>0$, we can apply Theorem (7.6) to obtain:

$$\mathfrak{L}(X(T)) = \int_0^T(\mathfrak{L}(\sigma(X(t)) - 1/2\sigma(X(t)))d\theta(t)$$
$$+ \int_0^T\mathfrak{L}(b(X(t)))dt$$

which becomes:

$$\mathfrak{L}(X(T)) = \int_0^T(1/2\sigma''(X(t))\langle X(t),X(t)\rangle_{\mathfrak{L}} + \sigma'(X(t))\mathfrak{L}(X(t)) - 1/2\sigma(X(t)))d\theta(t)$$
$$(7.12)\qquad + \int_0^T(1/2b''(X(t))\langle X(t),X(t)\rangle_{\mathfrak{L}} + b'(X(t))\mathfrak{L}(X(t)))dt\ .$$

Since we can now include $\langle X(\cdot),X(\cdot)\rangle_{\mathfrak{L}}$ among our known quantities, (7.12) uniquely determines $\mathfrak{L}(X(\cdot))$ and enables us to obtain estimates on $\|\sup_{0\le t\le T}\mathfrak{L}(X(t))\|_{L^q(\mathfrak{W})}$ in terms of $\|\sigma'\|_{C_b^1(R^1)}$, $\|b'\|_{C_b^1(R^1)}$, and $\|\sup_{0\le t\le T}\langle X(t),X(t)\rangle_{\mathfrak{L}}\|_{L^q(\mathfrak{W})}$. In other words, if we knew that $X(T)\in\mathcal{K}_{(q)}(\mathfrak{L})$, $T\ge 0$, and had minimal control over $\int_0^T|||X(t)|||^q_{(q)}dt$, we would have equations for $\langle X(\cdot),X(\cdot)\rangle_{\mathfrak{L}}$ and $\mathfrak{L}(X(\cdot))$ and could use these equatinos to get very good control over $|||X(\cdot)|||_{(q),T}$.

The unfortunate fact is that no such argument by itself can bring us to the desired result, because to justify such an argument requires our knowing a priori more than we do. In order to break out of the circle into which the above reasoning has led us, we return to Itô's original methodology for solving S.D.E.'s. Namely, define $X_0(\cdot) \equiv 0$ and for $\nu \geq 1$ define $X_\nu(\cdot)$ by

$$(7.13) \qquad X_\nu(T) = \int_0^T \sigma(X_{\nu-1}(t))d\theta(t) + \int_0^T b(X_{\nu-1}(t))dt \quad , \quad T \geq 0 \ .$$

As we already know, $\| \sup_{0\leq t\leq T} |X_\nu(t)-X(t)| \|_{L^q(\mathcal{W})} \to 0$ as $\nu \to \infty$ for each $q \in [2,\infty)$ and $T > 0$. Moreover, if we know that $|||X_{\nu-1}(\cdot)|||_{(q),T} < \infty$ for $q \in [4,\infty)$ and $T > 0$, then by Theorem (7.6) we know that $|||X_\nu(\cdot)|||_{(q),T} < \infty$ for $q \in [2,\infty)$ and $T > 0$. Since $|||X_0(\cdot)|||_{(q),T}$ obviously is finite for all $q \in [4,\infty)$ and $T > 0$, it follows by induction that $|||X_\nu(\cdot)|||_{(q),T} < \infty$ for all $q \in [2,\infty)$ and $T > 0$. In order to complete the proof that $X(T) \in \mathcal{K}_q(\mathcal{L})$ and that $\sup_{0\leq t\leq T} |||X(t)|||_{(q)} < \infty$ for all $q \in [4,\infty)$ and $T > 0$; it therefore suffices for us to show that $\sup_\nu |||X(\cdot)|||_{(q),T} < \infty$ for all $q \in [4,\infty)$ and $T > 0$. But we can for the $X_\nu(\cdot)$ justify the formal arguments given in the preceding paragraph for $X(\cdot)$ and from the resulting expressions obtain the needed estimate. In this way, one can prove that following theorem (cf. Theorem (4.11) and Corollary (4.13) in [S.,J.F.A.]).

(7.14) Theorem: Let $\sigma : R^N \to R^N \otimes R^d$ and $b : R^N \to R^N$ be thrice continuously differentiable functions satisfying $\|D^\alpha\sigma\|_{C_b(R^N)} \vee \|D^\alpha b\|_{C_b(R^N)} < \infty$ for all $1 \leq |\alpha| \leq 3$. For each $x \in R^N$, let $X(\cdot,x)$ be the unique progressively measurable solution to (2.13). Then $X(T,x) \in (\mathcal{K}_{(q)}(\mathcal{L}))^N$ for all $q \in [4,\infty)$; and, for each $q \in [4,\infty)$ and $T > 0$, $|||X(\cdot,x)|||_{(q),T}$ is finite and can be estimated in terms of

$\max_{1\leq|\alpha|\leq 2} \|D^\alpha\sigma\|_{C_b(R^N)} \vee \|D^\alpha b\|_{C_b(R^N)}$. Moreover, if $A(T,x) = \langle\langle X(T,x),X(T,x)\rangle\rangle$, then

$$(7.15) \qquad A(T,x) = \sum_{k=1}^{d}\int_0^T \{S_k(X(t,x)),A(t,x)\}d\theta_k(t)$$

$$+ \int_0^T(\{B(X(T,x)),A(t,x)\} + \sum_{k=1}^{d} S_k(X(t,x))A(t,x)S_k(X(t,x))^* + a(X(t,x)))dt$$

where

$$S_k(\cdot) = \left(\left(\frac{\partial\sigma_k^i}{\partial x_j}(\cdot)\right)\right)_{1\leq i,j\leq N} ,$$

$$(7.16) \qquad B(\cdot) = \left(\left(\frac{\partial b^i}{\partial x_j}(\cdot)\right)\right)_{1\leq i,j\leq N} ,$$

$$a(\cdot) = \sigma\sigma^*(\cdot) ,$$

and $\{M_1,M_2\} \equiv M_1M_2^* + M_2M_1^*$ for $M_1,M_2 \in R^N \otimes R^N$. Finally $A(T,x) \in \mathcal{K}_{(q)}(\mathcal{L})$ and $|||A(\cdot,x)|||_{(q),T}$ is finite for all $q \in [4,\infty)$ and $T > 0$.

In view of Corollary (5.17) , Theorem (7.14) allows us to conclude that the distribution of $X(T,x)$ under $\mathbb{W}$ admits a reasonably nice density as soon as we know that $1/\det(A(T,x)) \in \bigcap_1^\infty L^q(\mathbb{W})$. Thus the interesting question that remains is to find good criteria which guarantee that $1/\det(A(T,x)) \in \bigcap_1^\infty L^q(\mathbb{W})$. However, before we turn our attention to that problem and so long as we have the techniques of the present section clearly in mind, we would be well advised to investigate when we can apply higher order Malliavin operations to solutions of S.D.E.'s and what we can gain by doing so.

From what we have already seen, it is clear that the quantities resulting from the application of Malliavin's operations to solutions of S.D.E.'s again satisfy S.D.E.'s. In fact, if $X(\cdot,x)$ is the solution to (2.13) and we look at the vector

$$\Xi(T,x) \equiv \begin{pmatrix} X(T,x) \\ \langle\langle X(T,x),X(T,x)\rangle\rangle_{\mathcal{L}} \\ \mathcal{L}(X(T,x)) \end{pmatrix} \in R^N \times R^{N^2} \times R^N ,$$

then $\Xi(\cdot,x)$ is the solution to a S.D.E which is "lower triangular" in the sense that $X(\cdot,x)$ alone satisfies an equation which is autonomous, $\langle\langle X(\cdot,x),X(\cdot,x)\rangle\rangle_{\mathcal{L}}$ alone satisfies an equation which is autonomous given $X(\cdot,x)$, and $\mathcal{L}(X(\cdot,x))$ alone satisfies an equation which is autonomous given $X(\cdot,x)$ and $\langle\langle X(\cdot,x),X(\cdot,x)\rangle\rangle_{\mathcal{L}}$. Moreover, $\langle\langle X(\cdot,x),X(\cdot,x)\rangle\rangle_{\mathcal{L}}$ and $\mathcal{L}(X(\cdot,x))$ enter their respective equations linearly. These observations enable one to obtain estimates on these quantities by an inductive procedure and they motivate the following definition.

Let $D \geq 1$ and $V : R \to R$ be given. We say that $V(\cdot)$ is <u>lower triangular with respect to the grading</u> $\{D_\mu\}_{\mu=0}^M$ if $0 = D_0 < D_1 < \cdots < D_M = D$ and

$$V(X) = \begin{pmatrix} V_{(1)}(X_{(1)}) \\ \vdots \\ V_{(M)}(X_{(M)}) \end{pmatrix} \in R^{d_1} \times \cdots \times R^{d_M} , \quad X \in R^D ,$$

where $X_{(\mu)} = (X_1,\ldots,X_{D_\mu})$; $d_\mu = D_\mu - D_{\mu-1}$; and $V_{(\mu)} \in C^\infty(R^{D_\mu})$ has the property that for each $\alpha \in (\mathcal{N})^{D_\mu}$ there is a $C_\alpha < \infty$ and a $\gamma_\alpha \geq 0$ such that $\gamma_\alpha = 0$ when $\max\limits_{D_{\mu-1}<j\leq D_\mu} \alpha_j \geq 1$ and

$$\left|D^\alpha_{X_{(\mu)}} V_{(\mu)}(X_{(\mu)})\right| \leq C_\alpha(1 + |X_{(\mu)}|)^{\gamma_\alpha} , \quad X_{(\mu)} \in R^{D_\mu} .$$

Given functions $\sigma : R^D \to R^D \otimes R^d$ and $b : R^D \to R^D$, we say that $(\sigma(\cdot),b(\cdot))$ is a <u>lower trianguler system of coefficients</u> if the column vector fields $\dot{\sigma}_1(\cdot),\ldots,\dot{\sigma}_d(\cdot)$ and the vector field $\dot{b}(\cdot)$ are simultaneously lower triangular with respect to some common grading.

The importance to us of these notions is contained in the following theorem.

<u>(7.17)</u> <u>Theorem</u>: Let $(\sigma(\cdot),b(\cdot))$ be a lower triangular system of coefficients. Then for each $\Xi_0 \in R^D$ there is a unique solution $\Xi(\cdot)$ to

(7.18) $$\Xi(T) = \Xi_0 + \int_0^T \sigma(\Xi(t))d\theta(t) + \int_0^T b(\Xi(t))dt \ , \quad T \geq 0$$

Moreover, $\Xi(T) \in (K(\mathfrak{f}))^D$, $T \geq 0$, $T \to (\Xi(T),\langle\langle\Xi(T),\Xi(T)\rangle\rangle_{\mathfrak{L}},\mathfrak{L}(\Xi(T)))$ is continuous (a.s., $\mathfrak{W}$), and $|||\Xi(\cdot,X)|||_{(q),T} < \infty$ for all $q \in [4,\infty)$. Finally, there is a lower triangular system $(\tilde{\sigma}(\cdot),\tilde{b}(\cdot))$ and a $\tilde{\Xi} \in R^D \times R^{D^2} \times R^D$ such that

(7.19) $$\tilde{\Xi}(T) = \tilde{\Xi}_0 + \int_0^T \tilde{\sigma}(\tilde{\Xi}(t))d\theta(t) + \int_0^T \tilde{b}(\tilde{\Xi}(t))dt \ , \quad T \geq 0 \ ,$$

where

$$\tilde{\Xi}(\cdot) = \begin{pmatrix} \Xi(\cdot) \\ \langle\langle\Xi(\cdot),\Xi(\cdot)\rangle\rangle_{\mathfrak{L}} \\ \mathfrak{L}(\Xi(\cdot)) \end{pmatrix} .$$

<u>Proof</u>: The proof of this theorem is done by induction on the length M of the grading $\{D_\mu\}_{\mu=0}^M$. The case $M = 1$ is of course covered by Theorem (7.14) except for checking that $\Xi(\cdot)$ satisfies an equation having a lower triangular system of coefficients. However, a proof of this fact can be easily constructed out of our discussion preceding the statement of Theorem (7.14). To carry out the needed induction step requires us

to prove a result which generalizes Theorem (7.14) in a direction analogous to the sort of generalization which Lemma (2.9) represents. The interested reader is advised to consult Theorem (6.2) of [S.,J.F.A.] for more details. □

We can now show that there exists a rich class $\mathcal{S}_0(\mathcal{L})$ of $\Phi \in \mathcal{H}(\mathcal{L})$ such that $\mathcal{S}_0(\mathcal{L})$ has the following closure properties:

$$\text{(7.20)}\quad \begin{array}{l} \text{i) if } N \geq 1,\ \Phi = (\Phi_1,\ldots,\Phi_N) \in (\mathcal{S}_0(\mathcal{L}))^N, \text{ and } F \in C_\uparrow^\infty(R^N), \\ \quad \text{then } F \circ \Phi \in \mathcal{S}_0(\mathcal{L})\ ; \\ \text{ii) if } \Phi \in \mathcal{S}_0(\mathcal{L}), \text{ then } \mathcal{L}(\Phi) \in \mathcal{S}_0(\mathcal{L}) . \end{array}$$

Indeed, let $\mathcal{S}_0(\mathcal{L})$ consist of those Φ for which there exist a $D \geq 1$, a lower triangular system $(\sigma(\cdot),b(\cdot))$, a $\Xi_0 \in R^D$, and a $T \geq 0$ such that $\Phi = \Xi_D(T)$, where $\Xi(\cdot)$ is the solution to (7.18). By Itô's formula, $\mathcal{S}_0(\mathcal{L})$ has property (7.20) i) ; by Theorem (7.17), it has property (7.20) ii). Finally, we can now make the following extension of $\mathcal{S}_0(\mathcal{L})$. Namely, let $\mathcal{S}(\mathcal{L})$ denote the class of Φ for which there exists $\{\Phi_n\}_1^\infty \subseteq \mathcal{S}_0(\mathcal{L})$ such that if $q \in [4,\infty)$ then $\lim\limits_n \|\Phi_n - \Phi\|_{L^q(\mathcal{W})} = 0$ and $\lim\limits_{m\to\infty} \sup\limits_{n>m} \|\mathcal{L}^k(\Phi_n) - \mathcal{L}^k(\Phi_m)\|_{L^q(\mathcal{W})} = 0$ for all $k \geq 1$. It is easy to check that $\mathcal{S}(\mathcal{L})$ still enjoys the properties in (7.20). Moreover, if $\Phi \in \mathcal{S}(\mathcal{L})$ and $1/\Phi \in \bigcap_1^\infty L^q(\mathcal{W})$, then $1/\Phi \in \mathcal{S}(\mathcal{L})$.

(7.21) **Theorem:** Given $\Phi = (\Phi_1,\ldots,\Phi_N) \in (\mathcal{S}(\mathcal{L}))^N$, set $A = \langle\langle\Phi,\Phi\rangle\rangle_{\mathcal{L}}$ and $\Delta = \det(A)$. Assume that $1/\Delta \in \bigcap_1^\infty L^q(\mathcal{W})$. If $\mu = \mathcal{W} \circ \Phi^{-1}$, then μ admits a density $f \in \hat{C}^\infty(R^N)$.

Proof: Define $\mathcal{V}_k$, $1 \leq k \leq N$, as in (5.15) relative to Φ and A. Given $\Psi \in \mathcal{S}(\mathcal{L})$, define $\bar{\mathcal{V}}_k(\Psi) = \mathcal{V}_k(\Psi/\Delta)$. Then by repeated application of (5.14), we see that for $\alpha \in (\mathcal{N})^N$ and $F \in C_0^\infty(R^N)$:

$$(7.22) \qquad E^{\mathcal{W}}[(D^\alpha F\circ\Phi)\Psi] = (-1)^{|\alpha|}E^{\mathcal{W}}[(F\circ\Phi)\overline{\mathcal{V}}^\alpha(\Psi)]$$

where $\overline{\mathcal{V}}^\alpha \equiv \overline{\mathcal{V}}_N^{\alpha_N} \circ \cdots \circ \overline{\mathcal{V}}_1^{\alpha_1}$. In particular, with $\Psi = 1$, we see that for each $\alpha \in (\mathcal{N})^N$ there is a $\Psi^{(\alpha)} \in \mathcal{B}(\mathcal{L})$ such that

$$E^{\mathcal{W}}[D^\alpha F\circ\Phi] = (-1)^{|\alpha|}E^{\mathcal{W}}[(F\circ\Phi)\Psi^{(\alpha)}], \quad F \in C_0^\infty(R^N).$$

Proceeding as in the proof of Corollary (5.17), we now conclude that for each $\alpha \in (\mathcal{N})^N$ there is a $\psi^{(\alpha)} \in \bigcap_1^\infty L^q(\mu)$ such that

$$\int D^\alpha F d\mu = (-1)^{|\alpha|}\int F\psi^{(\alpha)}d\mu, \quad F \in C_0^\infty(R^N).$$

But, by Corollary (5.17), $d\mu = f dx$ where $f \in \hat{C}(R^N)$, and the preceding says that $D^\alpha f = \psi^{(\alpha)}f$. In particular, $\|D^\alpha f\|_{L^q(R^N)} \leq \|f\|_{C_b(R^N)}^{1-1/q}\|\psi^{(\alpha)}\|_{L^q(\mu)} < \infty$ for all $\alpha \in (\mathcal{N})^N$ and $q \in [0,\infty)$. It is easy to conclude from this that $f \in \hat{C}^\infty(R^N)$ (cf. the argument used to prove Lemma (3.2) and use the same argument to show that if $\max_{|\alpha|\leq k+1}\|D^\alpha f\|_{L^q(R^N)} < \infty$ for some $q \in (N,\infty)$ then $f \in \hat{C}^k(R^N)$). □

8. Criteria Guaranteeing Non-Degeneracy:

Let $\sigma : R^D \to R^D \otimes R^d$ and $b : R^D \to R^D$ be C^∞-functions having bounded derivatives of all orders greater than or equal to one. Given $x \in R^N$, denote by $X(\cdot,x)$ the solution to (2.13) and denote by $P(T,x,\cdot)$ the distribution of $X(\cdot,x)$ under $\mathcal{W}$. As we showed in section 2), $P(T,x,\cdot)$ is the fundamental solution to equation (1.1) with

$$L = 1/2 \sum_{i,j=1}^{D} \sigma\sigma^*(\cdot)^{ij} \frac{\partial^2}{\partial x_i \partial x_j} + \sum_{i=1}^{D} b^i(\cdot) \frac{\partial}{\partial x_i} \tag{8.1}$$

in the sense that if $\phi \in C_b^2(R^D)$ then $(t,x) \to u(t,x) \equiv \int \phi(y) P(T,x,dy) \in C^{1,2}([0,\infty) \times R^D) \cap C_b([0,\infty) \times R^D)$ and u is the unique bounded solution to (1.1) (cf. Theorem (2.21)).

Ever since section (2) , all our efforts have been directed toward the goal of using the representation $P(T,x,\cdot) = \mathbb{W} \circ X(T,x)^{-1}$ to say something about $P(T,x,dy)$ as a function of its "forward variable" y . In particular, we would like to know from this representation when we can assert that $P(T,x,dy) = p(T,x,y)dy$ where $P(T,x,\cdot)$ is smooth. Of course, just as soon as we know that $p(T,x,\cdot)$ is smooth, it follows from

$$\frac{\partial}{\partial t} \int \phi(y) p(t,x,y) dy = \int L\phi(y) p(t,x,y) dy \quad , \quad \phi \in C_0^\infty(R^D)$$

that, for each $x \in R^D$, $(t,y) \to p(t,x,y)$ satisfies the Fokker-Plank equation $\frac{\partial u}{\partial t} = L^* u$, $t > 0$, with initial data δ_x . (Here L^* is the formal adjoint of L . That is, $L^*\phi(y) = 1/2 \sum_{i,j=1}^{D} \frac{\partial^2}{\partial y_i \partial y_j}(\sigma\sigma^*(y)^{ij}\phi(y))$ $- \sum_{i=1}^{D} \frac{\partial}{\partial y}(b^i(y)\phi(y))$.)

Since $X(T,x) \in (\mathcal{S}(\mathcal{L}))^D$ for all $T \geq 0$ and $x \in R^D$, it follows from Theorem (7.21) that a smooth $p(T,x,\cdot)$ exists if $1/\Delta(T,x) \in \bigcap_1^\infty L^q(\mathbb{W})$ where $\Delta(T,x) = \det(A(T,x))$ and $A(T,x) = \langle\langle X(T,x), X(T,x)\rangle\rangle_{\mathcal{L}}$. That some condition besides smoothness must be imposed on $\sigma(\cdot)$ and $b(\cdot)$ in order to assure the existence of $p(T,x,\cdot)$ is clear. Indeed, in the extreme case when $\sigma(\cdot) \equiv 0$, $X(\cdot,x)$ is, $\mathbb{W}$-almost surely, just the integral curve of $b(\cdot)$ issuing from x. Thus, in this case, $P(T,x,\cdot)$ is the unit mass at the point reached by this integral curve at time T. Even in less extreme examples it is easy to give examples in which $P(T,x,\cdot)$ must be concentrated on a variety of dimension less than D and therfore cannot be absolutely continuous. (In general, $P(T,x,\cdot)$ has support equal to the closure of the set of points $\phi_\psi(T)$ where $\phi_\psi(t) = x + \int_0^t \sigma(\phi_\psi(s))\psi(s)ds + \int_0^t \tilde{b}(\phi_\psi(s))ds$, $t \geq 0$, with $\tilde{b}^i = b^i - 1/2 \sum_{k=1}^{d} \sum_{j=1}^{D} \sigma_k^j \frac{\partial \sigma_k^i}{\partial x_j}$ and $\psi \in C([0,\infty),R^d)$. The proof of this fact may be found in [S. & V., Berk. Symp.].) Thus some sort of "non-degeneracy" condition on $\sigma\sigma^*(\cdot)$ is required in order for $p(T,x,\cdot)$ to exist. What we have to do is find out how to translate non-degeneracy of $\sigma\sigma^*(\cdot)$ into the statement that $A(T,x)$ is non-degenerate (in the sense that $1/\Delta(T,x) \in \bigcap_1^\infty L^q(\mathbb{W})$).

In order to handle the problem outlined above, recall that $A(\cdot,x)$ satisfies (7.15). In order to solve (7.15), we use the method of variation of parameters. To this end, we need the following simple fact about linear S.D.E.'s.

(8.2) Lemma: Let $U_k : [0,\infty) \times \Theta \to R^D \otimes R^D$, $1 \leq k \leq d$, and $V : [0,\infty) \times \Theta \to R^D \otimes R^D$ be bounded progressively measurable functions. Then there is a precisely one progressively measurable $M : [0,\infty) \times \Theta \to R^D \otimes R^D$ satisfying

$$(8.3) \quad M(T) = I + \sum_{k=1}^{d} \int_0^T U_k(t)M(t)d\theta_k(t) + \int_0^T V(t)M(t)dt \quad , \quad T \geq 0 \quad .$$

In particular, for each $q \in [2,\infty)$ and $T > 0$, $E^{\mathbb{W}}[\sup_{0\leq t\leq T} \|M(t)\|^q_{H.S.}] < \infty$

$$(8.4)\qquad M(T)^* = I + \sum_{k=1}^{d}\int_0^T M(T)^* U_k(t)^* d\theta_k(t) + \int_0^T M(t)^* V(t)^* dt \quad , \quad T \geq 0 .$$

Finally, $M(\cdot)$ is non-singular (a.s., $\mathbb{W}$) and $M(\cdot)^{-1}$ is uniquely determined by

$$(8.5)\qquad M(T)^{-1} = I - \sum_{k=1}^{d}\int_0^T M(t)^{-1}U_k(t) + \int_0^T \Big(\sum_k M(t)^{-1}U_k(t)^2 - M(t)^{-1}V(t)\Big)dt$$

Proof: The existence and uniqueness of solutions to (8.3), (8.4), (8.5) is an easy consequence of Lemma (2.6). From the same lemma, we see that $M(\cdot)$ satisfies the stated $L^q(\mathbb{W})$-estimate. The fact that $M(\cdot)^*$ solves (7.18) requires no comment. Finally, to see that $M(\cdot)^{-1}$ exists and satisfies (8.5), denote by $Y(\cdot)$ the solution to (8.5) and use Itô's formula to show that $d(Y(T)M(T)) = 0$. □

(8.6) Remark: The equation (8.5) for $M(\cdot)^{-1}$ can be guessed in the following way. Assume that $dM(t)^{-1} = \sum_{k=1}^{d}\tilde{U}_k(t)d\theta_k(t) + \tilde{V}(t)dt$ for some $\tilde{U}_k(\cdot)$'s and $\tilde{V}(\cdot)$. Then $0 = M(t)^{-1}dM(t) + d(M(t)^{-1})M(t) + d(M(t)^{-1})dM(t) = M(t)^{-1}\sum U_k(t)M(t)d\theta_k(t) + \sum^{d}\tilde{U}_k(t)M(t)d\theta(t) + M(t)^{-1}V(t)M(t)dt + \tilde{V}(t)M(t)dt + \sum_{k=1}^{d}\tilde{U}_k(t)U_k(t)M(t)dt$. The only way in which this can hold is if:

$$M(t)^{-1}U_k(t)M(t) + \tilde{U}_k M(t) = 0 \quad , \quad 1 \leq k \leq d$$

and

$$M(t)^{-1}V(t)M(t) + \tilde{V}(t)M(t) + \sum_{k=1}^{d}\tilde{U}_k(t)U_k(t)M(t) = 0 \quad .$$

Hence: $\tilde{U}_k = -M(t)^{-1}U_k(t)$, $1 \leq k \leq d$, and $\tilde{V}(t) = -M(t)^{-1}V(t) + \sum_{k=1}^{d} M(t)^{-1}U_k(t)^2$.

(8.7) Lemma: Let $U_k(\cdot)$, $1 \leq d \leq d$, and $V(\cdot)$ be as in Lemma (8.2) and suppose that $a : [0,\infty) \times \Theta \to R^D \otimes R^D$ is a bounded progressively measurable function. Then there is one and only one progressively measurable $A : [0,\infty) \times \Theta \to R^N \otimes R^N$ such that

$$(8.8) \qquad A(T) = \sum_{k=1}^{d} \int_0^T \{U_k(t),A(t)\}d\theta_k(t)$$

$$+ \int_0^T (\{V(t),A(t)\} + \sum_{k=1}^{d} U_k(t)A(t)U_k(t)^* + a(t))dt \quad , \quad T \geq 0 \ .$$

(The notation $\{\cdot,\cdot\}$ is explained in the sentence following (7.16).) In fact, if $M(\cdot)$ is the solution to (8.3) and $M(s,t) \equiv M(t)M(s)^{-1}$, then the unique solution to (8.8) is

$$(8.9) \qquad A(T) = \int_0^T M(t,T)a(t)M(t,T)^* dt \quad , \quad T \geq 0 \ .$$

Proof: Existence and uniqueness of solutions to (8.8) is guaranteed by Lemma (2.6) . Thus, we will be done once we verify (8.9) . To this end, let $R(T)$, $T \geq 0$, denote the R.H.S. of (8.9) . Then

$$M(T)^{-1}R(T)M(T)^{-1*} = \int_0^T M(t)^{-1}a(t)M(t)^{-1*}dt \ .$$

Thus

$$M(t)^{-1}a(t)(M(t)^{-1})^* dt = -M(t)^{-1}(\sum_{k=1}^{d} \{U_k(t),R(t)\}d\theta_k(t))(M(t)^{-1})^*$$

$$- M(t)^{-1}(\{V(t),R(t)\} + \sum_{k=1}^{d} U_k(t)R(t)U_k(t)^*)dt(M(t)^{-1})^*$$

$$+ M(t)^{-1} dR(t)(M(t)^{-1})^* ;$$

and so

$$dR(t) = \sum_{k=1}^{d} \{U_k(t),R(t)\} d\theta_k(t)$$

$$+ (\{V(t),R(t)\} + \sum_{k=1}^{d} U_k(t)R(t)U_k(t)^* + a(t))dt \quad .$$

Since $R(0) = 0$, this proves that $R(\cdot)$ solves (8.8)

□

As a corollary of Lemma (8.7) we now see that if $J(\cdot;x)$ is the solution to

(8.10) $$J(T;x) = I + \sum_{k=1}^{d} \int_0^T S_k(X(t,x))J(t;x)d\theta_k(t)$$

$$+ \int_0^T B(X(T,x))J(t;x)dt \quad , \quad T \geq 0 \quad ,$$

where $S_k(\cdot)$, $1 \leq d \leq d$, and $B(\cdot)$ are given by (7.16) , and if

(8.11) $$J(s,t;x) = J(t;x)J(s;x)^{-1} \quad ,$$

then

(8.12) $$A(T,x) = \int_0^T J(t,T;x)a(X(t,x))J(t,T;x)^* dt \quad , \quad T \geq 0 \quad ,$$

where $a(\cdot) = \sigma\sigma(\cdot)^*$.

Clearly (8.12) goes a long way toward connecting non-degeneracy properties of $A(T,x)$ with those of $a(\cdot)$. For example, suppose that $a(\cdot) \geq \varepsilon I$ for some $\varepsilon > 0$. Then, by (8.12) :

$$(8.13)\qquad A(T,x) \geq \varepsilon\int_0^T J(t,T;x)J(t,T;x)^* dt \ .$$

In order to exploit (8.13) , we use the following simple variant of Jensen's inequaltiy.

(8.14) Lemma: Let $(E,\mathcal{F},P)$ be a probability space and $M : E \to R^N \otimes R^N$ an $\mathcal{F}$-measurable function such that $M(\cdot)$ is symmetric and positive definite a.s. Then $E[M]^{-1} \leq E[M^{-1}]$.

Proof: As usual, one need only show that if M_1 and M_2 are symmetric positive definite matrices and if $\lambda \in (0,1)$, then $(\lambda M_1 + (1-\lambda)M_2)^{-1} \leq \lambda M_1^{-1} + (1-\lambda)M_2^{-1}$. To this end, choose an orthogonal matrix $\mathcal{O}$ so that $\Lambda \equiv \mathcal{O}^* M_1^{-1/2} M_2 M_1^{-1/2}\mathcal{O}$ is diagonal. Then:

$$\begin{aligned}(\lambda M_1 + (1-\lambda)M_2)^{-1} &= M_1^{-1/2}(\lambda I + (1-\lambda)M_1^{-1/2}M_2M_1^{-1/2})M_1^{-1/2} \\ &= M_1^{-1/2}\mathcal{O}^*(\lambda I + (1-\lambda)\Lambda)^{-1}\mathcal{O}M_2^{-1/2} \\ &\leq M_1^{-1/2}\mathcal{O}^*(\lambda I + (1-\lambda)\Lambda^{-1})\mathcal{O}M_2^{-1/2} \\ &= \lambda M_1^{-1} + (1-\lambda)M_2^{-1} \ .\end{aligned}$$

□

Returning to (8.13) , we now see that

$$(8.15)\qquad A(T,x)^{-1} \leq \frac{1}{\varepsilon T^2}\int_0^T J(t,T;x)^{-1*} J(t,T;x)^{-1} dt \ .$$

But it follows easily from Lemma (8.2) that if $s \geq 0$, then

$$J(s,T;x)^{-1} = I - \sum_{k=1}^{d}\int_s^T J(s,t;x)^{-1} S_k(X(t,x))d\theta_k(t)$$

$$+ \int_s^T (\sum_{k=1}^{d} J(s,t;x)^{-1} S_k(X(t,x))^2 - J(s,t;x)^{-1} B(X(t,x)))dt$$

for $T \geq s$. Therefore, by Lemma (8.2) , for $q \in [2,\infty)$ and $T > 0$:

$$\sup_{0\leq s\leq T} E^{W}[\sup_{s\leq t\leq T} \|J(s,t;x)^{-1}\|^{q}_{H.S.}] < \infty .$$

We have therefore proved the following result.

(8.16) Theorem: If $a(\cdot) = \sigma\sigma^{*}(\cdot) \geq \varepsilon I$ for some $\varepsilon > 0$, then for each $q \in [1,\infty)$ and $T > 0$ there exists a $C_q(T)$, depending on $\max_{|\alpha|=1} \|D^{\alpha}\sigma(\cdot)\|_{C_b(R^D)} \vee \|D^{\alpha}b(\cdot)\|_{C_b(R^D)}$, such that $E^{W}[\sup_{0<t\leq T}|t^{N}/\Delta(t,x)|^{q}]^{1/q} \leq C_q(T)/\varepsilon^{D}$. In particular, $P(T,x,dy) = p(T,x,y)dy$ with $p(T,x,\cdot) \in \hat{C}^{\infty}(R^{D})$ for $T > 0$ and $x \in R^{D}$.

The conclusion drawn in the last part of Theorem (8.16) has been well known to students of P.D.E.'s for quite a long time. In fact, under the hypotheses of (8.16) the techniques used by experts in P.D.E. theory give one much more refined information about $p(T,x,y)$ than the simple statement that it exists (cf. [Friedman]). Nonetheless, Theorem (8.16) represents a victory for probabilists in that it provides a derivation which rests on the relation $P(T,x,\cdot) = W \circ (X(T,x))^{-1}$. Before Malliavin's calculus, no such derivation existed.

Having successfully recovered the classical result stated in the last part of (8.16) , we ought to try our hand at a Hörmander's renowned theorem on the hypoellipticity of second order operators [Hörmander]. To state Hörmander's theorem, define $V^{(k)}(\cdot)$ to be the vector field $\sum_{i=1}^{N} \sigma^{i}_{k}(\cdot)\frac{\partial}{\partial x}$, $1 \leq k \leq d$, and let $V^{(0)}(\cdot)$ be the vector field $\sum_{i=1}^{D} (b^{j}(\cdot) - 1/2 \sum_{k=1}^{d}\sum_{j=1}^{D} \sigma^{j}_{k}(\cdot)\frac{\partial\sigma^{i}_{k}}{\partial x_j}(\cdot))\frac{\partial}{\partial x_i}$. (Observe that $V^{(0)}(\cdot)$ is precisely the vector field associated with the map $\tilde{b}(\cdot)$ which arose in our discussion about $\mathrm{supp}(P(T,x,\cdot))$.) We now write L in "Hörmander's form":

(8.17) $$L = 1/2 \sum_{k=1}^{d} (V^{(k)})^2 + V^{(0)} ,$$

where $(V^{(k)})^2 f \equiv V^{(k)}(V^{(k)}f)$. (The importance of writing L is Hörmander's form is that whereas $\sigma(\cdot)$ and $b(\cdot)$ are not themselves differential geometric invariants the vector fields $V^{(0)},\ldots,V^{(d)}$ are. That is, if $\Phi : R^D \to R^D$ is a diffeomorphism and $(\tilde{L}f) \circ \Phi = L(f\circ\Phi)$, $f \in C_0^\infty(R^N)$, then $\tilde{L} = 1/2 \sum_{k=1}^{d} (\tilde{V}^{(k)})^2 + \tilde{V}^{(0)}$ where $(\tilde{V}^{(k)}f) \circ \Phi = V^{(k)}(f\circ\Phi)$ $0 \le k \le d$. This observation is, of course, the reason that $\tilde{b}(\cdot)$ enters in the description of $\mathrm{sup}(P(T,x,\cdot))$.) The theorem of Hörmander states that if $\mathrm{Lie}(\{\mathrm{ad}V^{(0)})^n V^{(k)} : n \ge 0$ and $1 \le k \le d\})$ has full rank at the point x , then, for $T > 0$, $P(T,x,dy) = p(T,x,y)dy$ with $p(T,x,\cdot) \in C^\infty(R^N)$. (Here $(\mathrm{ad}V^{(0)})^0 V^{(k)} \equiv V^{(k)}$ and $(\mathrm{ad}V^{(0)})^{n+1} V^{(k)} = [V^{(0)},(\mathrm{ad}V^{(0)})^n V^{(k)}] \equiv V^{(0)} \circ (\mathrm{ad}V^{(0)})^n V^{(k)} - (\mathrm{ad}V^{(0)})^n V^{(k)} \circ V^{(0)}$ for $n \ge 0$. The notation $\mathrm{Lie}(\{\cdot\})$ denotes the Lie algebra generated by the vector fields in $\{\cdot\}$.)

Hörmander's original proof of this result is quite intricate; and, in the sense that it provides some of the sharpest estimates, it is still the best. A far simpler derivation was found by J.J. Kohn [Kohn]. In fact, Kohn's proof is so straight-forward that it is doubtful if it can be substantially further simplified. Working with ideas similar to those of Kohn, Radekevich [Oleinik and Radekewich] made non-trivial extensions of Hörmander's theorem to include operators L which cannot be written in

Hörmadner's form. (The obstruction to writing an operator in Hörmander's form comes from the impossibility, in general, of finding a smooth square root of a smooth non-definite definite, symmetric matrix valued function.) More recently, Folland [Folland] and Rothschld and Stein [Roth. & Stein] have introduced new techniques which enable them to refine Hörmander's results.

Malliavin himself in [Malliavin] provided the critical link between Hörmander's condition and the non-degeneracy of $A(T,x)$. Malliavin's idea is as follows. From (8.12), we see that if

$$(8.18)\qquad \tilde{A}(T,x) \equiv J(T,x)^{-1}A(T,x)J(T,x)^{-1*},$$

then

$$(8.19)\qquad \tilde{A}(T,x) = \int_0^T J(t,x)^{-1}a(X(t,x))J(t,x)^{-1*}dt .$$

Since, as we have seen already, good estimates on $\|1/\det(J(T,x)J(t,x)^*)\|_{L^q(\mathfrak{w})}$ are readily available, it suffices for us to estimate $\|1/\det\tilde{A}(T,x)\|_{L^q(\mathfrak{w})}$ in order to get the desired bound on $\|1/\Delta(T,x)\|_{L^q(\mathfrak{w})}$. We next note that control on $\|1/\det(\tilde{A}(T,x))\|_{L^q(\mathfrak{w})}$ for all $q \in [1,\infty)$ can be obtained from control on $\sup_{|\eta|=1}\|1/(\eta,\tilde{A}(T,x)\eta)\|_{L^q(\mathfrak{w})}$ for all $q \in [1,\infty)$. We therefore choose and fix an $\eta \in R^N$ with $|\eta| = 1$ and consider $(\eta,\tilde{A}(T,x)\eta)$. Using (8.19), we see that

$$(\eta,\tilde{A}(T,x)\eta) = \int_0^T |J(t,x)^{-1}\sigma(X(t,x))^*\eta|^2 dt ,$$

which can be rewritten as:

$$(8.20)\qquad (\eta,\tilde{A}(T,x)\eta) = \sum_{k=1}^{d}\int_0^T (J(t,x)^{-1}V^{(k)}(X(t,x)),\eta)^2 dt .$$

(In (8.20) we have used $V^{(k)}(\cdot)$ to denote the vector $(\sigma_k^1(\cdot),\ldots,\sigma_k^D(\cdot))$ This abuse of notation is a familiar one.) It is clear from (8.20) that if $\text{rank}\{V^{(1)}(x),\ldots,V^{(d)}(x)\} = D$, then an estimate on $\|1/(\eta,A(T,x)\eta)\|_{L^q}$ will be forthcoming. Indeed, in this case there will exist an $\varepsilon > 0$ such that $\sum_{k=1}^{d} (V^{(k)}(y),\xi)^2 \geq \varepsilon$ for all y in a neighborhood U of x and all $\xi \in R^N$ with $|\xi| \geq 1/2$. Since one can easily estimate $\|1/\tau\|_{L^q(\mathfrak{W})}$, $q \in [1,\infty)$, where τ denotes the first time that either $X(\cdot,x)$ leaves U or that $\|J(\cdot,x)-I\|_{H.S.} \geq 1/2$, the required bound on $\|1/(\eta,A(T,x)\eta)\|_{L^q(\mathfrak{W})}$ follows quite quickly. Of course, the case when $\text{rank}\{(V^{(1)}(x),\ldots,V^{(d)}(x)\} = D$ does not really get us beyond the place where we arrived in Theorem (8.16)

Next, suppose that $\text{rank}\{V^{(1)}(x),\ldots,V^{(d)}(x)\} < N$. Obviously, it is still possible for the right hand side of (8.20) to be positive. In fact, the $V^{(k)}(\cdot)$ are "moved" in two different ways. In the first place, the path $X(\cdot,x)$ is going to move; one can even show that with positive probability, $X(\cdot,x)$ will stay arbitrarily close to the integral curve, starting at x, of any vector field $Z = Y + V^{(0)}$ where $Y \in \text{Lie}(\{(\text{ad}V^{(0)})^n V^{(k)} : n \geq 0 \text{ and } 1 \leq k \leq d\})$ (cf. [S. & V., Berk. Symp.]). Secondly, $V^{(k)}(\cdot)$ is getting "twisted" by the action of $J(\cdot,x)^{-1}$. Thus, even if $\sum_{k=1}^{d} (V^{(k)}(x),\eta)^2 = 0$, it is nonetheless possible for $\sum_{k=1}^{d} (J(\cdot,x)^{-1}V^{(k)}(X(\cdot,x)),\eta)^2$ to become positive immediately. Moreover, it is intuitively clear that Hörmander's condition is trying to say that the opportunity for $\sum_{k=1}^{d} (J(\cdot,x)^{-1}V^{(k)}(X(\cdot,x)),\eta)^2$ to become positive occurs. The problem is therefore one of making mathematics out of this intuition.

Malliavin's idea is to bring out the non-degeneracy contained in Hörmander's condition by computing the Itô differential of $(J(\cdot,x)^{-1}V^{(k)}(X(\cdot,x),\eta)$. To this end, let $V : R^D \to R^D$ be a $C_{\uparrow}^{\infty}(R^D)$ map. Then, by Itô's formula and (8.5) :

$$d(J(t,x)^{-1}V(X(t,x)))$$

$$= d(J(t,x)^{-1})V(X(t,x)) + J(t,x)^{-1}dV(X(t,x))$$

$$+ d(J(t,x)^{-1})d(V(X(t,x)))$$

$$= J(t,x)^{-1}\Big[\sum_{\ell=1}^{d}(-S_\ell(X(t,x))V(X(t,x)) + \sum_{i=1}^{N}\sigma_\ell^i(X(t,x))\frac{\partial V}{\partial x_i}(X(t,x)))d\theta_\ell(t)\Big]$$

$$+ J(t,x)^{-1}\Big[\sum_{\ell=1}^{d}S_\ell(X(t,x))^2V(X(t,x)) - B(X(T,x))V(X(t,x))$$

$$+ 1/2\sum_{i,j=1}^{D}a^{ij}(X(t,x))\frac{\partial^2 V}{\partial x_i\partial x_j}(X(t,x))$$

$$+ \sum_{i=1}^{D}b^i(X(t,x))\frac{\partial V}{\partial x_i}(X(t,x))$$

$$- \sum_{\ell=1}^{d}\sum_{i=1}^{D}S_\ell(X(t,x))(\sigma_\ell^i(X(t,x))\frac{\partial V}{\partial x_i}(X(t,x)))\Big]dt \quad .$$

Notice that

$$-S_\ell(X(t,x))V(X(t,x)) + \sum_{i=1}^{D}\sigma_\ell^i(X(t,x))\frac{\partial V}{\partial x_i}$$

$$= [V^{(\ell)},V](X(t,x)) \quad .$$

Next:

$$S_\ell(X(t,x))^2V(X(t,x)) = [(V(V^{(\ell)}))(V^{(\ell)})](X(t,x)) \quad ,$$

$$- B(X(t,x))V(X(t,x)) = -[V(V^{(0)} + 1/2V^{(\ell)}(V^{(\ell)}))](X(t,x)) \quad ,$$

$$1/2 \sum_{i,j=1}^{D} a^{ij}(X(t,x))\frac{\partial^2 V}{\partial x_i \partial x_j}(X(t,x)) + \sum_{i=1}^{D} b^i(X(t,x))\frac{\partial V}{\partial x_i}(X(t,x))$$

$$= 1/2\Big[\sum_{\ell=1}^{d} V^{(\ell)}(V^{(\ell)}(V)) + V^{(0)}(V)\Big](X(t,x)) ,$$

$$- S_\ell(X(t,x))\Big(\sum_{i=1}^{D} \sigma_\ell^i(X(t,x))\frac{\partial V}{\partial x_i}(X(t,x))\Big) = -\big[(V^{(\ell)}(V))(V^{(\ell)})\big](X(t,x))$$

Thus the bracketed quantity in the coefficient of dt can be written as:

$$1/2 \sum_{\ell=1}^{d} [V^{(\ell)}(V^{(\ell)}(V)) - V(V^{(\ell)}(V^{(\ell)}))$$

$$+ 2(V(V^{(\ell)}))(V^{(\ell)} - 2(V^{(\ell)}(V))(V^{(\ell)})](X(t,x))$$

$$= 1/2 \sum_{\ell=1}^{d} [V^{(\ell)}(V^{(\ell)}(V)) - V_i V_j^{(\ell)} \frac{\partial^2 V^{(\ell)}}{\partial x_i \partial x_j} + (V(V^{(\ell)}))(V^{(\ell)})$$

$$- 2(V^{(\ell)}(V))(V^{(\ell)})](X(t,x)) .$$

At the same time:

$$[V^{(\ell)},[V^{(\ell)},V]] = V^{(\ell)}(V^{(\ell)}(V)) - V^{(\ell)}(V(V^{(\ell)}))$$

$$- (V^{(\ell)}(V))(V^{(\ell)}) + (V(V^{(\ell)}))(V^{(\ell)})$$

$$= V^{(\ell)}(V^{(\ell)}(V)) - V_i^{(\ell)} V_j \frac{\partial^2 V^{(\ell)}}{\partial x_i \partial x_j} - 2(V^{(\ell)}(V))(V^{(\ell)})$$

$$+ (V(V^{(\ell)}))(V^{(\ell)}) .$$

Hence, we now see that:

(8.21) $$J(T,x)^{-1}V(X(T,x)) - V(x)$$

$$= \sum_{\ell=1}^{d} \int_0^T J(t,x)^{-1} [V^{(\ell)},V](X(t,x))d\theta_\ell(t)$$

$$+ \int_0^T J(t,x)^{-1}([V^{(0)},V] + 1/2 \sum_{\ell=1}^{d} [V^{(\ell)},[V^{(\ell)},V]])(X(t,x))dt \quad .$$

Exploiting (8.21) is not easy. The key to doing so is contained in the following observation. Given bounded progressively measurable functions $\alpha : [0,\infty) \times \Theta \to R^d$ and $\beta : [0,\infty) \times \Theta \to R^1$, consider the process $\xi(\cdot)$ given by:

(8.22) $$\xi(T) = \xi_0 + \int_0^T \alpha(t) \cdot d\theta(t) + \int_0^T \beta(t)dt \quad , \quad T \geq 0 \quad .$$

Then the variance of $\xi(\cdot)$ over a small time interval $[t_1,t_2]$ will be determined by $\int_{t_1}^{t_2} |\alpha(t)|^2 dt$ if $\int_{t_1}^{t_2} |\alpha(t)|^2 dt > 0$. The reason for this is that $\int_0^{\cdot} \alpha(t) \cdot d\theta(t)$ is a 1-dimensional Brownian motion run with the "clock" $\int_0^{\cdot} |\alpha(t)|^2 dt$ and the variance of a Brownian path over an interval is commensurate with the interval length (cf. Lemma (8.23) below for a precise statement). Thus, for short intervals, the variance of $\int_0^{\cdot} \alpha(t) \cdot d\theta(t)$ over $[t_1,t_2]$ will overwhelm that of $\int_0^{\cdot} \beta(t)dt$ and will be the principle contributor to the variance of $\xi(\cdot)$.

To make the ideas in the preceding paragraph precise, we introduce some notation. If I is a compact interval in R^1 and $f \in C(I)$, define $f_I = \frac{1}{|I|} \int_I f$ and $\sigma_I(f) = (\frac{1}{|I|} \int_I (f - f_I)^2)^{1/2}$. The following two results are adapted from theorems in Chapter (8) of [Ikeda & Watenabe] .

(8.23) Lemma: Let $B(\cdot)$ be a 1-dimensional Brownian motion and τ a stopping time. Then for $T > 0$ and $\varepsilon > 0$:

$$(8.24) \qquad P(\sigma_{[\tau,\tau+T]}(B(\cdot)) \le \varepsilon) \le 2^{1/2}\exp(-T/2^7\varepsilon^2) \ .$$

Proof: Since $\sigma_{[\tau,\tau+T]}(B(\cdot)) = \sigma_{[\tau,\tau+T]}(B(\cdot)-B(\tau))$ and because $B(\cdot+\tau) - B(\tau)$ has the same distribution as $B(\cdot)$, we may and will assume that $\tau \equiv 0$. Next, set $\overline{B}(t) = T^{-1/2}B(Tt)$, $t \ge 0$. Then $\overline{B}(\cdot)$ is again a 1-dimensional Brownian motion and $\sigma_{[0,T]}(B(\cdot))^2 = T\sigma_{[0,1]}(\overline{B}(\cdot))$. Thus we may and will assume that $T = 1$.

Now set $X(t) = B(t) - tB(1)$. Then $\{X(t) : 0 \le t \le 1\}$ is a Gaussian process with mean 0 and covariance $\rho(s,t) = s \wedge t - st$ . . Noting that:

$$\frac{\cos(2\pi ks)-1}{(2\pi k)^2} = \int_0^1 \rho(s,t)\cos(2\pi kt)dt$$

and

$$\frac{\sin 2\pi ks}{(2\pi k)^2} = \int_0^1 \rho(s,t)\sin(2\pi kt)dt$$

for $k \ge 1$, we see that the sequences $\{\xi_k\}_1^\infty$ and $\{\eta_k\}_1^\infty$ where

$$\xi_k = 2^{1/2}(2\pi k)\int_0^1 X(t)\cos(2\pi kt)dt$$

and

$$\eta_k = 2^{1/2}(2\pi k)\int_0^1 X(t)\sin(2\pi kt)dt$$

are independent of one another and consist of mutually independent $N(0,1)$-random varables. Moreover, since $B(1)$ is independent of $\{X(t) : 0 \le t \le 1\}$, $\xi_0 \equiv B(1)$ is an $N(0,1)$-random variable which is independent of $\{\xi_k\}_1^\infty \cup \{\eta_k\}_0^\infty$.

With these preliminaries, we will next prove that, almost surely,

(8.25) $$B(t) = t\xi_o + 2^{1/2}\sum_1^\infty \left[\frac{\xi_k}{2\pi k}(\cos 2\pi kt)-1 + \frac{\eta_k}{2\pi k}\sin(2\pi kt)\right]$$

where the convergence on the right is in the sense of $L^2([0,1])$. Clearly, it suffices to prove that the summation on the right converges almost surely (in $L^2([0,1])$) to $X(\cdot)$. To this end, let $f \in C([0,1])$ with $f(0) = f(1) = 0$ and define $f_r(t) = \int_0^1 f(t)dt + 2^{1/2}\sum_0^\infty r^k[a_k\cos(2\pi kt) + b_k\sin(2\pi kt)]$ for $0 < r < 1$, where

$$a_k = 2^{1/2}\int_0^1 f(t)\cos(2\pi kt)dt \quad \text{and} \quad b_k = 2^{1/2}\int_0^1 f(t)\sin(2\pi kt)dt \quad .$$

Then $f_r \in C^\infty([0,1])$ and is periodic for each $r \in (0,1)$. Moreover, $f_r \to f$ uniformly as $r \uparrow 1$. Next, using the relations mentioned above, we get:

$$f_r(s) - f_r(0) = -\int_0^1 \rho(s,t)f_r''(t)dt$$

$$= 2^{1/2}\sum_1^\infty r^k[a_k(\cos 2\pi kt) - 1) + b_k\sin(2\pi kt)]$$

In particular, if $\sum_1^\infty a_k$ converges, then

$$f(t) = 2^{1/2}\sum_1^\infty [a_k(\cos(2\pi kt) - 1) + b_k\sin(2\pi kt)]$$

where the convergence on the right is in $L^2([0,1])$. Since, by Kolmogorov's three series theorem, $\sum_1^\infty \frac{\xi_k}{k}$ converges almost surely, (8.25) now follows.

From (8.25) we see that

$$B(t) - \int_0^1 B(s)ds = (t-1/2)\xi_o + 2^{1/2}\sum_1^\infty\left[\frac{\xi_k}{2\pi k}\cos(2\pi kt) + \frac{\eta_k}{2\pi k}\sin(2\pi kt)\right] \quad .$$

Since $\int_0^1 (t-1/2)\cos(2\pi kt)dt = \int_0^1 \sin(2\pi \ell t)\cos(2\pi kt)dt = 0$, $k,\ell \geq 1$,

$\int_0^1 (B(t) - \int_0^1 B(s)ds)^2 ds \geq 2\int_0^1 (\sum_1^\infty \frac{\xi_k}{2\pi k}\cos(2\pi kt))^2 dt = \sum \frac{\xi_k^2}{(2\pi k)^2}$. Thus:

$$P(\sigma_{[0,1]}(B(\cdot)) \leq \varepsilon) \leq P(\sum_1^\infty \frac{\xi_k^2}{(2\pi k)^2} \leq \varepsilon)$$

$$\leq e^{\gamma^2\varepsilon^2/2} E[\exp(-\gamma^2/2 \sum_1^\infty \frac{\xi_k^2}{(2\pi k)^2})]$$

$$= e^{\gamma^2\varepsilon^2/2} \prod_1^\infty (1 + (\frac{\gamma}{2\pi k})^2)^{-1/2} = e^{\gamma^2\varepsilon^2/2} (\frac{\gamma/2}{\sinh(\gamma/2)})^{1/2}$$

for all $\gamma > 0$. Since

$$\frac{\gamma/2}{\sinh(\gamma/2)} = \frac{\gamma}{e^{\gamma/2} - e^{-\gamma/2}} \leq 2e^{-\gamma/4},$$

we obtain our estimate by taking $\gamma = \frac{1}{8\varepsilon^2}$. □

(8.26) **Theorem**: Let $\alpha : [0,\infty) \times \Theta \to R^d$ and $\gamma : [0,\Gamma) \times \Theta \to R^1$ be bounded progressively measurable functions and let $0 \leq \tau_1 \leq \tau_2 < \infty$ be stopping times. Given $\xi_0 \in R^1$, set

$$\xi(T) = \xi_0 + \int_0^T \alpha(t)\cdot d\theta(t) + \int_0^T \gamma(t)dt, \quad T \geq 0 .$$

If $M_1 \equiv \sup_\theta \sup_{\tau_1(\theta)\leq t\leq \tau_2(\theta)} |\alpha(t,\theta)|$ and $M_2 = \sup_\theta \sup_{\tau_1(\theta)\leq t\leq \tau_2(\theta)} |\gamma(t)|$, then for all $m \geq 5$, $Q > 0$, $R > 0$, and $N \geq 1$:

$$\mathbb{W}(\int_{\tau_1}^{\tau_2} \xi(t)^2 dt \leq Q/N^{4m-9}, \int_{\tau_1}^{\tau_2} |\alpha(t)|^2 dt \geq R/N^m, \tau_2 - \tau_1 = 1/N^3 | \mathcal{B}_{\tau_1})$$

$$\leq 2^{1/2} N^{m-5} \exp(-\frac{RN}{2^7((Q/R)^{1/2}M_1 + M_2)^2}) .$$

Proof: We begin by making a few simplifying observations. In the first place, because of the strong Markov property, there is no loss of generality if we assume that $\tau_1 \equiv 0$. Secondly, the desired estimate will follow in general once we prove it under the assumptions that $\tau_2 = \tau_1 + 1/N^3$ and that $|\alpha(\cdot)| \leq M_1$ and $|\gamma(\cdot)| \leq M_2$ everywhere. Finally, after introducing an extra dimension if necessary, we can easily construct for each $\varepsilon > 0$ an $\alpha_\varepsilon(\cdot)$ such that $|\alpha(\cdot)-\alpha_\varepsilon(\cdot)| \leq \varepsilon$ and $|\alpha_\varepsilon(\cdot)| \geq \varepsilon$ everywhere. Combining these remarks, we see that, without loss of generality, we may assume that $\tau_1 \equiv 0$, $\tau_2 \equiv 1/N$, $\varepsilon \leq |\alpha(\cdot)| \leq M_1$ everywhere for some $\varepsilon > 0$, and $|\gamma(\cdot)| \leq M_2$ everywhere. We will therefore make these assumptions.

Given $m \geq 5$, $Q > 0$, $R > 0$, and $N \geq 1$:

$$\mathfrak{w}\left(\int_0^{1/N^3} \xi(t)^2 dt \leq Q/N^{4m-9} , \int_0^{1/N^3} |\alpha(t)|^2 \geq R/N^m\right)$$

$$\leq \sum_0^{N^{m-5}-1} \mathfrak{w}\left(\int_{I(k)} \xi(t)^2 dt \leq Q/N^{4m-9} , \int_{I(k)} |\alpha(t)|^2 dt \geq R/N^{2m-5}\right)$$

where $I(k) = [k/N^{m-2}, k+1/N^{m-2})$. Next, define

$$A(T) = \int_0^T |\alpha(t)|^2 dt \quad , \quad T \geq 0 \ .$$

Then:

$$\int_{I(k)} \xi(t)^2 dt = \int_{A(k/N^{m-2})}^{A(k+1/N^{m-2})} (\xi\circ A^{-1}(t))^2 |\alpha\circ A^{-1}(t)|^{-2} dt$$

$$\geq \frac{1}{M_1^2} \int_{A(k/N^{m-2})}^{A(k+1/N^{m-2})} (\xi\circ A^{-1}(t))^2 dt \ .$$

Since $\int_{I(k)} |\alpha(t)|^2 dt \geq R/N^{2m-5}$ implies that $A(k+1/N^{m-2}) \geq A(k/N^{m-2}) + R/N^{2m-5}$, we now see that if $J(k) = [A(k/N^{m-2}), A(k/N^{m-2})+R/N^{2m-5}]$

$$\mathbb{W}\left(\int_{I(k)} \xi(t)^2 dt \leq Q/N^{4m-9}, \int_{I(k)} |\alpha(t)|^2 dt \geq R/N^{2m-5}\right)$$

$$\leq \mathbb{W}\left(\int_{J(k)} (\xi \circ A^{-1}(t))^2 dt \leq M_1^2 Q/N^{4m-9}, \int_{I(k)} |\alpha(t)|^2 dt \geq R/N^{2m-5}\right).$$

Set $B(t) = \int_0^{A^{-1}(t)} \alpha(s) \cdot d\theta(s)$, $t \geq 0$. Then $(B(t), \mathcal{B}_{A^{-1}(t)}, \mathbb{W})$ is a Brownian motion. Note that:

$$\left(\int_{J(k)} (\xi^2 \circ A^{-1}(t))^2 dt\right)^{1/2} = \left(\int_{J(k)} \left(\xi_0 + B(t) + \int_0^{A^{-1}(t)} \gamma(s) ds\right)^2 dt\right)^{1/2}$$

$$\geq \left(\int_{J(k)} \left(\xi_0 + \int_0^{k/N^{m-2}} \gamma(s) ds + B(t)\right)^2 dt\right)^{1/2}$$

$$- \left(\int_{J(k)} \left(\int_{k/N^{m-2}}^{A^{-1}(t)} \gamma(s) ds\right)^2 dt\right)^{1/2}$$

$$\geq \left(\frac{R}{N^{2m-5}}\right)^{1/2} \left(\sigma_{J(k)}(B(\cdot)) - \int_{k/N^{m-2}}^{A^{-1}(A(k/N^{m-2})+R/N^{2m-5})} |\gamma(s)| ds\right)$$

$$\geq \left(\frac{R}{N^{2m-5}}\right)^{1/2} \left(\sigma_{J(k)}(B(\cdot)) - M_2/N^{m-2}\right)$$

if $\int_{I(k)} |\alpha(t)|^2 dt \geq R/N^{2m-5}$. Thus

$$\mathbb{W}\left(\int_{J(k)} (\xi \circ A^{-1}(t))^2 dt \leq M_1^2 Q/N^{4m-9}, \int_{I(k)} |\alpha(t)|^2 dt \geq R/N^{2m-5}\right)$$
$$\leq \mathbb{W}\left(\sigma_{J(k)}(B(\cdot)) \leq (M_1 (Q/R)^{1/2} + M_2)/N^{m-2}\right).$$

Finally, $A(k/N^{m-2})$ is a stopping time relative to $\{\mathcal{B}_{A^{-1}(t)} : t \geq 0\}$, and therefore by Lemma (8.23) :

$$\mathbb{W}(\sigma_{J(k)}(B(\cdot)) \leq (M_1 (Q/R)^{1/2} + M_2)/N^{m-2})$$

$$\leq 2^{1/2}\exp\left(-(R/N^{2m-5})/(2^7(M_1(Q/R)^{1/2}+M_2)^2/N^{2m-4})\right)$$

$$= 2^{1/2}\exp\left(-\frac{RN}{2^7((Q/R)^{1/2}M_1+M_2)^2}\right) .$$

The result is immediate from this. □

We have now learned how to show that $\int_{\tau_1}^{\tau_2}\xi(t)^2dt$ cannot be too small by comparison with $\int_{\tau_1}^{\tau_2}|\alpha(t)|^2dt$. Referring back to (8.21) , we therefore see that for $w \in R^N$: $\int_{\tau_1}^{\tau_2}(J(t,x)^{-1}V(X(t,x)),w)^2dt$ can be estimated below in terms of $\int_{\tau_1}^{\tau_2}\sum_{k=1}^{d}(J(t,x)^{-1}[V^{(k)},V](X(t,x)),w)^2dt$. Obviously, this provides us with a "bootstrap" with which to obtain "non-degeneracy" of $n^{\underline{th}}$-order Lie products involving $V^{(1)},\ldots,V^{(d)}$ from "non-degeneracy" of $(n+1)^{st}$-order Lie products of these vector fields. In particular, we could use Theorem (8.26) to prove Hörmander's theorem in the case when Lie $(V^{(1)},\ldots,V^{(d)})$ has rank D at x . In order to take advantage of "non-degeneracy coming from Lie products involving $V^{(0)}$, we must learn how to estimate $\int_{\tau_1}^{\tau_2}\xi(t)^2dt$ from below in terms of $\int_{\tau_1}^{\tau_2}\gamma(t)^2dt$. In general, I do not know how to get such an estimate. However, in the situation with which we are dealing, we know that $\gamma(\cdot)$ itself is given by a stochastic integral. This observation is important because it allows us to get control over the regularity of $\gamma(\cdot)$ as a function of t .

(8.27) Lemma: The notation is the same as that in (8.26). For each $\alpha \in (0,1/2)$ there exist $A_\alpha < \infty$ and $0 < D_\alpha < \infty$ such that

$$\mathfrak{W}\Big(\sup_{\tau_1 \le s < t \le \tau_2} \frac{|\xi(t)-\xi(s)|}{|t-s|^\alpha} \ge L + M_2,\ \tau_2-\tau_1 \le T \Big| \mathcal{B}_{\tau_1}\Big)$$

$$\le A_\alpha \exp(-(L/D_\alpha M_1 T^{1/2-\alpha})^2)$$

for $0 < T \le 1$ and $L > 0$.

Proof: Just as in the proof of (8.26), we may and will assume that $\tau_1 \equiv 0$, $\tau_2 \equiv 1/N^3$, and $|\alpha(\cdot)| \ge \varepsilon > 0$. Define $A(T) = \int_0^T |\alpha(t)|^2 dt$ and $B(T) = \xi \circ A^{-1}(T)$. Then:

$$\sup_{0 \le s < t \le T} \frac{|\xi(t)-\xi(s)|}{|t-s|^\alpha} \le M_1^{2\alpha} \sup_{0 \le u < v \le M_1^2 T} \frac{|B(v)-B(u)|}{|v-u|^\alpha} + M_2 ;$$

and, because $B(\cdot)$ is a 1-dim Brownian motion under $\mathfrak{W}$,

$$\sup_{0 \le u < v \le M_1^2 T} \frac{|B(v)-B(u)|}{|v-u|^\alpha}$$

has the same distribution as

$$(M_1^2 T)^{1/2-\alpha} \sup_{0 \le u < v < 1} \frac{|B(v)-B(u)|}{|v-u|^\alpha} .$$

Thus

$$\mathfrak{W}\Big(\sup_{0 \le s < t \le T} \frac{|\xi(t)-\xi(s)|}{|t-s|^\alpha} \ge L + M_2\Big)$$

$$\le \mathfrak{W}\Big(\sup_{0 \le u < v \le 1} \frac{|B(v)-B(u)|}{|v-u|^\alpha} \ge L/M_1 T^{1/2-\alpha}\Big) .$$

Define $r > 2$ by $r-2/2r = \alpha$. Taking $\Psi(t) = (t/4)^{2r}$ and $p(t) = |t|^{1/2}$ in Theorem (2.13) of [S.&V.], we see that

$$|B(v)-B(u)| \leq 8\int_0^{|v-u|} \left(\frac{16K}{t^2}\right)^{1/2r} dt^{1/2}$$

$$= D_\alpha K^{\frac{1}{2r}} |v-u|^\alpha$$

whenever

$$\int_0^1\int \left(\frac{|B(v)-B(u)|}{2|v-u|^{1/2}}\right)^{2r} dudv \leq K \quad .$$

Hence

$$\mathbb{W}\left(\sup_{0\leq u<v\leq 1} \frac{|B(v)-B(u)|}{|v-u|^\alpha} \geq R\right)$$

$$\leq \mathbb{W}\left(\left(\int_0^1\int\left(\frac{|B(v)-B(u)|}{2|u-v|^{1/2}}\right)^{2r} dudv\right)^{1/2r} \geq R/D_\alpha\right)$$

$$\leq e^{-(R/D_\alpha)^2} E^{\mathbb{W}}\left[\exp\left(\int_0^1\int\left(\frac{|B(v)-B(u)|}{4|v-u|}\right)^r dudv\right)^{1/r}\right] \quad .$$

Finally, set $k_r = (r-1)^r$. Then $x \in [0,\infty) \to \exp((k_r+x)^{1/r})$ is convex. Hence, by Jensen's inequality:

$$E^{\mathbb{W}}\left[\exp\left(\left(\int_0^1\int\left(\frac{|B(v)-B(u)|^2}{4|v-u|}\right)^r dudv\right)^{1/r}\right)\right]$$

$$\leq E^{\mathbb{W}}\left[\exp\left(\left(k_r+\int_0^1\int\left(\frac{|B(v)-B(u)|^2}{4|v-u|}\right)^r dudv\right)^{1/r}\right)\right]$$

$$\leq E^{\mathbb{W}}\left[\int_0^1\int \exp\left(\left(k_r+\left(\frac{|B(v)-B(u)|^2}{4|v-u|}\right)^r\right)^{1/r}\right) dudv\right]$$

$$\leq e^{r-1}\int_0^1\int E^{\mathbb{W}}\left[\exp\left(\frac{|B(v)-B(u)|^2}{4|v-u|}\right)\right] dudv$$

$$= 2^{1/2} e^{r-1} \quad ,$$

since for all $0 \leq u \neq v \leq 1$, $\frac{B(v)-B(u)}{|v-u|^{1/2}}$ is distributed under $\mathcal{W}$ in the same way as $B(1)$ is.

□

We will use this regularity result in conjunction with the next very simple real-variables lemma.

(8.28) Lemma: Let $a < b$ and $f \in C([a,b])$. If $|f(\cdot)| \geq m$ on $[a,b]$, then $\sigma^2_{[a,b]}(\int_a^{\cdot} f(t)dt) \geq \frac{m^2}{12}(b-a)^2$. In particular, if $\sup_{a \leq s < t < b} \frac{|f(t)-f(s)|}{|t-s|^{\alpha}} \leq L$ for some $\alpha \in (0,1]$, then so long as $0 < \varepsilon \leq L(b-a)^{(1+2\alpha)/2}$:

$$\sigma^2_{[a,b]}(\int_a^{\bullet} f(t)dt) \geq \frac{1}{48\cdot 2^{3/\alpha}} \frac{\varepsilon^{(2\alpha-3)/\alpha}}{L^{3/\alpha}(b-a)^{(4\alpha+3)/2\alpha}}$$

whenever $\int_a^b f(t)^2 dt \geq \varepsilon^2$.

<u>Proof:</u> An easy translation and scaling argument shows that it suffices to prove the first part when $a = 0$ and $b = 1$. Set $F(t) = \int_0^1 f(s)ds$, $t \in [0,1]$. By the mean value theorem, $F(t_0) = \int_0^1 F(s)ds$ for some $t_0 \in (0,1)$. Thus $\left|F(t) - \int_0^1 F(s)ds\right| \geq m\left|t-t_0\right|$ and so $\sigma^2_{[0,1]}(F(\cdot)) \geq m^2\int_0^1 \left|t-t_0\right|^2 dt \geq m^2\int_0^1 \left|t-1/2\right|^2 dt = \frac{m^2}{12}$.

The second part is proved by an applicaton of the first. Namely, because $\int_a^b f(t)^2 dt \geq \varepsilon^2$, there is a $t_0 \in (a,b)$ such that $\left|f(t_0)\right| \geq \varepsilon/(b-a)^{1/2}$. Hence, $\left|f(t)\right| \geq \varepsilon/2(b-a)^{1/2}$ for $t \in [a,b]$ satisfying $L\left|t-t_0\right|^\alpha \leq \varepsilon/2(b-a)^{1/2}$. Since $(\varepsilon/2L(b-a)^{1/2})^{1/\alpha} \leq 1/2^{1/\alpha}(b-a) \leq (b-a)/2$, we now see that there is an interval $I \subseteq [a,b]$ with $|I| \geq (\varepsilon/2L(b-a)^{1/2})^{1/\alpha}$ such that $\left|f(t)\right| \geq \varepsilon/2(b-a)^{1/2}$ for $t \in I$ In particular:

$$(b-a)\sigma^2_{[a,b]}(\int_a^{\bullet} f(t)dt) \geq |I|\sigma^2_I(\int_a^{\bullet} f(t)dt)$$

$$\geq \frac{\varepsilon^2}{48(b-a)}|I|^3 \geq \frac{1}{48}\frac{\varepsilon^2}{b-a}\left(\frac{\varepsilon}{2L(b-a)^{1/2}}\right)^{1/\alpha} .$$ □

(<u>8.29</u>) <u>Theorem</u>: The notation is the same as that in (8.26) . Assume that $\gamma(T) = \gamma_0 + \int_0^T \tilde{\alpha}(t)\cdot d\theta(t) + \int_0^T \tilde{\gamma}(t)dt$, $T \geq 0$, where $\tilde{\alpha} : [0,\infty) \times \Theta \to R^d$ and $\tilde{\gamma} : [0,\infty) \times \Theta \to R^1$ are bounded progressively measurable functions. Set $\tilde{M}_1 = \sup_\Theta \sup_{\tau_1(\theta)\leq t\leq \tau_2(\theta)} \left|\tilde{\alpha}(t,\theta)\right|$ and

$\tilde{M}_2 = \sup_{\theta} \sup_{\tau_1(\theta)\le t\le\tau_2(q)} |\tilde{\gamma}(t,\theta)|$. There exist numbers $\lambda > 0$ and $K < \infty$, depending only on $M_1, \tilde{M}_1, M_2$, and $\tilde{M}_2$ such that for all $N \geq 1$ and $m \geq 6$:

$$\mathbb{W}\left(\int_{\tau_1}^{\tau_2} \xi(t)^2 dt \le 1/N^{20m-9} \;,\; \int_{\tau_1}^{\tau_2} \gamma(t)^2 dt \ge 1/N^m \;,\; \tau_2 - \tau_1 = 1/N^3 \,\middle|\, \mathcal{B}_{\tau_1}\right)$$

$$\le KN^{5m-5} \exp(-\lambda N^{1/8}) \;.$$

<u>Proof</u>: As in the proof of (8.26) , we may and will assume that $\tau_1 \equiv 0$, $\tau_2 \equiv 1/N^3$, and that $|\alpha(\cdot)| \geq \varepsilon$ for some $\varepsilon > 0$.

Set $E = \{\int_0^{1/N^3} \xi(t)^2 dt \le 1/N^{20m-9} \;,\; \int_0^{1/N^3} \gamma(t)^2 dt \ge 1/N^m\}$. We must estimate $\mathbb{W}(E)$. To this end, define:

$$E_1 = \{\int_0^{1/N^3} \xi(t)^2 dt \le 1/N^{20m-9} \;,\; \int_0^{1/N^3} |\alpha(t)|^2 dt \ge 1/N^{5m}\} \;,$$

$$E_2 = \{ \sup_{0\le s<t\le 1/N^3} \frac{|\gamma(t)-\gamma(s)|}{|t-s|^{3/8}} \ge \tilde{M}_2 + 1\} \;,$$

and

$$E_3 = E_2^c \cap \{\int_0^{1/N^3} \xi(t)^2 dt < 1/N^{20m-9} \;,\; \int_0^{1/N^3} |\alpha(t)|^2 dt < 1/N^{5m+1/3} \;,\; \int_0^{1/N^3} \gamma(t)^2 dt \ge 1/N^m\}$$

Clearly $E \subseteq E_1 \cup E_2 \cup E_3$.

Taking $Q = R = 1$ in (8.26) , we see that

$$\mathfrak{W}(E_1) \leq 2^{1/2} N^{5m-5} \exp\left(-\frac{N}{2^7 (M_1+M_2)^2}\right) .$$

By (8.27), with $\alpha = 3/8$ and $L = 1$:

$$\mathfrak{W}(E_2) \leq A_{3/8} \exp(-(1/D_{3/8} M_1 1/N^{1/16})^2) .$$

Thus it remains only to estimate $\mathfrak{W}(E_3)$.

Set $Y(T) = \int_0^T \alpha(t) \cdot d\theta(t)$, $T \geq 0$. Then, by the triangle inequality:

$$\left(\int_0^{1/N^3} \xi(t)^2 dt\right)^{1/2}$$

$$\geq \left(\int_0^{1/N^3} \left(\xi_0 + \int_0^t \gamma(s)ds\right)^2 dt\right)^{1/2} - \left(\int_0^{1/N^3} Y(t)^2 dt\right)^{1/2}$$

$$\geq N^{-3/2}\left(\sigma_{[0,1/N^3]}\left(\int_0^{\cdot} \gamma(s)ds\right) - \sup_{0 \leq t \leq 1/N^3} |Y(t)|\right) ;$$

and so:

$$\sup_{0 \leq t \leq 1/N^3} |Y(t)| \geq \sigma_{[0,1/N^3]}\left(\int_0^{\cdot} \gamma(s)ds\right) - \left(N^3 \int_0^{1/N^3} \xi(t)^2 dt\right)^{1/2} .$$

Note that $1/N^m \leq (\widetilde{M}_2+1)^2(1/N^3)^{7/4}$; and so, by (8.27) with $L = \widetilde{M}_2 + 1$ and $\varepsilon = 1/N^{m/2}$:

$$\sigma^2_{[0,1/N^3]}\left(\int_0^{\cdot} \gamma(s)ds\right) \geq \frac{1}{3 \cdot 2^{12}} \frac{(1/N^m)^5}{(\widetilde{M}_2+1)^8 (1/N^3)^6}$$

$$= \frac{1}{3 \cdot 2^{12} (\widetilde{M}_2+1)^8} N^{18-5m}$$

in E_3. At the same time, in E_3:

$$(N^3\int_0^{1/N^3}\xi(t)^2dt)^{1/2} \le N^{6-10m} .$$

Hence, if $H \equiv (3\cdot 2^{12}(\tilde{M}_2+1)^8)^{1/2}$, then

$$E_3 \subseteq \{ \sup_{0\le t\le 1/N^3} |Y(t)| \ge \frac{N^{9-5m/2}}{H}(1-HN^{-3-15m/2}), \int_0^{1/N^3}|\alpha(t)|^2dt \le 1/N^{5m+1/3}\}$$

Finally, set $B(T) = Y \circ A^{-1}(T)$, where $A(T) = \int_0^T|\alpha(t)|^2dt$. Then $B(\cdot)$ has the same distribution under $\mathcal{W}$ as does $\theta_1(\cdot)$. Hence, by Lemma (2.19),

$$\mathcal{W}(E_3) \le 2\exp(-\frac{N^{18-5m}}{2HN^{-5m}}(1-HN^{-3-15m/2})^2) .$$

□

We are at last ready to pick up where we left off after deriving (8.21). The notation is the same as that which we were using at the beginning of this section.

Given a smooth vector field $V : R^D \to R^D$, define $A(V) : R^D \to R^d \otimes R^D$ and $\Gamma(V) : R^D \to R^D$ by

$$A(V) = \begin{pmatrix} [V^{(1)},V] \\ \vdots \\ [V^{(d)},V] \end{pmatrix}$$

and

$$\Gamma(V) = ([V^{(0)},V] + 1/2 \sum_{k=1}^{d} [V^{(k)},[V^{(k)},V]]) .$$

(8.30) Theorem: Let $V : R^D \to R^D$ be a smooth vector field and let $\tau_1 \le \tau_2$ be finite stopping times. Assume that there is an $M < \infty$ such that:

$$\|J(t,x)^{-1}(A(V))(X(t,x))^*\|_{H.S.} \le M \quad ,$$

$$\max_{1\le k\le d} \|J(t,x)^{-1}(A([V^{(k)},V]))(X(t,x))^*\|_{H.S.} \le M \quad ,$$

$$\left|J(t,x)^{-1}(\Gamma(V))(X(t,x))\right| \le M \quad ,$$

$$\|J(t,x)^{-1}A(\Gamma(V))(X(t,x))^*\|_{H.S.} \le M \quad ,$$

and

$$\left|J(t,x)^{-1}\Gamma(\Gamma(V))(X(t,x))\right| \le M$$

for all $\tau_1 \le t \le \tau_2$. Then there exist $\lambda > 0$ and $K < \infty$, depending only on M , such that for any $\eta \quad S^{D-1}$, $N \ge 1$ and $m \ge 6$:

$$\mathbb{P}\Big(\int_{\tau_1}^{\tau_2}(J(t,x)^{-1}V(X(t,x)),\eta)^2 dt \le 1/N^{20m-9} \quad ,$$

$$\sum_{k=0}^{d} \int_{\tau_1}^{\tau_2}(J(t,x)^{-1}[V^{(k)},V](X(t,x)),\eta)^2 dt \ge 2/N^m \quad , \quad \tau_2 - \tau_1 = 1/N^3 \Big| \mathcal{B}_{\tau_1}\Big)$$

$$\le KN^{5m-5}\exp(-\lambda N^{1/8}) \quad .$$

<u>Proof</u>: Define

$$\alpha(t) = \chi_{[\tau_1,\tau_2]}(t)((A(V))(X(t,x)),J(t,x)^{-1*}\eta) \quad ,$$

$$\gamma(t) = \chi_{[\tau_1,\tau_2]}(t)(J(t,x)^{-1}(\Gamma(V))(X(t,x)),\eta) \quad ,$$

$$\xi_0 = (J(\tau_1,x)^{-1}V(X(\tau_1,x)),\eta) \quad ,$$

and

$$\xi(T) = \xi_0 + \int_0^T \alpha(t)\cdot d\theta(t) + \int_0^T \gamma(t)dt \quad , \quad T \geq 0 \quad .$$

By (8.21) , $\xi(t) = (J(t,x)^{-1}V(X(t,x)),\eta)$ for $t \in [\tau_1,\tau_2]$. Also

$$|\alpha(t)|^2 = \chi_{[\tau_1,\tau_2]}(t)\sum_1^d (J(t,x)^{-1}[V^{(k)},V](X(t,x)),\eta)^2$$

and

$$\gamma(t)^2 = \chi_{[\tau_1,\tau_2]}(t)((J(t,x)^{-1}([V^{(0)},V] + 1/2\sum_1^d [V^{(k)},[V^{(k)},V]])(X(t,x)),\eta))^2$$

Note that since ξ_0 is $\mathcal{B}_{\tau_1}$-measurable, Theorems (8.26) and (8.29) apply to $\xi(\cdot)$.

Next, define

$$\gamma^{(0)}(t) = \chi_{[\tau_1,\tau_2]}(t)(J(t,x)^{-1}[V^{(0)},V](X(t,x)),\eta)$$

and

$$\gamma^{(k)}(t) = \chi_{[\tau_1,\tau_2]}(t)(J(t,x)^{-1}[V^{(k)}[V^{(k)},V]](X(t,x)),\eta) \quad , \quad 1 \leq k \leq d \quad .$$

We have to estimate $\mathbb{W}(E)$, where:

$$E = \{\int_{\tau_1}^{\tau_2}\xi(t)^2dt \leq 1/N^{20m-9} \; , \; \int_{\tau_1}^{\tau_2}(|\alpha(t)|^2 + \gamma_0(t)^2)dt \geq 3/N^m \; , \; \tau_2 - \tau_1 = 1/N^3\}$$

Since $\gamma^{(0)}(t) = \gamma(t) - 1/2\sum_1^d \gamma^{(k)}(t)$ and therefore:

$$\gamma^{(0)}(t)^2 \leq 2\gamma(t)^2 + d/2 \sum_1^d \gamma^{(k)}(t)^2 \quad ,$$

we see that $E \subseteq F \cup G \cup H^{(1)} \cup \cdots \cup H^{(d)}$, where:

$$F = \{\int_{\tau_1}^{\tau_2} \xi(t)^2 dt \le 1/N^{20m-9} , \int_{\tau_1}^{\tau_2} |\alpha(t)|^2 dt \ge 1/N^3 , \tau_2 - \tau_1 = 1/N^3\} ,$$

$$G = \{\int_{\tau_1}^{\tau_2} \xi(t)^2 dt \le 1/N^{20m-9} , \int_{\tau_1}^{\tau_2} |\alpha(t)|^2 dt \ge 1/N^m , \tau_2 - \tau_1 = 1/N^3\}$$

and

$$H^{(k)} = \{\int_{\tau_1}^{\tau_2} \xi(t)^2 dt \le 1/N^{2m-9} , \int_{\tau_1}^{\tau_2} \gamma^{(k)}(t)^2 dt \ge 1/d^2N^m , \tau_2 - \tau_1 = 1/N^3\} .$$

Clearly Theorem (8.26) allows us to estimate $\mathbb{W}(F|\mathcal{B}_{\tau_1})$, and Theorem (8.29) provides us with the desired estimate for $\mathbb{W}(G|\mathcal{B}_{\tau_1})$. Thus it suffices to handle $\mathbb{W}(H^{(k)}|\mathcal{B}_{\tau_1})$, $1 \le k \le d$; and this will be done once we treat $\mathbb{W}(H^{(1)}|\mathcal{B}_{\tau_1})$.

Obviously:

$$H^{(1)} \subseteq \{\int_{\tau_1}^{\tau_2} \xi(t)^2 dt \le 1/N^{20m-9} , \int_{\tau_1}^{\tau_2} |\alpha(t)|^2 dt \le 1/N^{4m-9} , \tau_2 - \tau_1 = 1/N^3\}$$

$$\cup \{\int_{\tau_1}^{\tau_2} |\alpha(t)|^2 dt \le 1/N^{4m-9} , \int_{\tau_1}^{\tau_2} \gamma^{(1)}(t)^2 dt \ge 1/d^2N^m , \tau_2 - \tau_1 = 1/N^3\} .$$

Theorem (8.26) provides us immediately with the appropriate sort of estimate for the first of these sets. Moreover, since $|\alpha(t)|^2 \ge$

$(J(t,x)^{-1}[V^{(1)},V](X(t,x)),\eta)^2$, $\tau_1 \leq t \leq \tau_2$, and

$$\gamma^{(1)}(t)^2 \leq |(A([V^{(1)},V]))(X(t,x))J(t,x)^{-1*}\eta)^2 ,$$

equation (8.21) and Theorem (8.26) give us the desired sort of estimate for the second set. □

Theorem (8.30) provides us with the essential ingredient needed to prove our main result. In what follows, we will be using some new notation. Namely, define the sets C_ℓ , $\ell \geq 0$, by induction as follows:

$$C_0 = \{V^{(1)},\ldots,V^{(d)}\}$$

and

$$C_\ell = \{[V^{(k)},V] : 0 \leq k \leq d \text{ and } V \in C_{\ell-1}\} , \quad \ell \geq 1 .$$

(<u>8.31</u>) <u>Theorem</u>: Let $x \in R^D$ and $R > 0$ be given and suppose that there exist $\ell_0 \geq 0$ and $\varepsilon > 0$ such that

$$(8.32) \qquad \sum_{\ell=0}^{\ell_0} \sum_{V \in C_\ell} (V(y),\eta)^2 \geq \varepsilon , \quad y \in B(x,R) \text{ and } \eta \in R^D .$$

Then for each $q \in [1,\infty)$ there exists a $C_q < \infty$ depending only on d , D , R , ℓ_0 , $\|\sigma(\cdot)\|_{C_b^1(B(x,R))}$, $\|b(\cdot)\|_{C_b^1(B(x,R))}$, and $\max_{0\leq\ell\leq\ell_0+1} \max_{V\in C_\ell} \|V(\cdot)\|_{C_b(B(x,R))}$ such that for all $\eta \in S^{D-1}$:

$$(8.33) \qquad \|1/\langle\eta,\tilde{A}(t,x)\eta\rangle\|_{L^q(\mathbb{W})} \leq C_q(\varepsilon(t\wedge 1)^{2\ell_0+1})^{-\nu_{\ell_0}} , \quad 0 < t \leq T ,$$

where $\nu_{\ell_0} > 0$ depends on ℓ_0 alone (and $\tilde{A}(t,x)$ is given by (8.18)).

Proof: Since by an easy time-rescaling argument the estimate for small t can be derived from the estimate for $t = 1$ with ε replaced by $\varepsilon t^{2\ell_0+1}$, we will restrict our attention to $t = 1$.

Define σ_N, $N \geq 1$, by:

$$\sigma_N = (\inf\{t \geq 0 : |X(t,x)-x| \quad \|J(t,x)^{-1}-I\|_{H.S.} \geq R \wedge 1/2\}) \wedge 1/N^3 .$$

Then, by Lemma (2.19), there exist $K_1 < \infty$ and $\lambda_1 > 0$, depending only on d, D, R, $\|\sigma(\cdot)\|_{C_b^1(B(x,R))}$, and $\|b(\cdot)\|_{C_b^1(B(x,R))}$ such that

$$\text{(8.34)} \qquad \mathbb{W}(\sigma_N \neq 1/N^3) < K_1 e^{-\lambda_1 N^3} .$$

Moreover, for all $\eta \in S^{D-1}$:

$$\text{(8.35)} \qquad \int_0^{\sigma_N} \sum_{\ell=0}^{\ell_0} \sum_{V \in \mathcal{C}_\ell} (J(t,x)^{-1}V(X(t,x)),\eta)^2 dt \geq \varepsilon\sigma_N .$$

By Theorem (8.30), there exist $\rho \in (0,\infty)$, $\delta \in (0,\ell)$, $K_2 < \infty$ and $\lambda_2 > 0$ depending only on $\max\limits_{0 \leq \ell \leq \ell_0+1} \max\limits_{V \in \mathcal{C}_\ell} \|V(\cdot)\|_{C_b(B(x,R))}$ such that

$$\text{(8.36)} \qquad \mathbb{W}(\int_0^{\sigma_N} (J(t,x)^{-1}V(X(T,x)),\eta)^2 dt \leq \rho/N^{20m-9},$$

$$\sum_{k=0}^{d} \int_0^{\sigma_N} (J(t,x)^{-1}[V^{(k)},V](X(t,x)),\eta)^2 dt \geq \delta/N^m , \; \sigma_N = 1/N^3)$$

$$\leq K_2 N^{5m-5} \exp(-\lambda_2 N^\alpha)$$

for all $N \geq 1$, $m \geq 6$ and $V \in \bigcup_0^{\ell_0} \ell$. For $0 \leq \ell \leq \ell_0$ and $N \geq 1$, define

$$E_\ell(N) = \{ \sum_{\ell=0}^{\ell_0} \sum_{V \in \mathcal{C}_\ell} \int_0^{\sigma_N} (J(t,x)^{-1}V(X(t,x)),\eta)^2 dt \leq 2(d+1)^\ell \delta/N^{m_\ell} \}$$

where $m_\ell = 20^{\ell_0-\ell} \times 6$. Then

$$E_0(N) \supseteq \{\int_0^1 (\eta,\widetilde{A}(t,x)\eta)dt \le 2\delta/N^{m_{\ell_0}}\} . \tag{8.37}$$

Moreover,

$$E_0(N) \subseteq \Big(\bigcup_{\ell=1}^{\ell_0} (E_{\ell-1}(N) \cap E_\ell(N)^c)\Big) \cap E_{\ell_0}(N) \quad ; \tag{8.38}$$

and if

$$N^9 \ge 2(d+1)^{\ell_0}\delta/\rho \quad , \tag{8.39}$$

then

$$E_{\ell-1}(N) \cap E_\ell(N)^c \subseteq$$

$$\bigcup_{V\in\mathcal{C}_{\ell-1}} \{\int_0^{\sigma_N} (J(t,x)^{-1}V(X(t,x)),\eta)^2 \le \rho/N^{20m-9} \quad , \tag{8.40}$$

$$\sum_{k=0}^{d} \int_0^{\sigma_N} (J(t,x)^{-1}[V^{(k)},V](X(t,x)),\eta)^2 \ge \delta/N^{m_\ell}\} \quad .$$

Also, if

$$N^3 \ge 2(d+1)^{\ell_0}\delta/\varepsilon \quad , \tag{8.41}$$

Then, by (8.35)

$$E_{\ell_0}(N) \cap \{\sigma_N = 1/N^3\} = \emptyset \quad .$$

Combining (8.34), (8.36), (8.37), (8.38), and (8.40), it is now clear that so long as N satifies both (8.39) and (8.41):

$$(8.42)\qquad \mathbb{W}\left(\int_0^1 (\eta,A(t,x)\eta)dt \le 2\delta/N^{m_{\ell_0}}\right) \le K\exp(-\lambda N^{\beta}) ,$$

where $K < \infty$, $\lambda > 0$, and $\beta > 0$ depend only on d , D , R , ℓ_0 , $\|\sigma(\cdot)\|_{C_b^1(B(x,R))}$, $\|b(\cdot)\|_{C_b^1(B(x,R))}$, and $\max_{0\le\ell\le\ell_0+1}\max_{V\in C_\ell}\|V(\cdot)\|_{C_b(B(x,R))}$.
The desired estimate is obvious from (8.42) . □

With Theorem (8.31) we have completed the difficult part of our program. In fact, all that remains is for us to convert (8.33) into a statement about reciprocal moments of $\Delta(T,x)$. Since $A(t,x) = J(T,x)\tilde{A}(T,x)J(T,x)^*$ and we can easily control all moments of $1/(\det(J(t,x)))^2$, the required conversion will come easily from the next lemma.

(8.42) Lemma: Let $(E,\mathcal{F},P)$ be a probability space and let $A : E \to R^D \otimes R^D$ be an $\mathcal{F}$-measurable symmetric non-negative definite matrix valued function. Define the random variables $\lambda = \inf_{\eta\in S^{D-1}} (\eta,A\eta)$ and $\Lambda = \max_{\eta\in S^{D-1}} (\eta,A\eta)$. Then for each $p \in [1,\infty)$ there exist universal $C(p,D) < \infty$ such that:

$$E[(1/\lambda)^p] \le C(p,D) \sup_{\eta\in S^{D-1}} E[(\eta,A\eta)^{-2(D+p+1)}]^{1/2} E[(1+\Lambda)^{2D+6}]^{1/2} .$$

Proof: We first show that for any $\varepsilon > 0$ there exist $N = N_D(\varepsilon) \le C_D(1/(\varepsilon\wedge 1))^{D-1}$ points $\eta^1,\ldots,\eta^N \in S^{D-1}$ such that $S^{D-1} \subseteq \bigcup_1^N B(\eta^k,\varepsilon)$. Since this is trivial when $D = 1$ or $\varepsilon \ge 1$ we will assume that $D \ge 2$ and that $\varepsilon < 1$. Given $1 \le \ell \le D$, let A_ℓ be the set of $\eta \in S^{D-1}$ such that $\eta_j = m_j\varepsilon$ for $j \in \{1,\ldots,D\}\setminus\{\ell\}$ where the m_j's are integers from the interval $(-1/\varepsilon, 1/\varepsilon)$. Set $A = \bigcup_{1}^{D} A_\ell$. Clearly $\mathrm{card}(A) \le 2D(1/\varepsilon)^{D-1}$. Now suppose that $\xi \in S^{D-1}$. Choose $1 \le \ell \le D$ so that $|\xi_\ell| = \max_{1\le j\le D} |\xi_j|$. Clearly $|\xi_\ell| \ge D^{-1/2}$, and we may assume that $\ell = D$. If we choose $\eta \in A_D$ so that $\mathrm{sgn}\ \eta_D = \mathrm{sgn}\ \xi_D$ and $|\eta_j - \xi_j| < \varepsilon$, $1 \le j \le D-1$, then:

$$|\eta_D^2 - \xi_D^2| = \left|\sum_1^{D-1} \eta_j^2 - \sum_1^{D-1} \xi_j^2\right| = \sum_1^{D-1} |\eta_j - \xi_j||\eta_j + \xi_j|$$

$$\le 2^{1/2}\Big(\sum_1^{D-1} (\eta_j - \xi_j)^2\Big)^{1/2} \le (2D)^{1/2}\varepsilon .$$

Thus

$$|\eta_D - \xi_D| = |\eta_D^2 - \xi_D^2| / |\eta_D + \xi_D| \le 2^{1/2} D\varepsilon$$

since $|\eta_D + \xi_D| \ge |\xi_D| \ge D^{-1/2}$. We have therefore shown that $S^{D-1} \subseteq \bigcup_{\eta\in A} B(\eta,(2D^2 + D)^{1/2}\varepsilon)$. The assertion follows immediately from this.

Next suppose the A is a non-negative definite symmetric $D \times D$ matrix and define λ and Λ accordingly. Given $\varepsilon > 0$, choose $\eta^1,\ldots,\eta^N \in S^{D-1}$, with $N = N_D(\varepsilon/2)$, as in the preceding. Then $\min_{1\le k\le N} (\eta^k, A\eta^k) \ge 2\varepsilon\Lambda$ implies that $\lambda \ge \varepsilon\Lambda$. Indeed, for any $\xi \in S^{D-1}$ there is a $1 \le k \le N$ such that $|(\xi, A\xi) - (\eta^k, A\eta^k)| \le \varepsilon\Lambda$, and therefore $\lambda \ge \min_{1\le k\le N} (\eta^k, A\eta^k) - \varepsilon\Lambda$.

We now turn to the situation described in the statement of the lemma. Clearly:

$$E[(1/\lambda)^p] = \sum_{n=1}^{\infty} E[(1/\lambda)^p\ ,\ n-1 \le \Lambda < n]$$

$$= \sum_{n=1}^{\infty} \sum_{m=1}^{\infty} E[(1/\lambda)^p\ ,\ n-1 \le \Lambda < n \ \text{ and } \ \frac{n}{m+1} \le \lambda < \frac{n}{m}]$$

$$\leq \sum_{n=1}^{\infty} \sum_{m=1}^{\infty} \left(\frac{m+1}{n}\right)^p P(n-1 \leq \Lambda < n \text{ and } \frac{n}{m+1} \leq \lambda < \frac{n}{m}) .$$

For each $m \geq 1$, choose $K(m) \subseteq S^{d-1}$ so that $\text{card}(K(m)) \leq Cm^{d-1}$ and $\min_{\eta \in K(m)} (\eta^2, A\eta) \geq \frac{2}{m} \Lambda$ implies $\lambda \geq \frac{1}{m} \Lambda$. (This is possible by the preceding two paragraphs.) Then:

$$P(n-1 \leq \Lambda < n \text{ and } \frac{n}{m+1} \leq \lambda < \frac{n}{m})$$

$$\leq P(\Lambda \geq n-1 \text{ and } \min_{\eta \in K(m)} (\eta, A\eta) \leq \frac{2n}{m})$$

$$\leq \text{card}(K(m)) \sup_{\eta \in S^{D-1}} P(\Lambda \geq n-1 \text{ and } (\eta, A\eta) \leq \frac{2n}{m}) .$$

Hence

$$E[(1/\lambda)^p] \leq C \sum_{n=1}^{\infty} \sum_{m=1}^{\infty} \left(\frac{m+1}{n}\right)^p m^{D-1} \sup_{\eta \in S^{D-1}} P(\Lambda \geq n-1 \text{ and } (\eta, A\eta) \leq \frac{2n}{m})$$

$$\leq C \sum_{n=1}^{\infty} \frac{1}{n^p} (2n)^{p+D+1} \sum_{m=1}^{\infty} \frac{(m+1)^p m^{D-1}}{m^{p+D+1}} \sup_{\eta \in S^{D-1}} E[(\eta, A\eta)^{-(p+D+1)}, \Lambda \geq n-1]$$

$$\leq C_1(p,D) \sum_{n=1}^{\infty} n^{D+1} \sup_{\eta \in S^{D-1}} E[(\eta, A\eta)^{-(p+D+1)}, \Lambda \geq n-1]$$

$$\leq C_1(p,D) \sup_{\eta \in S^{D-1}} E[(\eta, A\eta)^{-2(p+D+1)}]^{1/2} \sum_{n=1}^{\infty} n^{D+1} P(\Lambda \geq n-1)^{1/2} .$$

Finally,

$$\sum_{n=1}^{\infty} n^{D+1} P(\Lambda + 1 \geq n)^{1/2} \leq \sum_{1}^{\infty} n^{D+1} \left(\frac{1}{n^{2D+6}} E[\Lambda^{2D+6}]\right)^{1/2}$$

$$\leq \left(\sum_{1}^{\infty} \frac{1}{n^2}\right) E[\Lambda^{2D+6}]^{1/2} . \qquad \square$$

(8.43) Theorem: Suppose that $x \in R^D$ and that $\text{Lie}(\{(\text{ad}^n V^{(0)})(V^{(k)}) : n \geq 0 \text{ and } 1 \leq k \leq d\})$ has full rank at x. Then there exist $\ell_0 \geq 0$, $\varepsilon > 0$, and $R > 0$ such that (8.32) holds. Moreover, for each $q \in [1,\infty)$ there is a $C_q(T) < \infty$ depending only on d, D, R, ℓ_0, $|x|$, $\max_{0 \leq \ell \leq \ell_0+1} \|V(\cdot)\|_{C_b(B(x,R))}$, and $(\max_{1 \leq k \leq d} \|S_k(\cdot)\|_{C_b(R^D)} \vee \|B(\cdot)\|_{C_b(R^D)})$ such that

$$(8.44) \qquad \|1/\Delta(t,x)\|_{L^q(\mathbb{W})} \leq C_q(T)(\varepsilon t^{2\ell_0+1})^{-\nu_{\ell_0} D}, \quad 0 < t \leq T,$$

where ν_{ℓ_0} depends on ℓ_0 alone. In particular, if $P(T,x,\cdot) = \mathbb{W} \circ X(T,x)^{-1}$, then $P(T,x,\cdot)$ admits a density function $p(T,x,\cdot) \in \hat{C}^\infty(R^D)$. Finally, for each $n \geq 0$ and $T > 0$ there exist $C_n(T) < \infty$ depending on $\|\sigma(\cdot)\|_{C_b^{n+2}(D^D)}$, $\|b(\cdot)\|_{C_b^{n+2}(R^D)}$ as well as the quantities on which the $C_q(T)$'s depend such that

$$(8.45) \qquad \|p(t,x,\cdot)\|_{C_b^n(R^D)} \leq C_n(T)(\varepsilon t^{2\ell_0+1})^{-\nu_{\ell_0} Dn}, \quad 0 < t \leq T.$$

Proof: The proof of this result is a matter of organizing facts which we already have in hand.

First note that the existence of ℓ_0, ε, and R is an immediate consequence of our assumptions about $\text{Lie}(\{(\text{ad}^n V^{(0)})V^{(k)} : n \geq 0 \text{ and } 1 \leq k \leq d\})$.

Next, by Lemma (8.2):

$$J(T,x)^{-1} = I - \sum_{k=1}^{d} \int_0^T J(t,x)^{-1} S_k(X(t,x))d\theta_k(t)$$
$$+ \int_0^T J(t,x)^{-1}\left(\sum_{k=1}^{d} S_k(X(t,x))^2 - B(X(t,x))\right)dt \quad .$$

Thus for each $q \in [1,\infty)$ there exist $K_q < \infty$ and $\lambda_q \geq 0$, depending only on d , D , and $\max_{1\leq k\leq d} \|S_k(\cdot)\|_{C_b(R^D)} \vee \|B(\cdot)\|_{C_b(R^D)}$, such that

(8.46) $$E^{\mathbb{W}}[\sup_{0\leq t\leq T} \|J(t,x)^{-1}\|^q_{H.S.}] \leq K_q e^{\lambda_q T} , \quad T \geq 0 .$$

At the same time, for each $q \in [1,\infty)$ there exist $K'_q < \infty$ and $\lambda'_q \geq 0$, with the same dependence, such that

(8.47) $$E^{\mathbb{W}}[\sup_{0\leq t\leq T} |X(t,x) - x|^q] \leq K'_q e^{\lambda'_q T} .$$

Since $\tilde{A}(t,x) = \int_0^T J(t,x)^{-1} a(X(t,x)) J(t,x)^{-1*} dt$ and $a(X(t,x)) \leq C(1 + |X(t,x)|)^2$, where C depends only on $\max_{1\leq k\leq d} \|S_k(\cdot)\|_{C_b(R^D)}$, we see that for each $q \in [1,\infty)$ there exist $K''_q < \infty$ and $\lambda''_q \geq 0$ depending only on K_q , K'_q , λ_q , and λ'_q such that

(8.48) $$E^{\mathbb{W}}[\sup_{0\leq t\leq T} \|\tilde{A}(t,x)\|^q_{H.S.}]^{1/q} \leq K''_q(1 + |x|)e^{\lambda''_q T} , \quad T \geq 0 .$$

Combining (8.48) with (8.33) and applying Lemma (8.42) , we arrive at

(8.49) $$E^{\mathbb{W}}[(1/\det(\tilde{A}(t,x)))^q]^{1/q} \leq K'''_q(1+|x|^2)^D e^{\lambda'''_q T}/(\varepsilon t^{2\ell_0+1})^{\nu_{\ell_0} D} , \quad 0 \leq t \leq T ,$$

where $K'''_q < \infty$ and $\lambda'''_q \geq 0$ depend only on K''_q , λ''_q , and the quantities upon which the constants in (8.33) depend.

Because $A(t,x) = J(t,x)\tilde{A}(t,x)J(t,x)$, $1/\Delta(t,x) = (\det(J(t,x))^{-1})^2/\det(\tilde{A}(t,x))$. Thus, from (8.46) and (8.49) , (8.44) is a consequence of Hölder's inequality.

Given (8.44) the remainder of the theorem is simply an application of Theorem (7.21) . The estimates on $\|p(t,x,\cdot)\|_{C_b^n(R^D)}$ come from an examination of the argument used to prove (7.21) . □

One can use (8.44) to derive more information about $p(t,x,y)$ than that given in Theorem (8.43) . For example, Theorem (8.43) deals with $p(t,x,y)$ as a function of its "forward" variable y , but (8.44) can be used to study $p(t,x,y)$ as a function of its "backwards" variable x as well. To do this, one may proceed as follows. From

$$\int f(y)p(T,x,y)dt = E^{\mathfrak{W}}[f(X(T,x))]$$

and the chain rule, it is clear that

$$(8.50) \qquad D_x^{\alpha}\int f(y)p(T,x,y)dy = \sum_{\beta\leq\alpha} E^{\mathfrak{W}}[(D^{\beta}f)(X(T,x))P_{\alpha,\beta}(\Xi_{(|\alpha|)}(T,x))]$$

where $\Xi_{(n)}(\cdot,x)$ denotes the vector of processes $D_x^{\gamma}X(\cdot,x)$, $|\gamma| \leq n$, and $P_{\alpha,\beta}$ is a polynomial. An examination of the proof of Theorem (2.12) reveals that if one grades the derivatives of $X(\cdot,x)$ according to the order of the derivatives, the resulting vector satisfies a lower triangular system. Hence $P_{\alpha,\beta}(\Xi_{(|\alpha|)}(T,x)) \in \mathscr{S}(\mathfrak{L})$. We can therefore make repeated use of Theorem (5.13) , with $\Psi = P_{\alpha,\beta}(\Xi_{(|\alpha|)}(T,x))/\Delta(T,x)$, to remove all the derivatives of f on the right hand side of (8.50) and thereby obtain the expression

$$(8.59) \qquad D_x^{\alpha}\int f(y)p(T,x,y)dt = E\,[f(X(T,x))\Psi_{(\alpha)}(T,x)]$$

where $\Psi_{(\alpha)}(T,x) \in \mathscr{S}(\mathfrak{L})$. From here it is a relatively simple matter to conclude that $p(T,x,y)$ is smooth as a function of x . See section (7) of [S, J. Fnal. Anal.] for more details.

9. Some Concluding Remarks and an Example:

In the preceding section we worked very hard to recover a result which was proved by Hörmander in 1967, was greatly simplified by Kohn in 1970, about the same time was extended by Radekevitch, and recently has been refined by Folland, Rothschild, Stein and others (cf. [R. & S.] for references). So why did we go to so much trouble? The answer is, at least in part, that we had developed a method and we simply wanted to see how well it worked in a somewhat delicate situation. In this sense, our efforts in section 8) have to be considered to have been worthwhile. At the same time, it is a little disappointing that nothing new came out of all our toil, and it may have occurred to some that struggling with the Malliavin calculus is not worth the rewards. Indeed, it is satisfying to devotées of probabilty theory that there now exist probabilistic proofs of heretofore purely analytic results, but is the satisfaction gained worth the time spent? In my view, the answer would have to be very uncertain were it not for the fact that the Malliavin calculus has been used to solve some problems which, as yet, the analysts cannot handle. Whether the solution of these problems justifies the time and effort involved is a matter that can be debated. In my opinion, it does.

All the examples of problems solved for the first time with Malliavin's calculus involve studying the regularity of quantities for which one has no differential equation. The simplest examples of this sort were provided by Shigekawa [Sh.] who dealt with the distribution of functionals defined in terms of iterated stochastic integrals. A second example was provided by D. Michael [Mich.] who applied Malliavin's calculus to the study of conditional distributions arising in non-linear filtering theory. Finally, I have applied Malliavin's calculus in order to get regularity results on the finite dimensional marginal distributions of infinite dimension diffusions (cf. [S., Sys. Th.] and [H. & S.]). In order to explain the sort of applications which I have in mind and which I consider to be a fruitful direction in which to persue this subject, I will close these lectures with a somewhat ad hoc but nonetheless intriguing example.

Let $d = 1$ and consider the stochastic integral equation:

$$\text{(9.1)} \qquad \xi(T) = \int_0^T \alpha\left(\int_0^t \rho(t-s)\xi(s)ds\right)d\theta(t) \quad , \quad T \geq 0 \quad ,$$

where $\alpha : R^1 \to R^1$ and $\rho : [0,\infty) \to R^1$ are smooth functions satisfying

(9.2)

i) $\alpha(\cdot) \geq \varepsilon$ for some $\varepsilon > 0$ and $\|\alpha(\cdot)\|_{C_b^n(R^1)} < \infty$ for all $n \geq 1$

ii) if $\rho_n \equiv (D^n\rho)(0)$, $n \geq 0$, then $|\rho_n| < K\lambda^n$, $n \geq 0$, for some $K < \infty$ and $\lambda > 0$.

What we want to do is to obtain regularity results about $\mu_T \equiv \mathbb{W} \circ \xi(T)^{-1}$ for $T > 0$. Notice that, by itself, $\xi(\cdot)$ is definitely not Markovian in general. Thus μ_T cannot be studied by representing it as the solution to one-dimensional Fokker-Planck equation. Indeed, there appears to be very little information about μ_T that comes from the usual analytic techniques; the only general result which applies is the beautiful theorem due to Krylov [Kr.] from which one can deduce that, for each $T > 0$, $\int_0^T \mu_t dt$ admits a density which is integrable to all orders.

We next point out where the difficulty lies in studying μ_T with Malliavin's calculus. Suppose that, by one means or another, we can show that $\xi(T) \in \mathcal{D}(\mathcal{L})$ (e.g. one could try to mimic the argument given in section (7)). Then the existence of a smooth density for μ_T would follow once we showed that $1/\langle\xi(T),\xi(T)\rangle_{\mathcal{L}} \in \bigcap^{\infty} L^q(\mathfrak{W})$. But where would such an estimate come from? Indeed, applying our rules for computing $\langle\xi(T),\xi(T)\rangle_{\mathcal{L}}$, we have:

$$\langle\xi(T),\xi(T)\rangle_{\mathcal{L}} = 2\int_0^T \langle\alpha(\int_0^t \rho(t-s)\xi(s)ds,\xi(t)\rangle_{\mathcal{L}}\, d\theta(t)$$

$$+ \int_0^T (\langle\alpha(\int_0^t \rho(t-s)\xi(s)ds),\alpha(\int_0^t \rho(t-s)\xi(s)ds)\rangle_{\mathcal{L}}$$

$$+ \alpha(\int_0^t \rho(t-s)\xi(s)ds)^2)dt \quad .$$

Clearly

$$\langle\alpha(\int_0^t \rho(t-s)\xi(s)ds),\xi(t)\rangle_{\mathcal{L}}$$

$$= \alpha'(\int_0^t \rho(t-s)\xi(s)ds)\int_0^t \rho(t-s)\langle\xi(s),\xi(t)\rangle_{\mathcal{L}} ds$$

and

$$\langle\alpha(\int_0^t \rho(t-s)\xi(s)ds),\alpha(\int_0^t \rho(t-s)\xi(s)ds\rangle$$

$$= \alpha'(\int_0^t \rho(t-s)\xi(s)ds)^2 \int_0^t\int_0^t \rho(t-s)\rho(t-\sigma)\langle\xi(s),\xi(\sigma)\rangle_{\mathcal{L}} ds d\sigma \quad .$$

Thus $\langle\xi(T),\xi(T)\rangle_{\mathcal{L}}$ does not satisfy an autonomous equation and one is forced to look at a two parameter equation describing the evolution of $(s,t) \to \langle\xi(s),\xi(t)\rangle_{\mathcal{L}}$. Although this two parameter equation is tractible, in the sense that it provides one with estimates on the moments of $\langle\xi(T),\xi(T)\rangle_{\mathcal{L}}$, I see no way of using it to get estimates on reciprocal moments. The origin of an inverse in the Markovian cases treated in section

8) is the linearity of the equation for $A(T)$. Indeed, this linearity prevents $A(T)$ from getting back to zero once the inhomogenious term has moved it away from zero. By contrast, in our present situation, it is not clear how small the coefficient of $d\theta(t)$ becomes when $\langle\xi(t),\xi(t)\rangle_{\mathfrak{L}}$ is small. The best that I can say is that $|\langle\xi(s),\xi(t)\rangle_{\mathfrak{L}}| \leq \langle\xi(s),\xi(s)\rangle_{\mathfrak{L}}^{1/2}\langle\xi(t),\xi(t)\rangle_{\mathfrak{L}}^{1/2}$, and therefore this coefficient is no worse than $\langle\xi(t),\xi(t)\rangle_{\mathfrak{L}}^{1/2}$. Unfortunately, as anyone familiar with S.D.E.'s will confirm, $\langle\xi(t),\xi(t)\rangle_{\mathfrak{L}}^{1/2}$ is not small enough in general to prevent a return to zero.

In order to curcumvent the problems raised in the preceding, we use an approximation procedure.

(9.3) Lemma: Suppose that $\bar{\rho} : [0,\infty) \to R^1$ is continuous and let $\bar{\xi}(\cdot)$ be the progressively measurable solution to

$$\bar{\xi}(T) = \int_0^T \alpha\left(\int_0^{t-} \bar{\rho}(t-s)\bar{\xi}(s)ds\right)d\theta(t) \quad .$$

Then for each $T > 0$:

$$E^{\mathcal{W}}\left[\sup_{0\leq t\leq T}\left|\xi(t)-\bar{\xi}(t)\right|^2\right] \leq A(T)\|\rho-\bar{\rho}\|_{L^1((0,T))}\exp(B(T)) \quad ,$$

where $A(T) = 4\|\alpha'\|^4_{C_b(R^1)}T^2$ and $B(T) = 8\|\alpha\|^2_{C_b^1(R^1)}\|\bar{\rho}\|^2$.

Proof:

$$E^{\mathcal{W}}\left[\sup_{0\leq t\leq T}\left|\xi(t)-\bar{\xi}(t)\right|^2\right]$$

$$\leq 4E^{\mathcal{W}}\left[\int_0^T \left(\alpha\left(\int_0^t \rho(t-s)\xi(s)ds\right)-\alpha\left(\int_0^{t-}\bar{\rho}(t-s)\bar{\xi}(s)ds\right)\right)^2 dt\right]$$

$$\leq 4\|\alpha'\|^2_{C_b(R^1)} \int_0^T \tilde{E}[(\int_0^t (\rho(t-s)\xi(s)-\bar{\rho}(t-s)\bar{\xi}(s))ds)^2]dt$$

$$\leq 8\|\alpha'\|^2_{C_b(R^1)} \int_0^T \tilde{E}[(\int_0^t (\rho(t-s)-\bar{\rho}(t-s))\xi(s)ds)^2]dt$$

$$+ 8\|\alpha'\|^2_{C_b(R^1)} \int_0^T \tilde{E}[(\int_0^t \bar{\rho}(t-s)(\xi(s)-\bar{\xi}(s))ds)^2]dt$$

$$\leq 8\|\alpha'\|^2_{C_b(R^1)} \|\rho-\bar{\rho}\|_{L^1(0,T)} \int_0^T \tilde{E}[\xi(s)^2]ds$$

$$+ 8\|\alpha'\|^2_{C_b(R^1)} \|\bar{\rho}\|^2_{L^2((0,T))} \int_0^T \tilde{E}[|\xi(t)-\bar{\xi}(t)|^2]dt$$

$$\leq 4\|\alpha\|^4_{C_b(R^1)} \|\rho-\bar{\rho}\|^2_{L^1(0,T)} T^2$$

$$+ 8\|\alpha'\|^2_{C_b(R^1)} \|\bar{\rho}\|^2_{L^1(0,T)} \int_0^T \tilde{E}[|\xi(s)-\bar{\xi}(s)|^2]ds \quad .$$

Thus the estimate follows from Gromwall's inequality. □

In the following we will be using the notation:

$$\rho^{(N)}(t) = \sum_0^{N-1} \rho_n t^n/n! \quad , \quad N \geq 1 \quad \text{and} \quad t \geq 0 \quad ,$$

$$\sigma_i^{(N)}(x) = \begin{cases} \alpha(x_1) & \text{if } i = 0 \\ 0 & \text{if } 1 \leq i \leq N \end{cases} , \quad N \geq 1 \quad \text{and} \quad x = (x_0,\dots,x_N) \in R^{N+1} \quad ,$$

$$b_i^{(N)}(x) = \begin{cases} 0 & \text{if } i = 0 \\ \rho_{i-1}x_0 + x_{i+1} & \text{if } 1 \leq k < N \text{ , } N \geq 1 \\ \rho_{N-1}x_0 & \text{if } i = N \text{ , } N \geq 1 \end{cases} \quad \text{and} \quad x = (x_0,\dots,x_N) \in R^{N+1} \quad ;$$

and $X^{(N)}(\cdot)$ is the solution to:

$$X^{(N)}(T) = \int_0^T \sigma^{(N)}(X(t))d\theta(t) + \int_0^T b^{(N)}(X(t))dt \quad , \quad T \geq 0 \quad .$$

(9.4) Lemma: For each $T > 0$, $\lim_{N\uparrow\infty} E^W[\sup_{0\leq t\leq T} |X_0^{(N)}(t)-\xi(t)|^2] = 0$.

Proof: The assertion follows from Lemma (9.4) once we show that for each $N \geq 0$:

$$X_0^{(N)}(T) = \int_0^T \alpha(\int_0^t \rho^{(N)}(t-s)X_0^{(N)}(s)ds)d\theta(t) \quad , \quad T \geq 0 \quad .$$

To this end note that

$$X_0^{(N)}(T) = \int_0^T \alpha(X_1(t))dt \quad , \quad T \geq 0 \quad .$$

Thus we need only check that $X_1(t) = \int_0^t \rho(t-s)X_0(s)ds$. But for $1 \leq i \leq N$:

$$X_i^{(N)}(t) = \int_0^t \rho_{i-1} X_0^{(N)}(s)ds + \int_0^t X_{i+1}^{(N)}(s)ds \quad ,$$

where $X_{N+1}^{(N)}(\cdot) \equiv 0$. Thus

$$\frac{(t-s)^{i-1}}{(i-1)!} \frac{d}{ds} X_i^{(N)}(s) - \frac{(t-s)^{i-1}}{(i-1)!} X_{i+1}^{(N)}(s) = \frac{(t-s)^{i-1}}{(i-1)!} \rho_{i-1} X_0^{(N)}(s) \quad , \quad 1 \leq i \leq N$$

Summing, we obtain:

$$\frac{d}{ds}\Big(\sum_1^N \frac{(t-s)^{i-1}}{(i-1)!} X_i^{(N)}(s)\Big) = \rho^{(N)}(t-s)X_0^{(N)}(s) \quad .$$

Since $X^{(N)}(0) = 0$, this proves that $X_1^{(N)}(t) = \int_0^t \rho^{(N)}(t-s)X_0(s)ds$. □

For each $N \geq 1$, $X^{(N)}(\cdot)$ is obviously Markovian and fits into the scheme studied in sectin (7) . In particular, $X^{(N)}(t) \in (\mathcal{B}(\mathfrak{L}))^{N+1}$ for all $T \geq 0$. Furthermore, if $A^{(N)}(T) = \langle\langle X^{(N)}(T),X^{(N)}(T)\rangle\rangle_{\mathfrak{L}}$, $T \geq 0$, then by (8.12) :

$$A^{(N)}(T) = \int_0^T J^{(N)}(t,T)a^{(N)}(X(t))J^{(N)}(t,T)^* dt \quad ,$$

where $a^{(N)}(x) = \sigma^{(N)}(x)\sigma^{(N)}(x)^*$ and for fixed $t \geq 0$:

$$J^{(N)}(t,T) = I + \int_t^T S^{(N)}(X(u))J^{(N)}(t,u)d\theta(u) + \int_t^T B^{(N)}(X(u))J^{(N)}(t,u)du \ , \ T \geq t \ ,$$

with

$$S_{ij}^{(N)}(x) = \begin{cases} \alpha'(x_1) & \text{if } i = 0 \text{ and } j = 1 \\ 0 & \text{otherwise} \end{cases}$$

and

$$B_{ij}^{(N)}(x) = \begin{cases} \rho_{i-1} & \text{if } 1 \leq i \leq N \text{ and } j = 0 \\ 1 & \text{if } 1 \leq i \leq N-1 \text{ and } j = i+1 \\ 0 & \text{otherwise.} \end{cases}$$

In particular:

$$(9.5) \qquad A_{00}^{(N)}(T) = \int_0^T \alpha(X_1(t))^2 J_{0,0}^{(N)}(t,T)^2 dt$$

$$\geq \varepsilon^2 \int_0^T J_{0,0}^{(N)}(t,T)^2 dt \ .$$

Therefore , from the discussion following (8.15) , it is clear that for each $N \geq 1$, $T > 0$, and $p \in [1,\infty)$:

$$E^{\mathbb{W}}[(1/A_{00}^{(N)}(T))^p] < \infty \ .$$

Before proceeding any farther, we now outline our plan. First, from the preceding remarks, we can show that for each $N \geq 1$ and $T > 0$, the measure $\mu_T^{(N)} = \mathbb{W} \circ X_0^{(N)}(T)^{-1}$ admits a density $f_T^{(N)} \in \hat{C}^\infty(R^1)$. Indeed (cf. the arguement given to prove Theorem (7.20)), $\|f_T^{(N)}\|_{C_b^n(R^1)}$ can be estimates in terms of

(9.6) $$\max_{1 \le m \le n+1} \| \mathcal{H}^m_{(N),T}(1) \|_{L^2(\mathbb{W})} ,$$

where

(9.7) $$\mathcal{H}_{(N),T}(\Psi) \equiv \langle X_0^{(N)}(T), \Psi / A_{00}^{(N)}(T) \rangle_{\mathcal{L}} + 2\Psi \mathcal{L}(X_0^{(N)}(T))) / A_{00}^{(N)}(T)$$

for $\Psi \in \mathcal{S}(\mathcal{L})$. Second, by Lemma (9.3) , $\mu_T^{(N)}$ tends weakly to μ_T as N^∞ . Thus, the desired conclusion about μ_T will follow immediately once we show that the quantity in (9.6) can be estimated independent of $N \geq 1$. There are two ingredients required in order to prove this : one is to estimate $X_0^{(N)}(T)$ and its Malliavin derivates independent of $N \geq 1$; the other is to estimate $\| 1/A_{00}^{(N)}(T) \|_{L^p(\mathbb{W})}$, for $p \in [1,\infty)$, independent of $N \geq 1$.

In the next lemma, the notation and terminology is that introduced in the discussion preceding Theorem (7.17) .

(9.8) Lemma: Let $V : R^D \to R^D$ and $W : R^D \to R^D$ be two smooth vector fields which are lower triangular with respect to the same grading $\{D_\mu\}_0^M$. Assume, in addition, that the quantities C_α and γ_α entering the definition of lower triangularity for V and W satisfy the conditions $\max_{|\alpha| \le n} C_\alpha \le C_n$ and $\max_{|\alpha| \le n} \gamma_\alpha \le \gamma_n$ for $n \geq 1$. Also, assume that $\max_{1 \le i \le D} \operatorname{card}(\{1 \le j \le D : \frac{\partial V^i}{\partial x_j} \not\equiv 0 \text{ or } \frac{\partial W^i}{\partial x_j} \not\equiv 0\}) \le L$. Let $Y(\cdot)$ be the solution to

$$Y(T) = \int_0^T V(Y(t))d\theta(t) + \int_0^T W(Y(t))dt \quad , \quad T \geq 0 \ .$$

Then for each $p \in [2,\infty)$ there exist $K_p < \infty$ and $\lambda_p \geq 0$ depending only on C_1, γ_1, M, and L (but not on $\{D_\mu\}_0^M$ or D) such that

$$\max_{1\le i\le D} E^{\mathfrak{W}}\Big[\sup_{0\le t\le T}|Y_i(t)|^p\Big]^{1/p} \le K_p e^{\lambda_p T}, \quad T \ge 0 .$$

Moreover, if $\Xi(\cdot) = \begin{pmatrix} Y(\cdot) \\ \langle\langle Y(\cdot),Y(\cdot)\rangle\rangle_{\mathcal{L}} \\ \mathcal{L}Y(\cdot) \end{pmatrix}$ and if V' and W' on $R^D \times R^{D^2} \times R^D$ into itself are the lower triangular vector fields such that

$$\Xi(T) = \int_0^T V'(\Xi(t))d\theta(t) + \int_0^T W'(\Xi(t))dt, \quad T \ge 0 ,$$

then the C_α''s and γ_α''s associated with $V'(\cdot)$ and $W'(\cdot)$ satisfy $\max_{|\alpha|\le n} C_\alpha' \le C_n'$ and $\max_{|\alpha|\le n} \gamma_\alpha' \le \gamma_n'$, $n \ge 1$, where C_n' and γ_n' depend only on L, C_{n+2}, and γ_{n+2}; and there is an L' depending only on L such that

$$\max_{1\le i'\le 2D+D^2} \operatorname{card}(\{1 \le j' \le 2D + D^2 : \frac{\partial V'}{\partial X_{j'}'} \not\equiv 0 \text{ or } \frac{\partial W'}{\partial X_{j'}'} \not\equiv 0\}) \le L' .$$

Proof: The second part of the lemma is simply a matter of book-keeping and is left to the reader. To prove the first part, we work by induction on M.

Assume that $M = 1$. Then for each $1 \le i \le D$ there is a set $J_i \subseteq \{1,\dots,D\}$ such that $\operatorname{card}(J_i) \le L$ and

$$|V_i(X)| \vee |W_i(X)| \le C_1(1 + \sum_{j\in J_i} X_j^2)^{1/2} .$$

Hence, by Burkholder's inequality

$$E^{\mathfrak{W}}\Big[\sup_{0\le t\le T}|X_i(t)|^p\Big]$$

$$\le A_p C_1^p(1 + T^{p-1})\int_0^T E^{\mathfrak{W}}\Big[(1 + \sum_{j\in J_i} X_j(t)^2)^{p/2}\Big]dt$$

$$\leq A'_p C_1^p (1 + T^p)$$

$$+ A'_p C_1^p (1 + T^{p-1}) \int_0^T \sum_{j \in J_i} E^{\mathcal{W}}[|X_j(t)|^p] dt$$

$$\leq A'_p C_1^p (1 + T^p)$$

$$+ A'_p C_1^p L (1 + T^{p-1}) \int_0^T \max_{1 \leq j \leq D} E^{\mathcal{W}}[|X_j(t)|^p] dt \quad ,$$

where A and A' depend only on $p \in [2,\infty)$. The desired estimate therefore follows from Gromwall's inequality. To handle the induction step, one proceeds in precisely the same manner, only now the estimate on $X_{(M-1)}(\cdot)$ in the induction hypothesis contributes on the right hand side of the integral inequality to which Gromwall is applied. □

(9.9) Lemma: Suppose that $|\rho_n| \leq C < \infty$ for all $n \geq 0$. Then there exist for each $m \geq 1$ and $p \in [1,\infty)$ a $K_{p,m} < \infty$ and a $\lambda_{p,m} \geq 0$ such that $\sup_{N\geq 1} \| A_{0,0}^{(N)}(T)^{2m} \mathcal{L}^m_{(N),T}(1) \|_{L^p(\mathcal{W})} \leq K_{p,m} e^{\lambda_{p,m} T}$, $T \geq 0$.

Proof: By a simple induction arguement it is possible to show that $A_{0,0}^{(N)}(T)^{2m} \mathcal{L}^m_{(N),T}(1)$ is a polynomial in $X_0^{(N)}(T)$ and its Malliavin derivates. Since, under the given hypotheses, $X^{(N)}(\cdot)$ satisfies an equation of the sort discussed in (9.8) with $C_n = C$, $n \geq 1$, and $L = 2$, it follows from (9.8) and induction that all order Malliavin derivates of $X_0^{(N)}(T)$ satisfy $L^p(\mathcal{W})$ estimates of the form asserted. Combining these observations, we get our result. □

In view of (9.9) and our earlier discussion, we will be finished, at least in the case when $\sup_n |\rho_n| < \infty$, once we show that, for all $T > 0$

and $p \in [1,\infty)$, $\sup_{N\geq 1} \|1/A^{(N)}_{0,0}(T)\|_{L^p(\mathbb{W})} < \infty$. By (9.5) , this reduces to estimating the reciprocal of $\int_0^T J^{(N)}_{0,0}(t,T)^2 dt$.

(9.10) **Lemma**: Let $U : [0,\infty) \times \Theta \to R^D \quad R^D$ and $V : [0,\infty) \times \Theta \to R^D \otimes R^D$ be progressively measurable functions such that
$\max_{1\leq i\leq D} \left(\sum_{j=1}^{D} \|U_{ij}(\cdot)\| \right) \vee \left(\sum_{j=1}^{D} \|V_{ij}(\cdot)\| \right) \leq C < \infty$ and
$\max_{1\leq j\leq D} \left(\sum_{i=1}^{D} \|U_{ij}(\cdot)\| \right) \vee \left(\sum_{i=1}^{D} \|V_{ij}(\cdot)\| \right) \leq C < \infty$. Let $M(\cdot)$ denote the solution to

$$M(T) = I + \int_0^T U(t)M(t)d\theta(t) + \int_0^T V(t)M(t)dt \quad , \quad T \geq 0 \quad ,$$

and set $M(t,T) = M(T)M(t)^{-1}$. Then for each $T > 0$ and $q \in [2,\infty)$ there is a $K_q(T) < \infty$, depending on C but not on D , such that

$$\max_{1\leq i,j\leq D} \Big\| \sup_{T-\delta\leq t\leq T} \left|M_{ij}(t,T)-\delta_{ij}\right| \Big\|_{L^q(\mathbb{W})} \leq K_q(T)\delta^{1/4} \quad , \quad 0 \leq \delta \leq T \quad .$$

Proof: Without loss of generality, we will assume that $\delta = T \leq 1$. Note that for $0 \leq t \leq \delta$:

$$\left|M_{ij}(t,\delta)-\delta_{ij}\right| \leq \left|M_{ij}(t,\delta)-M_{ij}(\delta)\right| + \left|M_{ij}(\delta)-\delta_{ij}\right|$$

and

$$\left|M_{ij}(t,\delta)-M_{i,j}(\delta)\right| = \left|\sum_{\nu} M_{i\nu}(\delta)(M^{-1}_{\nu j}(t)-\delta_{ij})\right|$$

$$\leq (MM^*)_{ii}(\delta)^{1/2}\left(\left|(M^{-1}M^{-1*})_{ii}(t)-1\right|^{1/2} + 2\left|M^{-1}_{ii}(t)-1\right|^{1/2}\right) \quad .$$

Thus it suffices for us to obtain estimates of the form:

$$\max_{1\le i,j\le D} E^{\mathfrak{W}}[\sup_{0\le t\le\delta}|M_{ij}(t)-\delta_{ij}|^q] \le K_q\delta^{q/2} ,$$

$$\max_{1\le i\le D} E^{\mathfrak{W}}[(MM^*)_{ii}(\delta)^q] \le K_q ,$$

$$\max_{1\le i\le D} E^{\mathfrak{W}}[\sup_{0\le t\le\delta}|(M^{-1}M^{-1*})_{ii}(t)-1|^q] \le K_q\delta^{q/2} ,$$

and

$$\max_{1\le i\le D} E^{\mathfrak{W}}[\sup_{0\le t\le\delta}|M^{-1}_{ii}(t)-1|^q] \le K_q\delta^{q/2} ,$$

for $0 \le \delta \le 1$, where K_q depends on C but not D .

To get these estimates, note that $M(\cdot)$, $MM^*(\cdot)$, $M^{-1}M^{-1*}(\cdot)$, and $M^{-1}(\cdot)$ all satisfy equations of the form:

$$H(T) = I + \int_0^T (B^{(1)}(t)H(t) + H(t)B^{(2)}(t))d\theta(t)$$

$$+ \int_0^T [B^{(3)}(t)H(t) + H(t)B^{(4)}(t) + B^{(5)}(t)H(t)B^{(6)}(t)]dt ,$$

where the $B^{(\ell)}(\cdot)$'s are progressively measurable functions satisfying:

$$\max_{1\le\ell\le 6}\left(\max_{1\le i\le D}\sum_{j=1}^{D}\|B^{(\ell)}_{ij}(\cdot)\| \vee \max_{1\le j\le D}\sum_{i=1}^{D}\|B^{(\ell)}_{ij}(\cdot)\|\right) \le K ,$$

with $K \le C(1 + C)$. Thus we need only check that

$$\max_{1\le i,j\le D} E^{\mathfrak{W}}[\sup_{0\le t\le\varepsilon}|H_{ij}(t)-\delta_{ij}|^q] \le K_q\delta^{q/2} ,$$

where K_q depends on K but not D . Set $\hat{H}(\cdot) = H(\cdot) - I$. Then:

$$\hat{H}(T) = \int_0^T (B^{(1)}(t) + B^{(2)}(t))d\theta(t) + \int_0^T (B^{(3)}(t) + B^{(4)}(t) + B^{(5)}(t)B^{(6)}(t))dt$$

$$+ \int_0^T (B^{(1)}(t)\hat{H}(t) + \hat{H}(t)B^{(2)}(t))d\theta(t)$$

$$+ \int_0^T [B^{(3)}(t)\hat{H}(t) + \hat{H}(t)B^{(4)}(t) + B^{(5)}(t)\hat{H}(t)B^{(6)}(t)]dt \ .$$

Thus, by Burkholder's inequality for each $q \in [2,\infty)$ there exists a universal $A_q < \infty$ such that:

$$E^{\mathfrak{W}}[\sup_{0\le t\le T} |\hat{H}_{ij}(t)|^q] \le$$

$$A_q((2K)^q T^{q/2} + (K(2+K))^q T^q)$$

$$+ A_q E^{\mathfrak{W}}[(\int_0^T |\sum_{\mu=1}^D (B_{i\mu}^{(1)}(t)\hat{H}_{\mu j}(t) + \hat{H}_{i\mu}(t)B_{\mu j}^{(2)}(t))|^2 dt)^{q/2}]$$

$$+ A_q E^{\mathfrak{W}}[|\int_0^T (\sum_{\mu=1}^D [(B_{i\mu}^{(3)}(t)\hat{H}_{\mu j}(t) + \hat{H}_{i\mu}(t)B_{\mu j}^{(4)}(t))$$

$$+ \sum_{\mu,\nu=1}^D B_{i\mu}^{(5)}(t)\hat{H}_{\mu\nu}(t)B_{\nu j}^{(6)}(t)])dt|^q]$$

Since

$$E^{\mathfrak{W}}[(\int_0^T |\sum_{\mu=1}^D (B_{i\mu}^{(1)}(t)\hat{H}_{\mu j}(t) + \hat{H}_{i\mu}(t)B_{\mu j}^{(2)}(t))|^2 dt)^{q/2}]$$

$$\le 2^{q-1}T^{q/2-1}\int_0^T (E^{\mathfrak{W}}[|\sum_{\mu=1}^D B_{i\mu}^{(1)}(t)\hat{H}_{\mu j}(t)|^q] + E^{\mathfrak{W}}[|\sum_{\mu=1}^D \hat{H}_{i\mu}(t)B_{\mu j}^{(2)}(t)|^q])dt$$

$$\le (2K)^{q-1}T^{q/2-1}\int_0^T (E^{\mathfrak{W}}[\sum_{\mu=1}^D |B_{i\mu}^{(1)}(t)||\hat{H}_{\mu j}(t)|^q]$$

$$+ E^{\mathfrak{W}}[\sum_{\mu=1}^D |B_{\mu j}^{(2)}(t)||\hat{H}_{i\mu}(t)|^q])dt$$

$$\le (2K)^q T^{q/2-1}\int_0^T \max_{1\le\mu,\nu\le D} E^{\mathfrak{W}}[|\hat{H}_{\mu\nu}(t)|^q]dt \quad ,$$

and similarly:

and, therefore, that the desired properties for μ_T follows (cf. the discussion preceding Lemma (9.8)). Thus, all that we have to do is eliminate the restriction on the range of λ .

If $\lambda \geq 1$, let $\bar{\rho}(t) = \rho(t/2\lambda)$, $\bar{\alpha}(x) = \frac{1}{(2\lambda)^{1/2}} \alpha(x/2\lambda)$, and define $\bar{\xi}(\cdot)$ by (9.1) with $\bar{\alpha}(\cdot)$ and $\bar{\rho}(\cdot)$ replacing $\alpha(\cdot)$ and $\rho(\cdot)$ Then it is not hard to show that the distribution of $\bar{\xi}(\cdot)$ under $\mathbb{W}$ coincides with that of $Y(\cdot/2\lambda)$. (For instance, one can use a martingale problem characterization of these distributions to check this equality.) In particular, $\mu_T = \bar{\mu}_{2\lambda T}$, $T > 0$, where $\bar{\mu}_T = \mathbb{W} \circ \bar{\xi}(T)^{-1}$. Since the preceding paragraph shows that $\bar{\mu}_T$ has the asserted properties for each $T > 0$, so must μ_T . ∎

$$E^{\mathbb{W}}\Big[\Big|\int_0^T\Big(\sum_{\mu=1}^{D}(B^{(3)}_{i\mu}(t)\hat{H}_{\mu j}(t) + \hat{H}_{i\mu}(t)B^{(4)}_{\mu j}(t))$$

$$+ \sum_{\mu,\nu=1}^{D} B^{(5)}_{i\mu}(t)\hat{H}_{\mu\nu}(t)B^{(6)}_{\nu j}(t)\Big)dt\Big|^q\Big]$$

$$\leq 3^{q-1}T^{q-1}(2K+K^2)\int_0^T \max_{1\leq\mu,\nu\leq D} E^{\mathbb{W}}[|\hat{H}_{\mu\nu}(t)|^q]dt \quad ,$$

the desired estimate follows easily from Gromwalls' inequality. □

Assume now that the λ in (9.2) ii) is strictly less than 1 . Then Lemma (9.10) applies to $J^{(N)}(t,T)$, and so for each $q \in [1,\infty)$ and $T > 0$ there exists a $K_q(t)$, depending only on the K and λ in (9.2) but not on N , such that

$$E^{\mathbb{W}}[\sup_{T-\delta\leq t\leq T} |J^{(N)}_{00}(t,T)-1|^q] \leq K_q(t)\delta^{q/4} \quad , \quad 0 \leq \delta \leq T \ .$$

Thus, if $\zeta^{(N)}_T = \sup\{0 \leq \delta \leq T : |J^{(N)}_{00}(t,T)-1| \leq 1/2 \text{ for } t \in [t-\delta,T]\}$, then $\mathbb{W}(\zeta^{(N)}_T \leq \delta) = \mathbb{W}(\sup_{T-\delta\leq t\leq T} |J^{(N)}_{00}(t,T)-1| \geq 1/2) \leq 2^q K_q(t)\delta^{q/4}$ for all $q \in [1,\infty)$. In particular, $E^{\mathbb{W}}[1/\zeta^{(N)}_T)^q] \leq B_q(T)$ where $B_q(T) < \infty$ depends only on the K and λ in (9.2) but not on N . Combining this with (9.5), we now arrive at:

$$\text{(9.11)} \qquad \sup_{N\geq 1} E^{\mathbb{W}}[(1/A^{(N)}_{00}(T))^q] \leq (2/\varepsilon)^2 B_q(T) \ .$$

(9.12) Theorem: Let $\xi(\cdot)$ be the solution to (9.1) , with $\alpha(\cdot)$ and $\rho(\cdot)$ satisfying (9.2) , and set $\mu_T = \mathbb{W}\circ\xi(T)^{-1}$. Then for each $T > 0$, μ_T admits a density $f_T \in \hat{C}^\infty(R^q)$.

Proof: Combining Lemma (9.8) and (9.9) , we see that, so long as $\lambda \in [0,1)$, the quantity in (9.6) can be estimated independent of $N \geq 1$

References

[Adams] Adams, R.A., Sobolev Spaces, Pure Appl. Math Series, Vol. 65, Academic Press (1975).

[Bismut] Bismut, J.M., "Martingales, the Malliavin Calculus and hypoellipticity under general Hörmander's conditions," Z. Wahr, 56, pp. 469-505 (1981).

[Folland] Folland, G.B., "Subelliptic estimates and function spaces on nilpotent Lie groups," Archiv f. Mat., 13, pp. 161-207 (1975).

[Friedman] Friedman, A., Partial Differential Equations of Parabolic Type, Prentice Hall (1964).

[H.&S.] Holley, R.A. and Stroock, D., "Diffusions on an infinite dimensional torus," J. Fnal. Anal., vol. 42 #1, pp. 29-63 (1981).

[Hörmander] Hörmander, L., "Hypoelliptic second order differential equations," Acta. Math. 119, pp. 147-171 (1967).

[Ikeda & Watenabe] Ikeda, N., and Watenabe, S., Stochastic Differential Equations and Diffusion Processes, North Holland Math. Library/Kodansha (1981).

[Kohn] Kohn, J.J., "Pseudo-differential operators and hypoellipticity," Proc. Symp. Pure. Math., 23, A.M.S., pp. 61-69 (1973).

[Krylov] Krylov, N.V., "Certain bounds on the density of the distribution of stochastic integrals," Izv. Akad. Nauk., 38, pp. 228-248 (1974).

[Kunita] Kunita, H., "On the decomposition of solutions of stochastic differential equations," Proc. of 1980 L.M.S. Conf. at Durham, Springer Lec. Notes in Math., ed. by D. Williams (1981).

[Malliavin] Malliavin, P., "Stochastic calculus of variation and hypoelliptic operators," Proc. Intern. Symp. S.D.E. Kyoto, ed. by K. Ito, Kinokuniya, Tokyo (1978).

[Mckean] Mckean, H.P., Jr., Stochastic Integrals, Academic Press (1969).

[Mich.] Michel, D., "Régularité des lois conditionelles en théorie du filtrage non linéaire et calcul des variations stochastique," J. Fnal. Anal., 41 #1, pp. 8-36 (1981).

[Rothschild & Stein] Rothschild, L. and Stein, E., "Hypoelliptic differential operators and nilpotent groups," Acta. Math., vol. 137, pp. 247-320 (1976).

[Shig.] Shigekawa, I., "Derivatives of Wiener functionals and aboluste continuity of induced measures," J. Math. Kyoto Univ., 20, pp. 263-289 (1980).

[S.] Stroock, D., "Topics in stochastic differential equations," to appear in Tata Inst. Lec. Notes Series, Springer Verlag.

[S.,Sys.Th.] Stroock, D., "The Malliavin calculus and its application to second order parabolic differential equations, Parts I and II," Math. Systems Th. 14, pp. 25-65 and pp. 141-171 (1981).

[S.,J. Fnal. Anal.] Stroock, D., "The Malliavin calculus, a functional analytic approach," to appear in J. Fnal. Anal.

[S.&V.] Stroock, D. and Varadhan, S.R.S., Multidimensional Diffusion Processes, Grund. Math. #233, Springer Verlag (1979).

[S.&V.,Berk. Symp.] Stroock, D. and Varadhan, S.R.S., "On the support of diffusion processes with applications to the strong maximum principle," Proc. 6th Berkeley Symp. on Math. Stat. and Prob., vol. III, pp. 333-360 (1970).

ANALYSE INFINITESIMALE DE FONCTIONS ALEATOIRES

PAR M. WEBER

Chapitre I : CLASSES SUPERIEURES ET INTEGRABILITE DE FONCTIONS ALEATOIRES.

1.0. Introduction. Dans ce chapitre nous avons cherché à présenter quelques méthodes élémentaires intervenant dans l'étude de la continuité des fonctions aléatoires non gaussiennes, le cas gaussien ayant, pour sa part, fait l'objet d'un cours de X. Fernique dans cette même Ecole d'Eté en 1974. Nous présentons ici quelques résultats très récents puisqu'ils ont été obtenus en 1980 et 1981. L'orientation générale de ce cours est la recherche de modules de continuité locale ou uniforme. On retiendra dans ce qui suit les contributions principales de N. Konô ,M. Hahn-Klass, G. Pisier, Nobelis-Nanopoulos, Ibragimov, X. Fernique, R.M. Dudley, Jain Marcus, nous-mêmes entre autres.

Le théorème suivant dû à Kolmogorov est classique et constitue une bonne entrée en matière.

THEOREME 1.0.1. Soit $(X_t)_{0\le t\le 1}$ un processus aléatoire. On suppose qu'il existe $0 < p < \infty$ et $\delta > 0$ tels que

$$\forall\ s,t \in [0,1]\ , \qquad \mathbb{E}\,|X_s - X_t|^p \le |t-s|^{1+\delta}\ .$$

Alors, le processus $(X_t)_{0\le t\le 1}$ admet une version à trajectoires continues.

Récemment (cf [5],[6],[7]) ce résultat a été raffiné.

THEOREME 1.0.2. Soit $\varphi : [0,1] \to \mathbb{R}_+$ une fonction croissante nulle en 0 telle que $\frac{\varphi(x)}{x}$ est décroissante et telle que

$$(I_p) \qquad \int_0^1 \frac{\varphi(t)}{t^{1+\frac{1}{p}}}\, dt < \infty\ .$$

Alors la condition intégrale (I_p) est équivalente à la condition suivante

$$K_p \quad \left\{ \begin{array}{l} \underline{\text{Tout processus } (X_t)_{0\le t\le 1} \text{ vérifiant}} \\ \forall\, t,s \in [0,1] \quad (E|X_s - X_t|^p)^{\frac{1}{p}} \le \varphi(|t-s|) \\ \underline{\text{possède nécessairement une version à trajectoires continues.}} \end{array} \right.$$

Dans une note parue aux Comptes Rendus ([5], 1979), Y. Ibragimov établit des conditions suffisantes pour que les trajectoires d'un processus aléatoire vérifiant la condition (K_p) soient lipschitziennes d'ordre α. Il montre qu'elles sont dans un certain sens aussi nécessaires. Plus précisément on a le

THEOREME 1.0.3. Soit k un entier. Soient aussi p tel que $1 < p < \infty$, $\alpha > 0$ et $\varphi : [0,1] \to \mathbb{R}_+$ une fonction croissante concave.
Alors les propriétés suivantes sont équivalentes

$$(I_{p,k,\alpha}) \qquad \int_0^1 \frac{\varphi(t)}{t^{1+\frac{k}{p}+\alpha}}\, dt < \infty ,$$

$$(K_p,k,\alpha) \quad \left\{ \begin{array}{l} \underline{\text{Tout processus } (X_t),\ t \in [0,1]^k \text{ tel que}} \\ \forall\, t,s \in [0,1]^k, \quad (E|X_t - X_s|^p)^{\frac{1}{p}} \le \varphi(\|t-s\|) \\ \underline{\text{admet une version à trajectoires lipschitziennes d'ordre } \alpha}. \end{array} \right.$$

On a posé ici $\|t\| = \sup_{i\le k} |t_i|$.

Ce théorème a été partiellement généralisé par Nobelis ([12]) par une méthode qui rappelle celle utilisée par H. Garcia et N. Konô. Elle fait intervenir une approximation intégrale du processus. Dans l'énoncé suivant, Φ est une fonction de Young ; on lui associe la fonction sur $\mathbb{R}^+$

$$Q(u) = \inf\{\alpha > 0 : \iint_{\{\|s-t\| < u\}} E\{\Phi(\frac{1}{\alpha}|X(s) - X(t)|)\}\, ds\, dt \le 1\}$$

THEOREME 1.0.4. Si $Q(u)$ est fini et $\Omega : [0,1] \to \mathbb{R}_+$ est croissante sous additive nulle en zéro telle que

$$\int_0^1 Q(u)\, \Phi^{-1}(u^{-2k})\, u^{-1}(\Omega(u))^{-1}\, du < \infty$$

alors $$E\{\sup_{0<\|s-t\|<\varepsilon} \frac{|X(s) - X(t)|}{\Omega(12\|s-t\|)}\} \le 6^{2k+3} \int_0^{\varepsilon} Q(u)\, \Phi^{-1}(u^{-2k})\, u^{-1}(\Omega(u))^{-1}\, du$$

et par conséquent les trajectoires de X sont p.s. Ω-lipschitziennes.

Les théorèmes précédents ne s'étendent au cas d'un espace métrique compact que dans certains cas. Nous allons en particulier démontrer le théorème suivant dû à G. Pisier ([13]).

THEOREME 1.0.5. Soit $p > 1$. Soit T un espace compact muni d'une pseudo-métrique d continue sur $T \times T$.

On définit $N_d(T,\epsilon)$ comme le plus petit nombre de d-boules ouvertes de rayon ϵ suffisant à recouvrir T.

On suppose que

$$(I(p)) \quad \int_0^1 N_d(T,\epsilon)^{\frac{1}{p}} d\epsilon < \infty .$$

Soit alors $(X_t)\ t \in T$ un processus aléatoire tel que

$$(E|X_t - X_s|^p)^{\frac{1}{p}} \leq d(s,t) .$$

On a alors

(i) $(X_t)_{t \in T}$ a une version à trajectoires continues

(ii) $\{E(\sup_{\delta(s,t)<\epsilon} |X_s - X_t|^p\}^{\frac{1}{p}} \leq K \int_0^{K\epsilon} N_d(T,u)^{\frac{1}{p}} du , \ \forall \epsilon > 0 .$

où $K > 0$, est une constante numérique.

(iii) De plus, pour toute fonction $\Phi : \mathbb{R}_+ \longrightarrow \mathbb{R}_+$ continue croissante nulle à l'origine et telle que

$$\int_0 \frac{N_d(T,u)^{\frac{1}{p}}}{\Phi(u)} du < \infty$$

on a :

$$(E \sup_{T \times T} \left| \frac{X_s - X_t}{\Phi(d(s,t))} \right|^p)^{\frac{1}{p}} \leq K_1 \int_0^{\mathrm{Diam}(T,d)} \frac{N_d(T,u)^{\frac{1}{p}}}{\Phi(u)} du ,$$

où K_1 est une constante numérique.

Remarque 1 : On notera qu'il existe toujours une fonction Φ de ce type vérifiant (I_Φ) dès que $\int_0 N_d(T,u)^{\frac{1}{p}} du < \infty$.

On ignore totalement si la conclusion de ii) reste vraie lorsque $P > 2$.

La démonstration utilise de manière cruciale le lemme suivant :

LEMME 1.0.1. <u>Soit</u> (Ω,μ) <u>un espace mesuré</u>. <u>Soit</u> $p \in [1,2]$ <u>et soit</u> $\{Y_i\ ,\ i \in I\} \subset L^p(\Omega,\mu)$. <u>Supposons que</u> $(Y_i)\ i \in I$ <u>engendre linéairement un sous-espace</u> E <u>de dimension</u> $n < \infty$ <u>de</u> $L^p(\Omega,\mu)$. <u>On a alors</u>

$$\| \sup_{i \in I} |Y_i| \|_p \leq n^{\frac{1}{p}} \sup_{i \in I} \|Y_i\|_p \ .$$

<u>Démonstration</u> : a) si $p = 2$; alors E admet une base orthonormale $(e_1,\dots,e_n)$ d'où l'écriture

$$Y_i = \sum_{k=1}^{n} \alpha_i^k e_k \quad \text{avec} \quad \sum_{k=1}^{n} |\alpha_i^k|^2 = \|Y_i\|_2^2 \ .$$

On a alors par Cauchy-Schwartz

$$\sup_{i \in I} |Y_i| \leq \sup_{i \in I} \left(\sum_1^n |\alpha_i^k|^2 \right)^{\frac{1}{2}} \left(\sum_1^n |e_k|^2 \right)^{\frac{1}{2}}$$

d'où ,

$$\| \sup_{i \in I} |Y_i| \|_2 \leq \sup_I \|Y_i\|_2 \ \| \left(\sum_1^n |e_k|^2 \right)^{\frac{1}{2}} \|_2$$

$$\leq \sqrt{n} \ \sup_I \|Y_i\|_2 \ .$$

ce qui établit le cas $p = 2$.

b) $1 \leq p < 2$, on pose

$$M = \sup_{i \in I} |Y_i| \quad \text{et} \quad \tilde{Y}_i = |Y_i| \, M^{\frac{p}{2} - 1} \ .$$

On note que $|\tilde{Y}_i| \leq |Y_i|^{\frac{p}{2}}$, donc

$$\sup_{i \in I} \|\tilde{Y}_i\|_2 \leq \sup_{i \in I} \|Y_i\|_p^{\frac{p}{2}}$$

La première étape de la démonstration montre

$$\| \sup_{i \in I} |\tilde{Y}_i| \|_2 \leq \sqrt{n} \ \sup_{i \in I} \|\tilde{Y}_i\|_2$$

d'où

$$\|M^{\frac{p}{2}}\|_2 \leq \sqrt{n} \ \sup_{i \in I} \|Y_i\|_p^{\frac{p}{2}} \ .$$

Soit aussi

$$\|M\|_p \leq n^{\frac{1}{p}} \sup_{i \in I} \|Y_i\|_p \ .$$

□

<u>Remarque</u> : Le lemme précédent devient faux si $p > 2$. On a seulement

$$\| \operatorname{Sup} |Y_i| \|_p \leq \sqrt{n} \ \sup \|Y_i\|_p \ .$$

<u>Démonstration du théorème 1.0.5.</u> : La démonstration suivra celle donnée en [13] :

On note D le diamètre de (T,d) $\delta_n = 2^{-n} D$, $N_n = N_d(T,\delta_n)$.

On note $(A_j^n, j \leq N_n)$ une partition de T subordonnée à un recouvrement minimal de T par des boules ouvertes de rayon δ_n, de centre $\{t_j^n, j \leq N_n\}$. On a donc

$$t_j^n \in A_j^n \qquad \text{et} \qquad A_j^n \subset B(t_j^n, \delta_n).$$

Posons alors

$$(1) \qquad \begin{aligned} &- X_t^n(\omega) = \sum_{j \leq N_n} \chi_{A_j^n}(t)\, X_{t_j^n}(\omega) \\ &- \Delta_t^n = X_t^n - X_t^{n-1} \ . \end{aligned}$$

On a donc par conséquent

$$(2) \qquad \|\Delta_t^n\|_p \leq \|X_t^n - X_t\|_p + \|X_t - X_t^{n-1}\|_p \leq \delta_n + \delta_{n-1} = 3\,\delta_n \ .$$

Fixons $k \in \mathbb{N}$; puisque $X_t = X_t^k + \sum_{n>k} \Delta_k^n$, on en déduit

$$(3) \qquad \| \sup_{d(t,s)<\delta_k} |X_t - X_s| \|_p \leq \sup_{d(t;s)<\delta_k} |X_t^k - X_s^k| \|_p + \sum_{n>k} \| \sup_{t,s} |\Delta_t^n - \Delta_s^n| \|_p \ .$$

D'une part

$$\|X_t^k - X_s^k\|_p \leq \|X_t^k - X_t\|_p + \|X_t - X_s\|_p + \|X_s - X_s^k\|_p$$

$$\leq 2\,\delta_k + d(t,s) \leq 3\,\delta_k \ .$$

On déduit du lemme (1.0.1)

$$(4)\qquad \Big\| \sup_{d(s,t)<\delta_k} |X^k_t - X^k_s| \Big\|_p \le N_k^{\frac{1}{p}} \sup_{d(s,t)<\delta_k} \|X^k_s - X^k_t\|_p$$

$$\le 3\,\delta_k\, N_k^{\frac{1}{p}} .$$

On a aussi, d'autre part,

$$\forall\, n,\ \forall\, s,t \in T \qquad \|\Delta^n_t - \Delta^n_s\|_p \le 6\,\delta_n .$$

Le lemme (1.0.1) montre aussi dans ce cas

$$(5)\qquad \Big\| \sup_{t,s} |\Delta^n_t - \Delta^n_s| \Big\|_p \le 6(N_n + N_{n-1})^{\frac{1}{p}}\, \delta_n .$$

En reportant ces majorations dans (3), on aboutit à

$$(6)\qquad \Big\| \sup_{d(s,t)<\delta_k} |X_t - X_s| \Big\|_p \le 3\,\delta_k\, N_k^{\frac{1}{p}} + 6\, 2^{\frac{1}{p}} \sum_{n>k} \delta_n\, N_n^{\frac{1}{p}}$$

$$\le 12\, 2^{\frac{1}{p}} \int_0^{\delta_k} N_d(T,u)^{\frac{1}{p}}\, du ,$$

ce qui permet d'établir i) et ii).

Démontrons iii) ; soit $\delta_{k+1} < d(s,t) \le \delta_k$, on constate

$$(7)\qquad \frac{|X_t - X_s|}{\Phi(\delta(s,t))} \le \frac{|X^k_t - X^k_s|}{\Phi(\delta_{k+1})} + \sum_{n>k} \frac{|\Delta^n_t - \Delta^n_s|}{\Phi(\delta_n)} .$$

Par conséquent, si on pose

$$S_1 = \Sigma_{k\ge 0} \sup_{d(s,t)\le\delta_k} \frac{|X^k_t - X^k_s|}{\Phi(\delta_{k+1})} ,$$

$$S_2 = \sum_{n\ge 1} \sup_{s,t} \frac{|\Delta^n_t - \Delta^n_s|}{\Phi(\delta_n)} ,$$

il suffit de majorer $\|S_1\|_p + \|S_2\|_p$ car (7) implique

$$\operatorname{Sup}_{s,t\in T} \frac{|X_t - X_s|}{\Phi(s(s,t))} \le S_1 + S_2 .$$

Les majorations (3) et (5) fournissent ici

$$\|S_1\|_p \leq 3 \sum_{k\geq 0} \delta_k \, N_k^{\frac{1}{p}} \, \Phi(\delta_{k+1})^{-1}$$

et

$$\|S_2\|_p \leq 6.2.^{\frac{1}{p}} \sum_{n\geq 1} \delta_n \, N_k^{\frac{1}{p}} \, \Phi(\delta_n)^{-1} .$$

On en déduit

$$\|S_1\|_p + \|S_2\|_p \leq 12(1 + 2^{\frac{1}{p}}) \int_0^D \frac{N_d(T,u)^{\frac{1}{p}}}{\Phi(u)} du ,$$

d'où le théorème (1.0.5) .

Avant de poursuivre introduisons quelques définitions.

1.1. Définitions, notations.

Soient (T,δ) et (F,d) deux espaces pseudométriques et $X : T \to F$ une fonction aléatoire sur T d'espaces d'épreuves (Ω,a,P) , la tribu a est supposée P-complète. On supposera toujours que la fonction aléatoire réelle $d(X(s),X(t))$ est δ-séparable.

DEFINITION 1.1.1. Soit φ un élément de $\mathcal{D}$; on dit que φ appartient à la classe locale supérieure $\mathcal{U}_\ell(X,d,t_o)$, (resp. inférieure $\mathcal{L}_\ell(X,d,t_o)$) par rapport à l'écart d , lorsque

(1.1.1.) $$P\{\omega : \exists\, \delta(\omega) > 0 : \forall\, s \in T,\ \delta(t_o,s) < \delta(\omega) \Rightarrow d(X(\omega,s), X(\omega,t_o)) \leq \delta(s,t_o)\ \varphi(\delta(s,t_o))\} = 1 \text{ (resp. 0)}$$

DEFINITION 1.1.2. Soit φ un élément de $\mathcal{D}$; on dit que φ appartient à la classe uniforme supérieure de X , $\mathcal{U}_u(X,d)$, (resp. inférieure $\mathcal{L}_u(X,d)$), par rapport à l'écart d , lorsque :

(1.1.2.) $$P\{\omega : \exists\, \delta(\omega) > 0 : \forall\, s \in T , \forall\, t \in T , \delta(s,t) < \delta(\omega) \Rightarrow d(X(\omega,s), X(\omega,t)) \leq \delta(s,t)\ \varphi(\delta(s,t))\} = 1 \text{ (resp. 0)} .$$

La définition suivante précise le sens des classes locales de la fonction alétoire X lorsque $(T,\delta) = (\overline{\mathbb{R}}, |.|)$, $t_o = \{\infty\}$ et (F,d) est un espace normé : $d(s,t) = \|s - t\|$.

DEFINITION 1.1.3. Soit φ un élément de $\mathcal{D}$; on dit que φ appartient à la classe à l'infini $\mathcal{U}_\infty(X)$, (resp. $(\mathcal{L}_\infty(X))$ lorsque :

(1.1.3.) $P\{\omega : \exists\, t_\omega < \infty : \forall\, t > t_\omega : \|X(\omega,t)\| \leq \varphi(\frac{1}{t})\} = 1$ (resp. 0) .

Ces définitions ont été introduites dans l'étude du Mouvement Brownien linéaire ([9]) . Elles servent aussi soit pour l'étude de processus gaussiens un peu plus généraux dont la covariance est par exemple du type suivant

$$E(X(s) - X(t))^2 \sim |s - t|^\beta \quad , \quad |s - t| \to 0 \, ,$$

avec $0 < \beta \leq 1$, (voir [20]), soit pour le comportement asymptotique de martingales ou de processus mélangeants ([1],[16]) .

Les théorèmes suivants sont classiques.

THEOREME LOCAL 1.1.1. ([15]) . Soit $(E, \|.\|)$ un espace euclidien de dimension N . Soit W un mouvement brownien sur E , c'est-à-dire un processus gaussien centré de covariance

$$E\{W(s).\, W(t)\} = \frac{-\|s - t\| + \|s\| + \|t\|}{2} \, .$$

Soit φ un élément de $\mathcal{D}$. On a les tests suivants

(1.1.4.) $\quad (\varphi \in \mathcal{U}_\ell(W,t_o) \quad \Leftrightarrow \quad (\int_{+0} \varphi^{2N-1}(u)\; e^{-\frac{1}{2}\varphi^2(u)}\, \frac{du}{u} < \infty)$,

(1.1.5.) $\quad (\varphi \in \mathcal{L}_\ell(W,t_o) \quad \Leftrightarrow \quad (\int_{+0} \varphi^{2N-1}(u)\; e^{-\frac{1}{2}\varphi^2(u)}\, \frac{du}{u} = \infty)$.

THEOREME UNIFORME 1.1.2. ([15]) . Soit φ un élément de $\mathcal{D}$. Sous les hypothèses précédentes on a aussi

(1.1.6.) $\quad (\varphi \in \mathcal{U}_u(W)) \quad \Leftrightarrow \quad (\int_{+0} \varphi^{4N-1}(u)\; e^{-\frac{1}{2}\varphi^2(u)}\, \frac{du}{u^{N+1}} < \infty)$,

$$(1.1.7.) \quad (\varphi \in \mathcal{L}_u(W)) \Leftrightarrow (\int_{+0} \varphi^{4N-1}(u) \; e^{-\frac{1}{2}\varphi^2(u)} \frac{du}{u^{N+1}} = \infty) .$$

Récemment ces résultats ont été renforcés ([17]).

Concernant les sommes partielles de variables aléatoires indépendantes, nous rappelons aussi le

THEOREME 1.1.3. ([1]). Soit $Y_1, Y_2, \ldots\ldots$ une suite de variables aléatoires indépendantes équidistribuées centrées et possédant un moment du second ordre. Soit $(\varphi(n))_{n\geq 1}$ une suite croissante. Notons $S_n = Y_1 + \ldots + Y_n$ et $\sigma = (E\, Y_1^2)^{\frac{1}{2}}$. Alors les assertions suivantes sont équivalentes.

$$(1.1.8.) \quad P\{ \frac{S_n}{\sigma\sqrt{n}} > \varphi(n) \text{ n.i.}\theta\} = 1 \quad (\text{resp.} = 0)$$

$$(1.1.9.) \quad \sum_n \frac{\varphi(n)}{n} \exp(-\frac{1}{2}\varphi^2(n)) = \infty \quad (< \infty) .$$

Les énoncés précédents mettent en évidence des lois du 0.1. On ne sait cependant pas répondre de façon satisfaisante à la question de l'existence de lois 0-1 pour les événements qui définissent les différentes notions des classes. Cependant, lorsque tel est le cas, cette notion apparaît comme un moyen de définir la meilleure famille de modules de continuité de la fonction aléatoire X. En relation avec (1.1.2) nous introduisons la .

DEFINITION 1.1.4. Soit $g : \mathbb{R}_+ \to \mathbb{R}_+$ une application sous-additive croissante s'annulant en 0 ; on dit que la fonction aléatoire est stable relativement à g, ou g-lipschitzienne lorsque

$$(1.1.10.) \quad E\{ \sup_{\substack{T \times T \\ \delta(s,t)\neq 0}} (\frac{d(X(s), X(t))}{g(\delta(s,t))})\} < \infty .$$

Remarque : Supposons que $g(t)$ s'écrive $g(t) = t\,\varphi(t)$, où φ est un élément de $\mathcal{D}$. Dans la suite, on verra que l'on peut avoir simultanément

$$\varphi \in \mathcal{U}_u(X,d) \quad \text{et} \quad E\{\sup_{\substack{T\times T \\ \delta(s,t)\neq 0}} \frac{d(X(s),\, X(t))}{g(\delta(s,t))}\} = \infty \,.$$

1.2. Majoration en loi.

Soient (T,δ) et (F,d) deux espaces pseudométriques. Soit $f : \mathbb{R}_+ \to \mathbb{R}_+$ une application croissante, nulle en zéro et sous-additive. Soit $X : T \to F$ une fonction aléatoire telle que la fonction aléatoire réelle $f(d(X(s),X(t)))$ soit δ-séparable et intégrable.

Dans [19] nous avions obtenu une bonne condition pour la majoration des trajectoires impliquant aussi leur continuité partielle presque sûre (en fait X. Fernique a montré depuis qu'elle implique aussi la continuité des trajectoires [4]). Cette condition s'exprime à l'aide des fonctions d'entropie de l'espace (T,δ). L'hypothèse "probabiliste" requise est du type suivant : on suppose qu'il existe une variable aléatoire réelle Λ_o telle que pour tous s,t éléments de T et tout $x > 0$ (tout x suffisamment grand suffit aussi)

(1.2.1.) $\quad P\{f(d(X(s),\, X(t)) > x\delta(s,t)\} \le P\{\Lambda_o > x\}$.

On remarquera à la lecture de la démonstration que la condition

(1.2.2.) $\quad E[f(d(X(s),\, X(t))) - x\delta(s,t)]^+ \le \delta(s,t)\, E(\Lambda_o - x)^+$

suffit aussi.

De la même façon, on se convaincra sans peine que la démonstration s'applique à toute fonction aléatoire réelle sur $T \times T$ définissant un écart aléatoire au sens de [4].

Pour des raisons de contexte, on s'en tiendra à ce qui a été dit plus haut.

Nous avions aussi obtenu dans [19] une évaluation de la loi de $\sup_{T\times T} d(X(s),X(t))$.

Celle-ci jouera un rôle important dans ce chapitre.

Soit $\Gamma(u)$ sur $[0,1]$ décroissante, telle que $P\{\Lambda_o > \Gamma(u)\} = u$. Alors, $R_{\Lambda_o}(x) = R(x) = x\, E[\Lambda_o\, \chi_{\Lambda_o > \Gamma(\frac{1}{x})}]$ est croissante sur $]0,\infty)$.

THEOREME 1.2.1. Soient $X : (T,\delta) \to (F,d)$ une fonction aléatoire, Λ_o une variable aléatoire positive intégrable et f sur $\mathbb{R}_+$ une fonction croissante sous-additive. On suppose que X, Λ_o et f sont liées par (1.2.1.) et que la fonction aléatoire réelle $f(d(X(s), X(t)))$ est δ-séparable. Dans ces conditions, pour que X soit presque sûrement majorée, il suffit que l'intégrale

$$(1.2.3.)\quad (C) \int_0^{\delta(T)} R[N(u)]\,du$$

soit convergente ; $\delta(T)$ est le diamètre de (T,δ) et $N(u) = N_\delta(T,u)$ définit comme d'habitude le cardinal minimal d'une famille de δ-boules ouvertes suffisant pour recouvrir T. On a alors pour tout t_o élément de T et tout a compris entre 0 et 1 strictement

$$(1.2.4.)\quad E[f(\sup_{t\in T} d(X(s), X(t_o)))] \leq 4 \int_0^{\delta(T)} R[N(u)]\,du \quad .$$

$$(1.2.5.)\quad P\{f(\sup_{t\in T} d(X(s), X(t_o))) > 4 \int_0^{\delta(T)} R[\frac{N(u)}{a}]\,du\} \leq a \quad .$$

Démonstration : Puisque (C) est vérifiée (T,δ) est préquasicompact donc séparable. En outre la fonction aléatoire réelle $f(d(X(s),X(t)))$ est non seulement δ-séparable, mais aussi δ-continue en probabilité ; elle est donc séparée par toute suite S dénombrable dense dans (T,δ) et nous avons

$$\text{p.s.} \quad \sup_T f(d(X(s), X(t_o))) = \sup_S f(d(X(s), X(t_o))) \; .$$

Introduisons deux suites de réels positifs

$$\underline{\varepsilon} = (\varepsilon_n)_{n\geq 1} \quad \text{décroissant vers zéro avec} \quad \varepsilon_o = \delta(T)$$

$$\underline{x} = (x_n)_{n\geq 0} \quad .$$

Nous notons pour tout entier $n \geq 0$, S_n une suite minimale de centres de δ-boules ouverts de rayon ε_n recouvrant T, et

$$(1)\qquad S^n = \bigcup_{k=0}^{n} S_k \; , \quad S^\infty = S \; , \quad S_o = \{t_o\} \text{ un point de } T \; .$$

Définissons pour tout élément s de S

$$\eta(s) = \inf\{k \geq 0 : s \in S_k\}$$

$$m(s) = x_o + \Sigma_{k=1}^{n(s)} \epsilon_{k-1} x_k \ .$$

Soit enfin $S \ni s \to \underline{s} \in S$ une application définie par $\underline{s} \in S_{\eta(s)-1}$ et $\delta(s,\underline{s}) < \epsilon_{\eta(s)-1}$.

LEMME 1.2.1. Pour tout entier $n \geq 1$,

$$(1.2.6.) \qquad z_n = \sup_{s \in S^n} f(d(X(s),X(t_o))) \leq \sum_{k=1}^{n} \epsilon_{k-1} \sup_{s \in S_k} \frac{f(d(X(s),X(\underline{s})))}{\delta(s,\underline{s})}$$

$$(1.2.7.) \qquad Z_n^+ = \sup_{s \in S^n} (f(d(X(s),X(t_o))) - m(s))^+$$

$$\leq \sum_{k=1}^{n} \epsilon_{k-1} \sup_{s \in S_k} \left(\frac{f(d(X(s),X(\underline{s})))}{\delta(s,\underline{s})} - x_k\right)^+ \chi_{\{f(d(X(s),X(t_o))) > x_o\}}$$

Démonstration du lemme 1.2.1. Pour tout entier $n \geq 1$,

$$(z_n - z_{n-1})^+ \leq [\operatorname{Sup}_{s \in S_n} f(d(X(s),X(t_o))) - \operatorname{Sup}_{s \in S_{n-1}} f(d(X(s),X(t_o)))]^+$$

$$(3) \qquad \leq \operatorname{Sup}_{s \in S_n} (f(d(X(s),X(t_o))) - f(d(X(\underline{s}),X(t_o))))^+$$

$$\leq \operatorname{Sup}_{s \in S_n} f(d(X(s),X(\underline{s}))) \leq \epsilon_{n-1} \operatorname{Sup}_{s \in S_{n-1}} \frac{f(d(X(s),X(\underline{s}))}{\delta(s,\underline{s})}$$

d'où (1.2.6.) par sommation.

De plus $(Z_n^+ - Z_{n-1}^+) \leq (Z_n^+ - Z_{n-1}^+)^+ \leq (Z_n - Z_{n-1})^+ \chi_{Z_n > 0}$

$$\leq \operatorname{Sup}_{s \in S_n} (f\circ d(X(s),X(t_o)) - f\circ d(X(\underline{s}),X(t_o)) - m(s) + m(\underline{s}))\chi_{\{f\circ d(X(s),X(t_o)) > m(s)\}} ;$$

ainsi,

$$(4)\qquad Z_n^+ - Z_{n-1}^+ \le \varepsilon_{n-1} \sup_{s\in S_n} \left(\frac{f\circ d(X(s),X(\underline{s}))}{\delta(s,\underline{s})} - x_n\right)^+ \chi_{\{f\circ d X(s),X(t_o)) > x_o\}} ,$$

d'où (1.2.7.) par sommation.

Retour à la démonstration du théorème 1.2.1. On déduit du lemme précédent

$$(5)\qquad (z_n - \sum_{k=1}^{n} \varepsilon_{k-1} x_k)^+ \le \sum_{k=1}^{n} \varepsilon_{k-1} \sup_{s\in S_k} \left(\frac{f\circ d(X(s),X(\underline{s}))}{\delta(s,\underline{s})} - x_k\right)^+ .$$

Posons alors $\varepsilon_k = 2^{-k}\varepsilon_o$, $x_k = \Gamma(\frac{1}{N(\varepsilon_k)})$ et

$$(6)\qquad Y_o = \sum_{k=1}^{\infty} \varepsilon_{k-1} \left(\sup_{s\in S_k} \frac{f\circ d(X(s),X(\underline{s}))}{\delta(s,\underline{s})} - x_k\right)^+ .$$

La condition (C) montre que Y_o est intégrable, la série de terme général $\varepsilon_{k-1} x_k$ converge aussi. Le théorème de convergence dominée ainsi que la séparabilité de $d(X(s),X(t))$ établissent (1.2.4.) par l'intermédiaire de la majoration (5). Si l'on détaille ce passage, on constate en posant $x_k = \Gamma(\frac{a}{N(\varepsilon_k)})$ où $0 < a < 1$,

$$(7)\qquad \begin{aligned} &E\{\sup_{s\in T} f\circ d(X(s),X(t_o)) - \sum_{k=1}^{\infty} \varepsilon_{k-1} \Gamma(\frac{a}{N(\varepsilon_k)})\}^+ \\ &\le a \sum_{k=1}^{\infty} \varepsilon_{k-1} [R(\frac{N(\varepsilon_k)}{a}) - \Gamma(\frac{a}{N(\varepsilon_k)})] . \end{aligned}$$

Or pour toute variable aléatoire réelle U et tous réels u et $\varepsilon > 0$, on sait que

$$E(U-u)^+ \ge \varepsilon\, P\{U > u+\varepsilon\} .$$

Nous en déduisons à l'aide de (7) en posant

$$u = \sum_{k=1}^{\infty} \varepsilon_{k-1} \Gamma(\frac{a}{N(\varepsilon_k)})$$

$$\varepsilon = \sum_{k=1}^{\infty} \varepsilon_{k-1} [R(\frac{N(\varepsilon_k)}{a})) - \Gamma(\frac{a}{N(\varepsilon_k)})] .$$

$$(8)\qquad P\{\sup_{s\in T} f\circ d(X(s),X(t_o)) > \sum_{k=1}^{\infty} \varepsilon_{k-1} R(\frac{N(\varepsilon_k)}{a})\} \le a .$$

ce qui permet d'établir (1.2.5.). □

Remarques : 1- Un cas particulier intéressant est $R(x) = O(\Gamma(\frac{1}{x}))$ $x \to \infty$, ce qui se produit dès que $t^r P\{\Lambda_o > t\}$ est décroissant pour un $r > 1$. Dans cette situation, on peut montrer

$$(1.2.8.)\quad P\{\sup_{s\in T} f\circ d(X(s),X(t_o)) > 5\int_0^{\delta(T)} \Gamma(\frac{a}{N(T,u)})\,du\} \le Ka$$

où $K > 0$ dépend de R.

2- Posons pour tout $0 < a < 1$ et tout $\lambda > 0$

$$\epsilon(a,\lambda) = \sup\{\epsilon > 0 : \sup_{t_o\in T}\int_0^{\epsilon} R(\frac{N(B(t_o,\epsilon)u)}{a})\,du \le \frac{\lambda}{4}\}.$$

On déduit facilement de (1.2.5.) le corollaire suivant

COROLLAIRE 1.2.1. <u>Sous les hypothèses du théorème 1.2.1, si la condition (C) est satisfaite alors</u>

$$(1.2.9.)\quad P\{\sup_{B(t_o,\epsilon(a,\lambda))} f\circ d(X(s),X(t_o)) > \lambda\} \le a$$

<u>si de plus l'écart δ est induit par une semi-norme, alors</u>

$$(1.2.10.)\quad \forall Y > 0,\quad P\{\sup_{t\in T} f(\|X(t)\|) > \lambda+Y\} \le N(T,\epsilon(a,\lambda))\,[a+\sup_{t_o\in T} P\{\|X(t)\|>Y\}].$$

La dernière majoration est intéressante lorsque $P\{f(\|X(t)\|) > x\} \le P\{\Lambda_o > x\}$. Dans ce cas, on a en posant $a = P\{\Lambda_o > Y\}$,

$$(1.2.11.)\quad \frac{P\{\sup f(\|X(t)\|) > \lambda + Y\}}{P\{\Lambda_o > Y\}\,N(T,\epsilon(P\{\Lambda_o>Y\},\lambda))} \le 2.$$

3- Soit U une partie de $T\times T$, notons Δ l'écart produit $\Delta((s,t),(s',t')) = \mathrm{Max}(\delta(s,s'),\delta(t,t'))$. Soit $U(\epsilon)$ une suite minimale dans $T\times T$ de centres de Δ-boules ou de rayon ϵ recouvrant U. Le corollaire (1.2.1.) montre alors

$$(1.2.11.)\quad P\{\sup_{(s,t)\in U} f\circ d(X(s),X(t)) > 2\lambda + Y\}$$

$$\leq N_\Delta(U,\epsilon(a,\lambda)) \sup_{(s,t)\in U(\epsilon(a,\lambda))} \{P\{f\circ d(X(s),X(t)) > Y\} + 2a\} .$$

Cette majoration interviendra dans l'étude des classes uniformes.

4- La méthode employée reste bien entendu valable lorsque l'on change de densité. Soit $A \in \mathcal{C}$, si $\mathcal{C}$ est la tribu de l'espace d'épreuves, supposons que $P(A) > 0$. En travaillant à partir de la probabilité $Q_A(.) = P(./A)$ on majoré sous les hypothèses adaptées $E\{\sup_T f\circ d(X(s),X(t))\ \chi_A\}$ ou encore

$$P\{\sup_T f\circ d(X(s),X(t)) > x,\ A\} .$$

5- L'intérêt de f se trouve lorsque $d(X(s),X(t)) \in L^p(\Omega)$ avec $p \leq 1$. C'est un des points développés par X. Fernique dans le cours suivant. Dans tous les autres cas, on a avantage à poser $f(x) = x$.

1.3. <u>Intégrabilité de fonctions aléatoires.</u> Le théorème 1.2.1. montre que lorsque la condition (C) est satisfaite, la fonction aléatoire X possède des propriétés d'intégrabilité interessantes. On montre en effet

THEOREME 1.3.1. <u>Soit</u> $X : (T,\delta) \to (F,d)$ <u>une fonction aléatoire. Soit</u> $f : \mathbb{R}_+$ $f : \mathbb{R}_+ \to \mathbb{R}_+$ <u>croissante sous-additive. Soit</u> $\Lambda_o \geq 0$ <u>une variable aléatoire intégrable. On suppose que</u> X,Λ_o <u>et</u> f <u>sont liés par</u> (1.2.1.) <u>et que la condition</u> (C) <u>est satisfaite. On pose pour tout</u> $0 < a < 1$

$$\theta_T(a) = \theta(a) = 4 \int_0^{\delta(T)} R\left(\frac{N(T,u)}{a}\right) du ,$$

<u>et soit</u> $G : \mathbb{R}_+ \to \mathbb{R}_+$ <u>croissante. Dans ces conditions on a l'implication suivante</u>

$$(1.3.1.)\quad \left(\int_1^\infty \theta_T^{-1}(u)\,d\,G(u) < \infty\right) \Rightarrow \left(E\,G\left(\sup_{T\times T} f\circ d(X(s),X(t))\right) < \infty\right) .$$

Démonstration : C'est une conséquence facile de la majoration (1.2.5.)

On en déduit les corollaires suivants qui sont des cas particuliers.

COROLLAIRE 1.3.1. (Cas puissance). Soient $p > 1$ et $X : (T,\delta) \to (F,d)$ une fonction aléatoire telle que la fonction aléatoire réelle $d(X(s),X(t))$ soit δ-séparable et vérifie l'une des deux conditions suivantes.

$$(1.3.2.) \quad \sup_{T \times T} E\left(\frac{d(X(s),X(t))}{\delta(s,t)}\right)^p < \infty$$

$$(1.3.3.) \quad \limsup_{x \to \infty} \left(x^p \sup_{T\times T} P\{d(X(s),X(t)) > x\,\delta(s,t)\}\right) < \infty \;.$$

Si de plus l'intégrale $\int_0^{\delta(T)} N(T,u)^{\frac{1}{p}} du$ est convergente, pour toute application $G : \mathbb{R}_+ \to \mathbb{R}_+$ croissante on a

$$(1.3.4.) \quad \left(\int_1^\infty y^{-p}\, dG(y) < \infty\right) \Rightarrow E\left[G\left(\sup_{T\times T} d(X(s),X(t))\right)\right] < \infty \;.$$

En particulier,

$$(1.3.5.) \quad \forall\, \varepsilon > 0,\; E\left\{\frac{\sup_{T\times T} d(X(s),X(t))^p}{\left[\log \sup_{T\times T} d(X(s),X(t))\right]^{1+\varepsilon}}\right\} < \infty \;.$$

Dans le corollaire suivant on a posé

$$\forall\, x > 0,\; \forall\, R > 0,\; \forall\, \tau > 0, \qquad \Phi_{R,\tau}(x) = \exp(-Rx^\tau) \;.$$

COROLLAIRE 1.3.2. (Cas exponentiel). Soit $X : (T,\delta) \to (F,d)$ une fonction aléatoire telle que la fonction aléatoire réelle $d(X(s),X(t))$ soit δ-séparable et vérifie l'une des deux conditions suivantes :

$$(1.3.6.) \quad \sup_{T\times T} E\left\{\Phi_{R,\tau}\left(\frac{d(X(s),X(t))}{\delta(s,t)}\right)\right\} < \infty$$

(1.3.7.) $\limsup_{x \to \infty} \left(\sup_{T \times T} \frac{P\{d(X(s),X(t)) > x\delta(s,t)\}}{\Phi_{R,\tau}(x)} \right) < \infty$.

Si de plus l'intégrale $\int_0^{\delta(T)} (\log N(T,u))^{\frac{1}{\tau}}$ est convergente, alors pour toute application $G : \mathbb{R}_+ \to \mathbb{R}_+$ croissante, on a l'implication

(1.3.8.) $\left(\int_0^\infty e^{Rk_o y^\tau} dG(y) < \infty\right) \Rightarrow \left(E[G[\sup_{T \times T} d(X(s),X(t))]] < \infty\right)$

où $k_o = R^\tau \left(9 \int_0^{\delta(T)} (\log N(T,u))^{\frac{1}{\tau}} du\right)^{-\tau}$.

En particulier,

(1.3.9.) $\forall \alpha < R k_o, \quad E\{\exp(\alpha(\sup_{T \times T} d(X(s),X(t)))^\tau)\} < \infty$.

COROLLAIRE 1.3.3. (Cas non intégrable) . Soient $0 < q < p < 1$ et $X : (T,\delta) \to (F,d)$ une fonction aléatoire telle que la fonction aléatoire réelle $d(X(s),X(t))$ soit δ-séparable et vérifie l'une des deux conditions suivantes :

(1.3.10) $\sup_{T \times T} E\{[\frac{d(X(s),X(t))}{\delta(s,t)}]^p\} < \infty$

(1.3.11) $\limsup_{x \to \infty} (x^p \sup_{T \times T} P\{d(X(s),X(t)) > x\,\delta(s,t)\}) < \infty$.

Si de plus l'intégrale $\int_0^{\delta(T)} N(T,v)^{\frac{q}{p}} v^{q-1} dv$ est convergente, alors pour toute application $G : \mathbb{R}_+ \to \mathbb{R}_+$ croissante, on a

(1.3.12) $\left(\int_1^\infty y^{-p/q} dG(y)\right) \Rightarrow \left(E[G(\sup_{T \times T} d(X(s),X(t)))] < \infty\right)$.

En particulier,

(1.3.13) $\forall \varepsilon > 0, \quad E\left\{ \frac{\sup_{T \times T} d(X(s),X(t))^p}{(\log \sup_{T \times T} d(X(s),X(t)))^{1+\varepsilon}} \right\} < \infty$.

1.4. <u>Continuité locale.</u> La majoration (1.2.5) du théorème (1.2.1) implique facilement la

PROPOSITION 1.4.1. 1) <u>Soit</u> $X : (T,\delta) \to (F,d)$ <u>une fonction aléatoire telle que</u> $d(X(s),X(t))$ <u>soit séparable. On suppose que les conditions</u> (1.1.2) <u>et</u> (C) <u>sont réalisées et on note</u> $\Phi(x) = P\{\Lambda_o > x\}$. <u>Alors on a aussi</u>

(1.4.1.) $$\forall\, \varepsilon > 0,\ \forall\, 0 < a < 1, \quad P\left\{ \frac{\sup\limits_{s\in B(t_o,\varepsilon)} d(X(s),X(t_o))}{4\int_0^{\frac{\varepsilon}{2}} R\left(\frac{N(B(t_o,\varepsilon,u))}{a}\right)du} > 1 \right\} \leq a .$$

2) <u>Si</u> $\Phi(x) = O(x^{-p})$ <u>avec</u> $p > 1$, (1.4.1) <u>est précisée par</u>

(1.4.2.) $\forall\, \varepsilon > 0$, $\forall\, x$ <u>assez grand,</u>

$$P\left\{ \frac{\sup\limits_{B(t_o,\varepsilon)} d(X(s),X(t_o))}{\int_0^{\varepsilon} N(B(t_o,\varepsilon),u)^{\frac{1}{p}}\, du} > x \right\} \leq \frac{5^p}{2}\, x^{-p} .$$

3) <u>Lorsque</u> $\Phi(x) = O(\exp - Rx^{\tau})$ <u>avec</u> $R > 0$, $\tau > 0$, <u>on a, de même</u>

(1.4.3.) $\forall\, \varepsilon > 0$, $\forall\, a$ <u>suffisamment petit</u>

$$P\{ \sup_{s\in B(t_o,\varepsilon)} d(X(s),X(t_o)) > 5 \int_0^{\varepsilon} (R \log \frac{N(B(t_o,\varepsilon),u)}{a})^{\frac{1}{\tau}} du\} \leq 4a .$$

4) <u>En outre si</u> $\Phi(x) = O(x^{-\alpha})$ <u>avec</u> $0 < \alpha < 1$, <u>supposons qu'il existe</u> $0 < \beta < \alpha$ <u>tel que l'intégrale</u> $\int_0^{\delta(T)} N(T,u)^{\frac{\beta}{\alpha}} u^{\beta-1}\, du$ <u>converge ; dans ce cas on a aussi</u>

(1.4.4.) $\forall\ \varepsilon > 0$, $\forall\ x$ <u>assez grand</u>,

$$P\left\{ \frac{\sup\limits_{s\in B(t_o,\varepsilon)} d(X(s),X(t_o))}{\left(\int_0^{\varepsilon} N(B(t_o,\varepsilon),u)^{\beta/\alpha}\, u^{\beta-1} du\right)^{\frac{1}{\beta}}} > x \right\} \leq (5\beta)^{\frac{\alpha}{\beta}}\ x^{-\alpha} \quad .$$

5) <u>Enfin lorsque</u> $\Phi(x) = O(x(\log x)^{-\nu})$ <u>avec</u> $\nu > 1$, <u>on a de même</u>

(1.4.5.) $\forall\ \varepsilon > 0$, $\forall\ 0 < a < 1$

$$P\{ \sup_{s\in B(t_o,\varepsilon)} d(X(s),X(t_o)) > \frac{8}{\nu-1} \int_0^{\varepsilon} \frac{N(B(t_o,\varepsilon),u)}{a} \left[\log \frac{N(B(t_o,\varepsilon),u)}{a} \right]^{1-\nu} du \} \leq 4a$$

<u>si l'intégrale</u> $\int_0^1 N(T,u)\ (\log N(T,u))^{-\nu}\ du$ <u>est convergente</u>.

En liaison avec la proposition précédente, nous démontrons le

THEOREME 1.4.1. <u>Soit</u> $X : (T,\delta) \to (F,d)$ <u>une fonction aléatoire telle que</u> $d(X(s),X(t))$ <u>soit séparable. Soit</u> Λ_o <u>une v.a. intégrable ; soit</u> $f : \mathbb{R}_+ \to \mathbb{R}_+$ <u>sous-additive croissante nulle en zéro. On suppose que</u> X,Λ_o <u>et</u> f <u>sont liés par la relation (1.1.2.) et que la condition (C) soit satisfaite. On pose</u>

$$\forall\ \varepsilon > 0,\ \forall\ 0 < a < 1 \quad \theta(a,\varepsilon) = 4 \int_0^{\frac{\varepsilon}{2}} R\left(\frac{N(B(t_o,\varepsilon),u)}{a}\right) du \quad .$$

<u>Soit</u> $a(t) \in L^1(]0,1], \frac{dt}{t})$ <u>croissante, à valeurs positives. On lui associe l'application croissante</u>

$$\forall\, 0 < x < 1, \quad \varepsilon_a(x) = \theta^{-1}(a(x),.)\,(x) \quad .$$

<u>Dans ces conditions</u>

$$P\{\omega : \exists\, \delta(\omega) > 0 : \delta(s,t_o) < \delta(\omega) \Rightarrow d(X(s,\omega),X(t_o,\omega)) \le 2\,\varepsilon_a^{-1}(\delta(s,t_o))\} = 1 \quad .$$

<u>Démonstration</u> : Pour tout entier n , on déduit de (1.2.5.)

$$(1) \qquad P\{\sup_{s\in B(t_o,\varepsilon)} f\circ d(X(s),X(t_o)) > \theta(\varepsilon,a(2^{-n}))\} \le a(2^{-n})$$

et donc en posant $\varepsilon = \varepsilon_a(2^{-n})$ dans (1), on déduit la convergence de la série de terme général

$$(2) \qquad P\{\sup_{s\in B(t_o,\varepsilon_a(2^{-n}))} f\circ d(X(s),X(t_o)) > 2^{-n}\} \quad .$$

Le lemme de Borel-Cantelli montre donc

$$\text{p.s.}\ \exists\, N(\omega) < \infty : \forall\, n > N(\omega)$$

$$(3) \qquad \sup_{\delta(s,t_o) < \varepsilon_a(2^{-n})} f\circ d(X_\omega(s),X_\omega(t_o)) \le 2^{-n} \quad .$$

Soit $\delta(s,t_o) < \varepsilon_a(2^{-N(\omega)})$ et n l'entier tel que

$$\varepsilon_a(2^{-n-1}) \leq \delta(s,t_o) < \varepsilon_a(2^{-n}) .$$

On déduit de (3)

$$\text{fod } (X(\omega,s),X(\omega,t_o)) \leq 2\ \varepsilon_a^{-1}\ (\delta(s,t_o))$$

d'où le théorème

1.5. Continuité uniforme.

1.5.1. Les données restent celles du paragraphe 1.2. : on considère une fonction aléatoire $X : (T,d) \to (F,d)$ telle que $d(X(s),X(t))$ soit δ-séparable, on lui associe f et Λ_o qu'on suppose liés à X par (1.1.2.) et la condition (C) .

On a vu que (C) assure la continuité en chaque point de X p.s., (en fait la continuité des trajectoires) . Cependant on ne peut actuellement répondre de manière définitive à la question de l'existence de modules de continuité uniforme. On se heurte à une difficulté de fond qui ne peut être levée que lorsque la structure géométrique de (T,δ) est particulière. Cependant la majoration (1.2.5.) permet d'écrire des conditions suffisantes d'existence de modules de continuité uniforme sous forme intégrale qui rappelle les tests classiques sur les classes du Mouvement Brownien.

Nous posons à cet effet

$\underline{\varepsilon} = (\varepsilon_n)$ une suite décroissant vers zéro.

- $\forall\, n \geq 1, \quad C_n = \{(s,t) \in T \times T : \varepsilon_n \leq \delta(s,t) < \varepsilon_{n-1}\}$.

- $\forall (s,t),(s',t') \in T \times T \qquad \Delta((s,t),(s',t')) = \delta(s,s') \vee \delta(t,')$.

- $\forall\, \varphi \in \mathcal{D},\ \forall \lambda > 0,\ \forall\, n \in \mathbb{N}^*$,

$$\eta(\lambda,\varphi,\varepsilon_n) = \eta_n = \operatorname{Sup}\ \{\eta > 0 : \sup_{t_o \in T} 4\int_0^{\eta} R\Big(\frac{N(B(t_o,\eta),u)}{P\{\Lambda_o > \varphi(\varepsilon_n)\}}\Big) \leq \lambda\,\varepsilon_{n-1}\} .$$

THEOREME 1.5.1. Sous les hypothèses ci-dessus, si la série

(1.5.1.) $\sum_n P\{\Lambda_o > \varphi(\varepsilon_n)\} N_\Delta(C_n,\eta_n)$

est convergente, alors

(1.5.2.) $P\{\omega : \exists\, \delta(\omega) > 0 : \forall\, (s,t) \in T \times T,$

$\delta(s,t) < \delta(\omega) \Rightarrow f \circ d(X(\psi,s),X(\omega,t)) \le \delta(s,t)[2\lambda + \varphi(\delta(s,t))]\} = 1$.

Démonstration : Notons $S_n = \{(s_n,t_n)\ 1 \le n \le N_\Delta(C_n,\eta_n)\}$ une famille minimale dans $T \times T$ de centre de Δ-boules ouvertes de rayon η_n recouvrant C_n .

Soit (s,t) dans C_n et n' un entier tel que

$$\Delta((s_{n'},t_{n'}),(s,t)) < \eta_n .$$

On a :

$$(1)\quad f\circ d(X(s),X(t)) \le f\circ d(X(s),X(s_{n'})) + f\circ d(X(t),X(t_{n'})) + f\circ d(X(s_{n'}),X(t_{n'}))$$

donc

$$(2)\quad P\{\sup_{C_n} f\circ d(X(s),X(t)) > t + 2h\}$$

$$\le N_\Delta(C_n,\eta_n) \sup_{(s_{n'},t_{n'})\in S_n} \{2\,P\{\sup_{B_\delta(s_{n'},\eta_n)} f\circ d(X(s),X(s_{n'})) > h\} + P\{f\circ d(X(s_{n'}'),X(t_{n'}')) > t\}\} .$$

On pose $t = \varepsilon_{n-1}\, \varphi(\varepsilon_n)$ $\quad h = \lambda\, \varepsilon_{n-1}$.

On a donc

$$(4)\quad P\{\sup_{C_n} f\circ d(X(s),X(t)) > \varepsilon_{n-1}(2\lambda + \varphi(\varepsilon_n))\} \le 3N_\Delta(C_n,\eta_n)\, P\{\Lambda_o > \varphi(\varepsilon_n)\} .$$

Puisque l'on a supposé que la série de terme général $N_\Delta(C_n,\eta_n)\ P\{\Lambda_o > \varphi(\varepsilon_n)\}$ était convergente, on conclut de la manière habituelle en appliquant le lemme de Borel-Cantelli.

Dans certains cas $N_\Delta(C_n,\eta_n)$ se simplifie.

LEMME 1.3.1. Soit $(E,\|\|)$ un espace normé de dimension finie ; soit T une partie compacte de E ; soit $\varepsilon > \varepsilon' \geq 2\eta > 0$. Notons pour tout $(s,t),(s',t')$ dans $E \times E$

$$\Delta((s,t),(s',t')) = \sup\{\|s-s'\|,\|t-t'\|\}$$

et soit $C = C(T,\varepsilon',\varepsilon) = \{(s,t) \in T \times T : \varepsilon' < \|s-t\| < \varepsilon\}$.

Alors on a : $N_\Delta(C,2n) \leq N(B(0,\varepsilon),\eta).N(T,\eta)$.

Démonstration : Soient $N = N(B(0,\varepsilon),\eta)$, $M = N(T,\eta)$, $u_1,\dots,u_N$ une suite minimale dans $B(0,\varepsilon)$ au sens de N, $t_1,\dots,t_M$ une suite dans E minimale au sens de M. Alors la famille de Δ-boules $\{B_\Delta((u_i+t_j,t_j),2\eta)\ 1\leq i\leq N,\ 1\leq j\leq M\}$ forme un recouvrement de C qui répond au problème.

1.5.2. Cas des f.a. de type puissance.

THEOREME 1.5.2. Soient $X : T \to F$ une fonction aléatoire, $f : \mathbb{R}_+ \to \mathbb{R}_+$ croissante sous additive, s'annulant en 0. On suppose que $d(X(s),X(t))$ est séparable et que pour un nombre $p \geq 1$ l'une des deux conditions suivantes est réalisée

(1.5.3.) $$\limsup_{x\to\infty}\ (x^p\ P\{f\circ d(X(s),X(t)) > x\,\delta(s,t)\}) < \infty .$$

(1.5.4.) $$E\{f^p(d(X(s),X(t)))\}^{\frac{1}{p}} \leq \delta(s,t) .$$

Posons pour tout $\varepsilon > 0$, $D(\varepsilon) = \sup_{t_o\in T}\int_0^\varepsilon N_\delta(B(t_o,\varepsilon),u)^{\frac{1}{p}}\,du$ si

(1.5.5.) $$\lim_{\varepsilon\to 0} D(\varepsilon) = 0 ,$$

<u>pour toute suite</u> (ϵ_n) <u>décroissant vers zéro, tout élément</u> φ <u>de</u> $\mathcal{D}$ <u>, on a</u>

$$(1.5.6.) \quad \sum_n \varphi(\epsilon_n)^{-p} N_\Delta\left(C_n, D^{-1}\left(\frac{\epsilon_n}{\varphi(\epsilon_n)}\right)\right) < \infty$$

$$\Rightarrow \varphi \in \mathcal{U}_u(f \circ X) .$$

Lorsque $T = [0,1]^N$, $\delta(s,t) = \delta(\|s-t\|)$, avec δ croissante, $\delta(0) = 0$, $\|.\|$ la norme euclidienne. Le théorème précédent se traduit par le

COROLLAIRE 1.5.1. a) <u>On a l'implication</u>

$$(1.5.7.) \quad \int_{+0} \varphi(u)^{-p} \left(\frac{\sqrt{\delta^{-1}(u)}}{\delta^{-1} \circ D^{-1}\left(\frac{u}{\varphi(u)}\right)} \right)^N \frac{du}{u} < \infty$$

$$\Rightarrow \varphi \in \mathcal{U}_u(f \circ X) .$$

b) <u>De même,</u>

$$(1.5.8.) \quad \int_{+0} \varphi(u)^{-p} \left(\frac{\sqrt{\delta^{-1}(u)}}{\delta^{-1} \circ D^{-1}(u)} \right)^N \frac{du}{u} < \infty$$

$$\Rightarrow \text{p.s.} \quad \limsup \frac{f \circ d(X(s),X(t))}{\delta(\|s-t\|\varphi \circ \delta(\|s-t\|)} \leq 1 .$$

c) Si $\delta(u) = u^\alpha$ $0 < \alpha \leq 1$, $N < p\,\alpha$, <u>alors pour tout</u> $\epsilon > 0$

$$(1.5.9.) \quad \text{p.s.} \quad \limsup_{\|s-t\| \to 0} \frac{f \circ d(X(s),X(t))}{\|s-t\|^{1-\frac{N}{\alpha p}} \left|\log \|s-t\|\right|^{\frac{1+\epsilon}{p}}} = 0 .$$

n.b. Ces résultats sont des applications faciles du théorème 1.5.2. Ils sont à comparer avec ceux de G. Pisier (th. 1.0.5), Y. Ibragimov (th. 1.0.3) et P. Nobelis (th. 1.0.4) .

On déduit notamment des énoncés (1.0.3) et (1.5.1 c) qu'il existe une fonction aléatoire réelle définie sur $[0,1]^N$ de type puissance avec $p > 1$ et $\delta(\|s-t\|) = \|s-t\|^{\alpha}$ telle que si $N < p\,\alpha$, on ait simultanément

(1.5.10) $$P\{\sup_{s \neq t} \frac{|X(s) - X(t)|}{\|s-t\|^{\alpha - \frac{N}{p}}} = \infty\} = 1$$

(1.5.11) $\forall\, \varepsilon > 0$,

$$P\{\limsup_{\|s-t\| \to 0} \frac{|X(s) - X(t)|}{\|s-t\|^{1-\frac{N}{\alpha p}} \log\|s-t\|^{1+\varepsilon}} = 0\} = 1 \quad .$$

1.6. Le cas ultramétrique.

Soit (T,δ) un espace métrique. Lorsque δ vérifie l'inégalité forte du triangle

$$(*)\ \forall\, s,t,u \in T\ ,\quad \delta(s,t) \leq \max(\delta(s,u),\delta(u,t))\quad ,$$

on dit que la métrique est non archimédienne ou aussi ultra-métrique (par extension (T,δ) est un espace ultramétrique).

La caractéristique principale de ces espaces tient dans le fait que deux boules de même rayon sont ou bien disjointes ou bien confondues.

Dans ce qui suit, (T,δ) est séparable ; on a noté $\varepsilon_n = \rho^n \varepsilon_o$ avec $0 < \rho < 1$, S_n une suite m F minimale finie ou non de centres de boules ouvertes de rayon ε_n suffisant pour recouvrir T. On introduit les applications suivantes

- $\theta_n : T \to S_n$ définie par $\delta(s,\theta_n(s)) < \varepsilon_n$

- $\pi_{n,n-1} : S_n \to S_{n-1}$ définie par $\delta(t,\pi_{n,n-1}(t)) < \varepsilon_{n-1}$,

- $\pi_{n,k} = \pi_{k+1,k} \circ \dots \circ \pi_{n,n-1}$, de S_n dans S_k, $n \geq k \geq 0$

- $\pi_{n,m} = Id(S_n)$, $n \geq 0$.

Le lemme (facile) qui suit donne une représentation des espaces ultramétriques séparables.

LEMME 1.6.1. <u>Soit</u> (T,δ) <u>un espace ultramétrique séparable. Le complet</u> $((S_n), (\pi_{n,k}))$ <u>forme un système projectif d'ensembles et on a</u> $\theta_k = \pi_{n,k} \circ \theta_n$ <u>pour tous les entiers</u> $n \geq k \geq 0$.
<u>Notons</u> $L = \lim (S_n, \pi_{n,k})$ <u>sa limite projective,</u> $G = \prod_{n=1}^{\infty} S_n$ <u>et pour tout entier</u> $k \geq 0$, <u>soit</u> π_k <u>la restriction à</u> L <u>de la projection de</u> G <u>sur</u> S_k.
Posons aussi pour tous $\underline{s},\underline{t}$ éléments de L,

$$\Delta(\underline{s},\underline{t}) = \varepsilon_{n(\underline{s},\underline{t})} \quad \text{avec} \quad n(\underline{s},\underline{t}) = \operatorname{Sup}\{n : \pi_n(\underline{s}) = \pi_n(\underline{t})\} .$$

Alors (L,Δ) est un espace ultramétrique compact. De plus l'application $\ell : (T,\delta) \to (L,\Delta)$ définie par $\ell(s) = \{\theta_n(s), n \geq 0\}$ détermine un plongement continu de (T,δ) dans (L,Δ) et on a les relations

$$\forall\ s,t \in T , \quad \rho\, \Delta(\ell(s),\ell(t)) \leq \delta(s,t) \leq \Delta(\ell(s),\ell(t)) .$$

Enfin lorsque (T,δ) est compact, ℓ est un isomorphisme de structures uniformes.

<u>Démonstration</u> : Pour tous $s \in T$, $n \geq k \geq 0$ entiers, on a

$$\delta(\theta_k(s),\pi_{n,k}(\theta_n(s))) \leq \operatorname{Max}\{\delta(\theta_k(s),s),\ \delta(s,\pi_{n,k}(\theta_n(s)))\}$$

$$\leq \operatorname{Max}\{\delta(s,\theta_k(s)),\delta(s,\theta_n(s)),\delta(\theta_n(s),\pi_{n,n-1}(\theta_n(s))),\operatorname{Max}\{\delta(\pi_{n,j}(\theta_n(s)),\pi_{n,j-1}(\theta_n(s))), j \in]k,n]\}\}$$

$$< \varepsilon_k .$$

Par conséquent, pour tout $s \in T$, $\delta(\theta_k(s),\pi_{n,k}(\theta_n(s))) = 0$.
En outre, pour tous $n \geq j \geq k \geq 0$, on a par construction,

$$(1) \qquad \pi_{n,k} = \pi_{j,k} \circ \pi_{n,j} ,$$

d'où le système projectif que forment ces applications. (L,Δ) est de toute évidence un espace ultramétrique compact ; cela tient à ce que pour tout

$q \leq n(\underline{s},\underline{t})$, $\pi_q(\underline{s}) = \pi_q(\underline{t})$ et pour tout $q > n(\underline{s},\underline{t})$, $\pi_q(\underline{s}) \neq \pi_q(\underline{t})$. L'application ℓ est injective. En effet si $\ell(s) = \ell(s')$, alors pour tout entier n, $\theta_n(s) = \theta_n(s')$. Par conséquent

(2) $$\delta(s,s') \leq \inf_{n\geq 1} \mathrm{Max}\{\delta(s,\theta_n(s)),\delta(\theta_n(s),s')\} = 0$$

ce qui montre que $s = s'$.

On remarque par ailleurs que $\delta(s,t) \leq \epsilon_{n(\ell(s),\ell(t))} = \Delta(\ell(s),\ell(t))$, et, en outre, si p est l'entier déterminé par

$$\epsilon_{p+1} \leq \delta(s,t) < \epsilon_p \; ,$$

on vérifie que $\theta_p(s) = \theta_p(t)$, donc $p \leq n|\ell(s),\ell(t))$. Par conséquent

(3) $$\Delta(\ell(s),\ell(t)) \leq \epsilon_p \leq \rho^{-1}\,\delta(s,t) \; .$$

Enfin lorsque (T,δ) est compact ; il reste à montrer que ℓ est surjective pour établir le dernier point du lemme.

Soit s un élément de L. Pour tout entier n, il existe un point s_n vérifiant les relations

$$\forall\, k \leq n\,,\quad \theta_k|s_n) = \pi_k(s) \; .$$

La suite (s_n) obtenue est une suite de Cauchy dans (T,δ) puisque

$$\delta(S_p,S_q) \leq \mathrm{Max}\{\delta(S_q,\theta_{q\wedge p}(S_q)),\delta(S_p,\theta_{q\wedge p}(S_q)\}$$
$$< \epsilon_{q\wedge p} \; .$$

Elle est donc convergente et on montre facilement que sa limite s^o vérifie, grâce au caractère ultramétrique de Δ, pour chaque n, la relation $\theta_n(s^o) = \pi_n(s)$. Donc $\ell(s^o) = s$, et le lemme est établi.

THEOREME 1.6.1. Soient (T,δ) un espace ultramétrique, D un écart aléatoire sur $T \times T$, δ-séparable, φ une fonction de Young telle que

$$\sup_{T \times T} E\{ \varphi(\frac{D_{s,t}}{\delta(s,t)})\} \leq 1 \ .$$

a) <u>Dans ces conditions pour que les trajectoires de</u> D <u>soient presque sûrement continues, il suffit que l'intégrale</u>

(1.6.1.) $\int_{+0} \varphi^{-1}(N\ \delta(T,u))\, du$ <u>converge.</u>

b) <u>Soit</u> $\Omega(u) \geq 0$ <u>un module de continuité ; pour que les trajectoires de</u> D <u>soient p.s. Ω-Lipschitziennes, il suffit que l'intégrale</u>

(1.6.2.) $\int_{+0} \frac{\varphi^{-1}(N(T,u))du}{\Omega(u)}$ <u>converge.</u>

<u>On a alors</u>

(1.6.3.) $E\{ \operatorname{Sup}_{T \times T} \frac{D_{s,t}}{\Omega(\delta(s,t))} \} \leq \text{Const.} \int_0^{\text{Diam}\,(T,\delta)} \frac{\varphi^{-1}(N(T,u))}{\Omega(u)} du \ .$

<u>Démonstration</u> : On conserve les notations habituelles du paragraphe en posant $\epsilon_o = \text{Diam}\,(T,\delta)$.

On notera ici pour tout entier N

(1) $S(N) = \bigcup_{k=N} S_k$.

Comme D est δ-séparable, δ continu en probabilité, pour tout entier n , il existe un négligeable N_n tel que

$$\forall \omega \in N_n , \ \forall E \subset T \times T, \ \operatorname{Sup}_E D_{s,t}(\omega) = \operatorname{Sup}_{E \cap S(n)^2} D_{s,t}(\omega) \ .$$

Notons $N = \bigcup_n N_n$, $\Omega_o = \Omega \backslash N$.

Soit $\omega \in \Omega_o$, soient N entier et (s,t) dans $S(N)^2$ tel que

(1') $\delta(s,t) < \epsilon_N$.

Il existe $p \geq N$, $q \geq N$ tels que $s \in S_p$, $t \in S_q$; donc $\theta_p(s) = s$ et $\theta_q(t) = t$.

Par ailleurs $\delta(s,t) < \epsilon_N$ implique $\theta_N(t) = s_o \in S_N$.

Ainsi : $D_{s,t} = D_{\theta_p(s),\theta_q(t)}$

$$\leq \Sigma_{k=N+1}^{p} D_{\theta_k(s),\theta_{k-1}(s)} + \Sigma_{k=N+1}^{q} D_{\theta_k(t),\theta_{k-1}(t)}$$

(2)
$$\leq 2 \sum_{k=N+1}^{p\vee q} \sup_{s\in S_k} D_{s,\pi_{k,k-1}(s)} .$$

Notons alors

(3)
$$Y_N = 2 \Sigma_{k\geq N} \epsilon_{k-1} \sup_{s\in S_k} D_{s,\pi_{k,k-1}(s)}/\delta(s,\pi_{k,k-1}(s)) .$$

On a
$$E\, Y_N \leq 2 \sum_{k\geq N} \epsilon_{k-1} \; E\, \varphi^{-1} \circ \varphi\Big(\sup_{s\in S_k} \frac{D_{s,\pi_{k,k-1}(s)}}{\delta(s,\pi_{k,k-1}(s))} \Big)$$

(4)
$$\leq 2 \sum_{k\geq N} \epsilon_{k-1} \varphi^{-1}(N(T,\epsilon_k)) .$$

Notre hypothèse implique donc que la suite de variables aléatoires Y_n décroit p.s. vers zéro.

Comme on a

(5)
$$\sup_{\substack{T\times T \\ \delta(s,t)<\epsilon_N}} D_{s,t} \leq 2 Y_N ,$$

on en déduit la continuité des trajectoires de D.

Ce dernier point montre aussi pour tout entier N et tout $\omega \in \Omega_o$

$$\sup_{\epsilon_{N+1}\leq\delta(s,t)<\epsilon_N} \frac{D_{s,t}(\omega)}{\Omega\circ\delta(s,t)}$$

(6)
$$\leq 2 \sum_{k=N+1}^{\infty} \frac{\epsilon_{k-1}}{\Omega(\epsilon_k)} \sup_{s\in S_k} \frac{D_{s,\pi_{k-1,k}(s)}(\omega)}{\delta(s,\pi_{k-1,k}(s))} .$$

Par conséquent

$$(7)\quad \text{p.s.} \quad \sup_{\delta(s,t)<\varepsilon_N} \frac{D_{s,t}}{\Omega\circ\delta(s,t)} \leq 2 \sum_{k=N+1}^{\infty} \frac{\varepsilon_{k-1}}{\Omega(\varepsilon_k)} \sup_{s\in S_k} \frac{D_{s,\pi_{k-1,k}(s)}}{\delta(s,\pi_{k-1,k}(s))}$$

d'où, en intégrant chaque membre de (7)

$$E\{ \sup_{T\times T} \frac{D_{s,t}}{\Omega(\delta(s,t))} \} \leq \text{Const.} \int_0^{\text{Diam}\,(T,\delta)} \frac{\varphi^{-1}(N(T,u))}{\Omega(u)}\, du \ .$$

ce qui montre que

$$\text{p.s.} \quad D \in \text{Lip}(\Omega_o\delta) \ .$$

□

Remarque : (cas euclidien)

Soit r un entier positif, soient D un écart aléatoire séparable sur le cube $[0,1]^r$, φ une fonction de Young et $d : [0,1] \to R^+$ une fonction croissante sous-additive nulle en 0. On suppose que D vérifie la condition de moment

$$(1.6.4)\quad \sup_{s,t\in[0,1]^r} E\{\varphi(\frac{D_{s,t}}{d(\|s-t\|)})\} \leq 1 \ .$$

où $\|.\|$ est la norme euclidienne.

On a alors de manière analogue le résultat suivant

THEOREME 1.6.2 :

Sous les hypothèses précédentes, pour que D soit presque sûrement continu il suffit que l'intégrale

$$(1.6.5)\quad \int_0^1 \frac{d(u)}{u}\, \varphi^{-1}(u^{-r})\, du$$

soit convergente. Soit $\Omega : [0,1] \to R_+$ une fonction croissante sous additive nulle en 0. Pour que D appartienne à l'espace $\text{Lip}(\Omega)$ presque sûrement il suffit que l'intégrale

1.6.6 $$\int_0^1 \frac{d(u)\ \varphi^{-1}(u^{-r})}{u\ \Omega(u)}\ du$$

converge et l'on a alors

1.6.7 $$E\{ \sup_{T\times T} \frac{D_{s,t}}{\Omega(\|s-t\|)} \} \leq \text{Const.} \int_0^1 \frac{d(u)\ \varphi^{-1}(u^{-r})}{u\ \Omega(u)}\ du \ .$$

Indication de démonstration

On pose pour tout entier p

(1) $$S_p^r = \{\frac{j_i}{2^p}\ ,\ 1 \leq i \leq r)\ ,\ j_i \in \{0, 2^p\}\ i = 1,\ldots,r\}$$

Soit k un entier fixé, soit n un entier supérieur à k et soient s,t deux éléments de S_n^r. On peut définir par induction une suite double $\{(s^q,t^q),\ q = k,\ldots,n\}$ vérifiant pour tout q

(2)
a) $s^q, t^q \in S_q^r$ et $s^n = s\ ,\ t^n = t\ ,$

b) $\text{Max}(\|s^q - s^{q-1}\|, \|t^q - t^{q-1}\|) \leq 2^{-q}$

c) $\|s^q - t^q\| \leq \|s^{q+1} - t^{q+1}\|$

En particulier si $\|s-t\| < 2^{-k}$, 2-c) implique $s^k = t^k$. On suppose maintenant $\|s-t\| < 2^{-k}$. Alors on a immédiatement

(3) $$D_{s,t} \leq \sum_{q=k+1}^{n} (D_{s^q,s^{q-1}} + D_{t^q,t^{q-1}})$$

Notons formellement $s \to \underline{s}$ l'application qui détermine s^{q-1} connaissant s^q

De (3) , on tire

(4) $$D_{s,t} \leq 2 \sum_{q=k+1}^{\infty} \sup_{s \in S_q^r} D_{s,\underline{s}}$$

A partir de là, il suffit de reconduire le raisonnement utilisé plus haut.

REFERENCES DU CHAPITRE I

[1] L. BREIMAN : Probability (1968). Addison-Wesley.

[2] X. FERNIQUE : Régularité des trajectoires des fonctions aléatoires gaussiennes. Ecole d'Eté de Probabilité de Saint-Flour (1974). Lect. Notes Math. 480, (1975) p. 295.

[3] X. FERNIQUE : Caractérisation de processus à trajectoires majorées ou continues, Lectures Notes Springer, 649, 691-706.

[4] X. FERNIQUE : Continuité des trajectoires des écarts aléatoires, application aux fonctions aléatoires, + cours suivant.

[5] I. IBRAGIMOV : Sur la régularité des trajectoires des fonctions aléatoires, C.R. Acad. Sc. Paris A 289 (1979) 545-547.

[6] M.J. KLASS - M.G. HAHN : Sample continuity of square intégrable processes, Annals of Prob. 5 (1977) 361-377.

[7] N. KONO : Best Possibility of an intégral Test for sample continuity of L_p-Processes ($p \geq 2$). Proc. Japan Acad. 54 ser A (1978) 197-201.

[8] N. KONO : Sample path. properties of stochastic processes, J. Math. Kyoto Univ. 20-2 1980, 295-313.

[9] P. LEVY : Processus stochastiques et Mouvement Brownien (1965). Paris Gauthier-Villars .

[10] K. MENGER : Statistical Metrics Proc. Nat. Acad. Sc. U.S.A. (1942), 28, 535-537.

[11] C. NANOPOULOS - P. NOBELIS : Etude de la régularité des fonctions aléatoires et leurs propriétés limites. Thèse de 3e cycle (1977) Strasb.

[12] P. NOBELIS : Fonctions aléatoires lipschitziennes, à paraître.

[12 a] O. ONICESCU : Nombres et systèmes aléatoires (1964), Ed. Eyrolles, Paris.

[13] G. PISIER : Conditions d'entropie assurant la continuité de certains processus ... Séminaire d'Analyse fonctionnelle exposés n° 13-14. Ecole Polytechnique, Palaiseau 1980.

[14] B. SCHWEITZER: Sur la possibilité de distinguer les points dans un espace métrique aléatoire.
C.R. Acad. Sc. Paris t. 280 (1975), 459-461.

[15] T. SIRAO : On the continuity of Brownian Motion with a Multidimensional Parameter. Nagoya Math. Journal V.16, 1960, 135-156.

[16] W.F. STOUT, K. JOGDOA,N.C. JAIN : Upper and lower functions for Martingales and mixing Processes.
Ann. of Prob. (1975) V3, n°1, 119-145.

[17] TAKASHIMA : Frequency of exceptionnal growth of the N-parameter Wiener Process. à paraître.

[18] M. WEBER : Analyse asymptotique de processus gaussiens stationnaires, thèse d'Etat (1980), Strasbourg.

[19] M. WEBER : Une méthode élémentaire pour l'étude de la régularité d'une large classe de fonctions aléatoires. C.R. Acad. Sc. Paris t. 292 (1981), 599-602.

Chapitre II : ANALYSE DES TRAJECTOIRES DE CERTAINS PROCESSUS GAUSSIENS STATIONNAIRES.

2.0. Dans ce chapitre nous étudions des expressions telles que $P\{\exists\, t \in I : X(t) > \varphi(t)\}$ lorsque cette probabilité est petite ; X est en général un processus gaussien stationnaire dont la régularité de la covariance à l'origine et à l'infini est précisée.

Dans un premier temps nous mettons en place les résultats classiques de Ito, Cramer, Ylvisaker et autres sur les zéros d'un processus gaussien stationnaire dans un intervalle borné : c'est l'objet du théorème 2.1.1. Dans cette partie nous nous sommes largement inspirés de [4] chap. X et [13].

Puis nous énonçons deux résultats remarquables sur le caractère poissonnien des zéros : l'un dû à H. Cramer [4], l'autre plus récent est l'oeuvre de S.M. Berman. Pour des raisons faciles à comprendre (longueur du texte, répétitions) nous nous sommes limités à la démonstration du résultat le plus récent : celui de S.M. Berman ; nous en donnons la démonstration originale.

Enfin, dans la dernière partie nous nous intéressons à l'étude de $P\{\exists\, I \text{ intervalle} : \forall\, t \in I \quad X(t) > \varphi(t)\}$ sous plusieurs aspects (temps de séjour, instants de grande amplitude).

On remarquera que dans la première partie les techniques d'entropie sont absentes, de même que des arguments d'intégrabilité forte. La raison en est bien simple : ils sont dans cette situation inefficaces.

2.1. Espérance du nombre de zéros dans un intervalle borné.

2.1.0. Dans ce chapitre nous étudions certaines v.a. particulières associées à un processus gaussien stationnaire $\xi(t)$, t réel , par exemple le nombre de fois que ses trajectoires rencontrent une droite donnée durant l'intervalle de temps $[0,T]$. Des problèmes très concrets sont à l'origine de ces questions. Nous en donnons un exemple.

Problème : Certaines perturbations intervenant dans les émissions radios peuvent être interprétées de la façon suivante : une onde émise d'équation $A\,e^{i\omega t}$ est perturbée par un bruit représenté sous la forme d'un processus complexe gaussien stationnaire $\xi^*(t)$; d'où l'équation modifiée

$$A\,e^{i\omega t} + \xi^*(t) = e^{i\omega t}[A + \xi^*(t)\,e^{-i\omega t}] = K(t)\,e^{i\omega t + i\theta(t)} ,$$

donnant lieu à

$$\begin{cases} K(t)\cos\theta(t) = A + \xi_1(t) , \\ K(t)\sin\theta(t) = \xi_2(t) , \\ \text{si } \xi^*(t)\,e^{-i\omega t} = \xi_1(t) + i\xi_2(t) . \end{cases}$$

Les perturbations, c'est-à-dire les variations brusques de phase, correspondent aux évènements : $\xi_2(t)$ change de signe pendant que $A + \xi_1(t) < 0$. C'est un problème de zéros de processus gaussiens.

2.1.1. Introduisons quelques définitions utiles.

Soit $\mathcal{C}$ l'ensemble des applications $f : [0,1] \to \mathbb{R}$ continues ne s'annulant ni en 0 ni en 1 et telles que $\mathrm{Int}(f^{-1}(0)) = \emptyset$.

Par extension, quel que soit $u : [0,1] \to \mathbb{R}$ continue, on notera

$$\mathcal{C}_u = u + \mathcal{C} .$$

Soit f appartenant à $\mathcal{C}_u$:

D_1 - On dit que f a une montée par rapport à u en t_o, s'il existe $\epsilon > 0$ tel que $f - u \leq 0$ sur $(t_o - \epsilon, t_o)$ et $f - u \geq 0$ sur $(t_o, t_o + \epsilon)$. On notera M_u le nombre de montées de f par rapport à u_o.

D_2 - On dit que f descend par rapport à u en t_o, si $-f$ monte par rapport à $-u$ en t_o. On note D_u le nombre de descentes.

D_3 - f croise u en t_o s'il existe une suite $s_n < t_o < t_n$ avec $\lim_{n\to\infty} t_n - s_n = 0$, vérifiant pour tout entier n, $[f(t_n) - u(t_n)][f(s_n) - u(s_n)] < 0$. On note C_u leur nombre.

Clairement $C_u \geq D_u + M_u$; si f croise u en t_o, alors $f(t_o) = u(t_o)$ par continuité ; la réciproque est évidemment fausse car f peut être tangente à u. Notons T_u le nombre de points de contacts de f à u, N_u le nombre de zéros de $f-u$, ($N_u = C_u + T_u$). On a cependant la proposition suivante :

PROPOSITION 2.1.1. (Ylvisaker, [13]). Soit $Y(t)$, $0 \leq t \leq 1$, un processus gaussien à trajectoires continues. Si pour tout $0 \leq t \leq 1$, $\sigma^2(t) = E[Y(t)-EY(t)]^2 > 0$, alors pour toute application u sur $[0,1]$ continue,

$$P\{\omega : Y(\omega,\cdot) \text{ n'a pas de points de contact avec } u\} = 1 .$$

Démonstration : Notons $m(t) = EY(t)$ et considérons le processus auxiliaire

$$X(t) = \frac{Y(t) - u(t)}{\sigma(t)} + \lambda \quad , \qquad \lambda > 0 .$$

Alors $\bar{m}(t) = E\,X(t) = \dfrac{m(t) - u(t)}{\sigma(t)} + \lambda$ et $\bar{\sigma}(t) = (E[X(t) - \bar{m}(t)]^2)^{\frac{1}{2}} = 1$.

Choisissons λ assez grand pour que $\bar{m}$ soit positif sur $[0,1]$; évidemment les points de contacts de Y avec u sont ceux de X avec λ. Nous montrons que X est presque sûrement sans points de contact avec λ. Un argument de dénombrabilité d'événements et la symétrie de la loi de X réduisent le problème à établir pour tout intervalle $I \subset [0,1]$ fermé,

$$P\{\forall t \in I,\ X(t) \leq \lambda \text{ et } \exists\, t_o \in I : X(t_o) = \lambda\} = 0 .$$

Or une telle éventualité impliquerait $\sup_I X(\cdot) = \lambda$ et nous avons le

LEMME 2.1.1. La v.a. $\sup_I X$ a une distribution absolument continue.

Démonstration du lemme (esquisse). Notons $\varphi(x)$ la densité de la loi gaussienne centrée réduite. Soit $T = \{t_1,\dots,t_n\}$ un sous-ensemble de $[0,1]$. La loi de $\max_{j=1}^{n} X(t_j)$ a pour densité

$$(2\pi)^{-\frac{1}{2}} \sum_{j=1}^{n} e^{-\frac{1}{2}(x - \bar{m}(t_j))^2} \; P\{\sup_{j \neq i} X(t_i) \leq x \mid X(t_j) = x\} ,$$

$$= \varphi(x) \sum_{j=1}^{n} e^{x\overline{m}(t_j) - \frac{1}{2} m^2(t_j)} G_{t_j}(x) .$$

On va montrer que les fonctions $G_{t_j}(x)$ sont croissantes. Il suffit de montrer que toute fonction du type $P\{\max(X_1,\dots,X_k) \le u \mid X_0 = u\}$ est croissante. Montrons-le lorsque les variables aléatoires $X_0,X_1,\dots,X_k$ sont centrées réduites;; cela ne restreint pas la généralité. Soit $r \le k$ le rang de la famille $[X_0,\dots,X_k]$. On peut supposer que la sous famille $[X_0,\dots,X_r]$ est maximale, les autres variables aléatoires s'écrivant en fonction des précédentes ; on a

$$\forall\ i \in]r,k] \qquad X_i = \Sigma_{j=0}^{r} \theta_{i,j} X_j .$$

Soit $\alpha = (\alpha_{i,j})$ $i,j = 0,\dots,r$ la matrice inverse de corrélation du vecteur $(X_0,\dots,X_r)$.

La densité conditionnelle du vecteur $(X_1,\dots,X_r)$ par rapport à $\{X_0=u\}$ s'écrit sous la forme

$$\frac{\sqrt{\det \alpha}}{(2\pi)^{\frac{r}{2}}} \exp\Big(-\frac{1}{2} \Sigma_{i,j=1}^{r} \alpha_{i,j} x_i x_j - \big(\sum_{j=1}^{r} \alpha_{0,j}\, u x_j\big)\Big) .$$

La probabilité conditionnelle $P\{\mathrm{Max}(X_1,\dots,X_k) \le u \mid X_0 = u\}$ s'écrit en fonction de la densité précédente, après recentrage de celle-ci

$$P\{\mathrm{Max}\,(X_1,\dots,X_k) \le u \mid X_0 = u\}$$

$$= \int_{\substack{\forall\ j \in [1,r],\ y_i \le u(1-E[X_0 X_i]) \\ \forall\ i \in]r,k],\ \Sigma_{j=1}^{r} \theta_{i,j} y_j \le u(1-E[X_0 X_i])}} e^{-\frac{1}{2}\Sigma_{i,j=1}^{r} \alpha_{i,j} y_i y_j} \frac{d_y}{\sqrt{\det\alpha^{-1}}(2\pi)^{\frac{r}{2}}}$$

ce qui montre bien la croissance de ces fonctions.

Comme $\overline{m} \ge 0$, la densité de la loi de $\max_T X$ s'écrit sous la forme $\varphi . G_T$ où G_T est croissante. Soit (T_n) une suite croissante d'ensembles finis de réunion partout dense dans $[0,1]$. Supposons que la loi de $\sup_T X$ ne soit pas absolument continue ; alors il existerait au moins un point x_0 et un réel $\rho > 0$ tels que l'on ait

$$P\{\sup_{[0,1]} X = x_o\} \geq \rho > 0$$

et, a fortiori,

$$\rho \leq P\{\forall\, k \geq 1,\ \exists\, N_k < \infty : \forall\, n \geq N_k \quad \sup_{T_n} X \in [x_o - \tfrac{1}{k}, x_o + \tfrac{1}{k}]\}$$

donc

$$\forall\, k \geq 0,\ \exists\, N_k < \infty \text{ tel que } \frac{\rho}{2} \leq \inf_{n \geq N_k} P\{\sup_{T_n} X \in [x_o - \tfrac{1}{k}, x_o + \tfrac{1}{k}]\}$$

et

$$\frac{\rho}{2} \leq \inf_{n \geq N_k} G_{T_n}(x_o + \varepsilon) \int_{x_o - \varepsilon}^{x_o + \varepsilon} \varphi(u)\, du$$

ce qui implique

$$\forall\, \varepsilon > 0, \quad \lim_{n \to \infty} G_{T_n}(x_o + \varepsilon) = \infty .$$

Or, pour tout réel y, on a :

$$G_{T_n}(y) \left(\int_y^{\infty} \varphi(u)\, du\right) \leq \int_y^{\infty} \varphi(u)\, G_{T_n}(u)\, du \leq 1 ,$$

ce qui montre $\sup_n G_{T_n}(y) < \infty$, d'où la contradiction ; le théorème est démontré.

□

2.1.2. Dans cette partie, on approche C_u à l'aide de variables aléatoires étagées du même type. On ne suppose pas que le processus ξ soit gaussien. Le lemme qui va suivre concerne le cas où u est une application constante prenant la valeur u. Il peut être adapté au cas d'une fonction continue (voir remarques en fin de démonstration).

Nous notons auparavant pour tous entiers n,r positifs avec $u \le 2^n$,

$t_{n,r} = r.2^{-n}$ et

$$\xi_n(t) = \xi(t_{n,r}) + 2^n(t - t_{n,r})\,[\xi(t_{n,r+1}) - \xi(t_{n,r})] ,$$

$$C_n^* = \text{le nombre de croisements de } \xi_n \text{ avec } u .$$

LEMME 2.1.2. Soit $\xi(t)$, $0 \le t \le 1$, un processus à trajectoires continues vérifiant de plus que

1) les lois des variables aléatoires $\xi(t)$ sont toutes diffuses,
2) pour tous s,t, $P\{\xi(s) = \xi(t)\} = 0$,
3) pour tout réel x, $P\{\xi$ a un point de contact avec $x\} = 0$.

Dans ces conditions, et avec les notations du haut, presque sûrement C_n^* croît vers C_u.

La proposition 2.1.1. et le lemme 2.1.2. établissent immédiatement le

COROLLAIRE 2.1.1. Soit $\xi(t)$, $0 \le t \le 1$, un processus gaussien stationnaire centré à trajectoires continues tel que $E\{\xi^2(0)\} > 0$. Alors, presque sûrement C_n^* croît vers C_u.

Démonstration du lemme 2.1.2. •) Montrons tout d'abord que presque sûrement tous les C_n^* sont finis. En effet, dans le cas contraire, il faudrait (et il suffirait) que $\xi(t_{n,r}) = \xi(t_{n,r+1})$ pour au moins un couple (n,r). Or la réunion de ces éventualités est P-négligeable.

•) Montrons maintenant que presque sûrement les C_n^* forment une suite croissante.

Comme ξ_n ne rencontre u qu'au plus une fois dans chaque subdivision $[t_{n,r}, t_{n,r+1}]$, il suffit de montrer que si ξ_n rencontre u en t_o avec $t_{n,r_1} \leq t_o \leq t_{n,r_1+1}$, alors ξ_{n+1} rencontre u au moins une fois dans $[t_{n,r_1}, t_{n,r_1+1}]$ (on peut même exclure l'éventualité $\{\xi(t_{n,r}) = u$ pour au moins un $t_{n,r}\}$ qui est négligeable). Mais ceci résulte immédiatement du théorème des valeurs intermédiaires appliqué au processus continu ξ_{n+1}, puisque u appartient à

$$]\xi_{n+1}(t_{n,r_1}) \wedge \xi_{n+1}(t_{n,r_1+1}), \xi_{n+1}(t_{n,r_1}) \vee \xi_{n+1}(t_{n,r_1+1})[,$$

et $\xi_{n+1} = \xi_n$ sur les $t_{n,r}$.

•) Notons maintenant

$$(1) \quad \begin{aligned} &\Omega_1 = \{\text{les } C_n^* \text{ sont finis et croissants}\} \\ &C^* = \lim_{n\to\infty} C_n^* \cdot \chi_{\Omega_1} . \end{aligned}$$

Le théorème des valeurs intermédiaires montre que presque sûrement pour chaque n, on a $C_n^* \leq C_u$ et donc presque sûrement $C^* \leq C_u$.

•) Si C_n est fini, on déduit de l'hypothèse 3) qu'en dehors d'une négligeable, le processus ne comprend qu'un nombre fini de montées et de descentes qui sont alors les seuls croisements de ξ avec u. On peut définir un entier n suffisamment grand tel qu'en chaque point t_c de croisement, on ait

$$(2) \quad \begin{aligned} &\xi(t) - u < 0 \quad \text{sur }]t_c - \tfrac{1}{2^n}, t_c[\\ &\text{et } \xi(t) - u > 0 \quad \text{sur }]t_c, t_c + \tfrac{1}{2^n}[\end{aligned}$$

ou l'inverse. En particulier les intervalles $]t_c - \frac{1}{2^n}, t_c + \frac{1}{2^n}[$ sont disjoints.

A tout point de croisement t_c, associons l'entier r_c tel que

(3) $$t_{n+1,r_c} < t_c \le t_{n+1,r_c+1} .$$

On en déduit que le processus ξ_{n+1} a une montée ou une descente dans chaque subdivision

$$]t_{n+1,r_c}, t_{n+1,r_c+1}]$$

d'où $C_{n+1} = C_u$, et donc $C_u = C^*$.

On en déduit qu'il existe un ensemble mesurable Ω_2 avec $P(\Omega_2) = 1$ tel que

(4) $$\{C_n < \infty\} \cap \Omega_2 \subset \{C_u = C^*\} .$$

•) Enfin, lorsque $C_u = \infty$, presque sûrement $C_u \ge n_1$ pour tout n_1. Soient $t^1, \dots, t^{n_1}$, n_1 points de croisements distincts, ordonnés.

La définition d'un point de croisement implique qu'il existe pour tout $i \le n_1$ une suite $s_k^i < t^i < t_k^i$ telle que pour tout entier k

$$(\xi(s_k^i) - u)\,(\xi(t_k^i) - u) < 0 .$$

On peut donc déterminer un entier k_1 assez grand tel que

(5) $$t_{k_1}^1 - \frac{1}{k_1^2} < s_{k_1}^1 < t_{k_1}^1 < s_{k_1}^2 < t_{k_1}^2 < \dots\dots$$

La continuité de ξ montre aussi qu'il existe $k_2 > k_1$ tel que

(6) $$(\xi(s) - u)\,(\xi(t) - u) < 0$$

pour $s \in \,]s_{k_1}^i - \frac{1}{k_2^2}, s_{k_1}^i + \frac{1}{k_2^2}[$,

$t \in \,]t_{k_1}^i - \frac{1}{k_2^2}, t_{k_1}^i + \frac{1}{k_2^2}[$,

i variant de 1 à n_1.

Considérons le processus ξ_{k_2+1} .

Pour tout $i \le n_1$, soient $t_{k_2,r_i'}$, $t_{k_2,r_i''}$ les points de subdivision vérifiant

$$t_{k_2+1,r_i'} \le s^i_{k_1} < t_{k_2+1,r_i'+1} ,$$

(7)

$$t_{k_2+1,r_i''} \le t^i_{k_1} < t_{k_2+1,r_i''+1} .$$

Alors $t_{k_2+1,r_i'} \in \left]s^i_{k_1} - \frac{1}{2^{k_2}} , s^i_{k_1} + \frac{1}{2^{k_2}}\right[$,

de même $t_{k_2+1,r_i''} \in \left]t^i_{k_1} - \frac{1}{2^{k_2}} , t^i_{k_1} + \frac{1}{2^{k_2}}\right[$,

donc

(8) $[\xi(t_{k_2+1,r_i'}) - u]\,[\xi(t_{k_2+1,r_i''}) + u] < 0$.

Ainsi, en vertu du théorème des valeurs intermédiaires, ξ_{k_2+1} rencontre u dans la subdivision

$$[t_{k_2+1,r_i'} , t_{k_2+1,r_i''}] .$$

Ces intervalles sont deux à deux disjoints ( ) .

Il en résulte que ξ_{k_2+1} rencontre au moins n_1 fois u , ce qui implique

(9) $C^* \ge C_{k_2+1} \ge n_1$.

Comme n_1 est arbitraire, on en déduit que $C^* = \infty$. En conclusion, nous venons de montrer qu'il existe un ensemble mesurable Ω_3 avec $P(\Omega_3) = 1$, tel que

(10) $\{C_n = \infty\} \cap \Omega_3 \subset \{C_u = C^*\}$.

Nous avions aussi vu plus haut que

(11) $\{C_u < \infty\} \cap \Omega_2 \subset \{C_u = C^*\}$

où Ω_2 est mesurable de probabilité 1 .

Ces relations ensemblistes montrent donc

(12) $\quad \Omega \cap \Omega_3 \cap \Omega_2 \subset \{C_u = C^*\} \subset \Omega$.

L'ensemble $\{C_u = C^*\}$ est donc mesurable de probabilité 1 . Nous avons montré que presque sûrement C_n^* croit vers C_u et en prime la mesurabilité de C_u .

Remarques : -) La dernière étape aurait été facilitée si on savait a priori que

$$(C_u = \infty) \Rightarrow (\text{Max}\,(M_u , D_u) = \infty) \; .$$

Or ce point est vrai, mais il résulte de la démonstration. Notons M_n (resp. D_n) le nombre de montées (resp. le nombre de descentes) du processus ξ_n . On peut montrer de façon similaire que $M_n \uparrow M_u$, $D_n \uparrow D_u$ et que M_u et D_u sont mesurables. On en déduit

$$M_n + D_n = C_n^* \uparrow M_u + D_u = C_u (= N_u) \; .$$

-) Pour traiter le cas où $u(t)$ est simplement continu, il suffit d'adapter les hypothèses du lemme au processus $\xi(t) - u(t)$.

THEOREME 2.1.1. (Cramer [4], Ito [7], ...). Soient $\xi(t)$, t réel un processus gaussien centré stationnaire à trajectoires continues, $r(t) = \int_0^\infty \cos \lambda t \, dF(\lambda)$ sa covariance. Notons $\lambda_o = r(0)$ et pour tout réel u et $T > 0$, $N_u(T)$ le nombre de zéros de $\xi - u$ dans l'intervalle $[0,T]$.

2.1.1. Si $\lambda_2 = \int_0^\infty \lambda^2 \, dF(\lambda) < \infty$, alors $E[N_u(T)] = \frac{T}{\pi} \left(\frac{\lambda_2}{\lambda_o}\right)^{1/2} e^{-u^2/2\lambda_o}$.

2.1.2. Si $\lambda_2 = \infty$, alors $E[N_u(T)] = \infty$.

Démonstration : La continuité et la stationnarité de ξ ainsi que l'additivité de $N(.)$ montrent en premier lieu $E[N_u(T)] = T\, E[N_u(1)]$.

Le lemme précédent établit par convergence monotone

$$(1) \qquad E\, N_u(1) = E\, C_u(1) = \lim_{n\to\infty} E\, C_n^* .$$

Or,

$$E\, C_n^* = 2^n [P\{\xi(0) > u > \xi(2^{-n})\} + P\{\xi(0) < u < \xi(2^{-n})\}] .$$

Notons $\gamma_n = 2^n(\xi(2^{-n}) - \xi(0))$, $p_n(x,z)$ la densité u loi du couple $(\xi(0),\gamma_n)$ et $\Phi(x)$ la densité de la gaussienne centrée réduite.

$$2^n\, P\{\xi(0) > u > \xi(2^{-n})\} = 2^n\, P\{ u < \xi(0) < u - 2^{-n}\gamma_n\}$$

$$(2) \qquad = \int_{-\infty}^{0} d_z \int_u^{u-2^{-n}z} 2^n p_n(x,z)\,dx = \int_{-\infty}^{0} d_z \int_0^{-z} 2^n p_n(u+2^{-n}x,z)\, dx .$$

Si $\lambda_2 < \infty$, on voit que

$$\lim_{n\to\infty} 2^n p_n(u+2^{-n}x,z) = \lambda_2^{-\frac{1}{2}}\, \Phi(u)\, \Phi\left(\frac{z}{\lambda_2^{\frac{1}{2}}}\right)$$

et il existe des constantes $K_1 > 0$ et K_2 telles que

$$2^n\, p_n(u + 2^{-n}\, x,z) \le K_1\, e^{-K_2 z^2} .$$

Le théorème de convergence dominé montre

$$(3) \qquad \lim_{n\to\infty} 2^n\, P\{\xi(0) > u > \xi(2^{-n})\} = \lambda_2^{-\frac{1}{2}}\, \Phi(u) \int_{-\infty}^{0} |z|\, \Phi\left(\frac{z}{\lambda_2^{\frac{1}{2}}}\right) dz = \frac{\lambda_2^{\frac{1}{2}}}{\pi}\, e^{-u^2/2}$$

ce qui suffit pour conclure.

Si $\lambda_2 = \infty$, le changement de variables $z = 2^n(2(1 - r(2^{-n})))^{\frac{1}{2}}\, x$ dans (2) permet de conclure.

□

nb : Ce théorème donne aussi des indications sur le comportement de $N_u(U,T)$ quand $T \to \infty$. Plus précisément

2.1.3. Si ξ est ergodique alors,

$$\text{p.s.} \lim_{T \to \infty} \frac{N_u(0,T)}{T} = \left[\frac{\lambda_2}{\lambda_0}\right]^{\frac{1}{2}} e^{-u^2/2\lambda_0}.$$

Ce point est une conséquence directe du théorème de Birkhoff. Nous rappelons que ξ est ergodique si et seulement si la mesure spectrale de ξ est complètement diffuse ; dans ce cas le processus est aussi faiblement mélangeant ([8]).

2.2. Caractère Poissonnien des zéros.

2.2.0. Lorsque la covariance du processus est assez régulière tant à l'origine qu'asymptotiquement, les zéros de $\xi - u$ ont des propriétés limites remarquables bien que relativement délicates à établir. Nous en présentons deux de démonstrations voisines. Le premier théorème établit une convergence en loi des marges du processus $N_u([s,t])$ $0 < s < t$; il fut d'abord obtenu par Volkonski et Rozanov [9], puis amélioré par d'autres auteurs dont H. Cramer [4]. L'énoncé que nous donnons est dû à H. Cramer. Le second théorème, plus récent (1971), est dû à S.M. Berman. Nous le démontrons complètement car il nous a semblé que sa démonstration, par la richesse de son argumentation, illustre bien les différents procédés utilisés pour ces questions.

THEOREME 2.2.1. (H. Cramer [4]). Soit ξ un processus gaussien stationnaire centré. On suppose que les hypothèses suivantes sont réalisées

(2.2.1.) $r(t) = 1 - \frac{\lambda_2}{2!} t^2 + \frac{\lambda_4}{4!} t^4 + o(t^4)$ $\quad t \to 0$

(2.2.2.) $r(t) = O(t^{-\alpha})$ $t \to \infty$, pour un $\alpha > 0$.

Notons $\epsilon(u) = (\lambda_2)^{\frac{1}{2}} e^{-u^2/2}$.

Soient $I_1, \dots, I_j$, j intervalles disjoints de longueurs respectives $\frac{\tau_1}{\epsilon(u)}, \dots$ $\dots, \frac{\tau_j}{\epsilon(u)}$, $\tau_1 > 0, \dots, \tau_j > 0$ étant des constantes. Soient aussi $k_1, \dots, k_j$ des entiers positifs; alors :

(2.2.3.) $\lim_{u\to\infty} P\{M_u(I_i) = k_i,\ i = 1,\dots,j\} = \prod_{i=1}^{j} e^{-\tau_i} \frac{\tau_i^{k_i}}{k_i!}$.

Remarque : Choisissons $j = 1$, $I_1 = [0,T]$, $k = 0$ et $T = (\varepsilon(u))^{-1} e^{-z}$ dans (2.2.3.) . Alors,

$$\lim_{n\to\infty} P\{M_u(T) = 0\} = e^{-e^{-z}}, \text{ grâce à la proposition 2.1.1. Mais}$$

$$P\{M_u(T) = 0\} = P\{\xi(0) < u,\ M_u(T) = 0\} + P\{\xi(0) > u,\ M_u(T) = 0\}$$

$$= P\{\sup_{[0,T]} \xi < u\} + O(1),\quad u \to \infty$$

donc $\lim_{n\to\infty} P\{\sup_{[0,T]} \xi < u\} = e^{-e^{-z}}$,

ou encore, (cf. th. 2.1.1.)

(2.2.4.) $\lim_{T\to\infty} P\{\sup_{0\le t\le T} \xi(t) < \sqrt{2\log T} + \frac{z + \log\frac{\sqrt{\lambda_2}}{2\pi}}{\sqrt{2\log T}}\} = e^{-e^{-z}}$.

THEOREME 2.2.2. (S.M. Berman, [1]). Soit $X(t)$, t réel un processus gaussien stationnaire à trajectoires continues. On suppose que la covariance $r(t)$ de X vérifie les conditions suivantes :

(2.2.4) $1 - r(t) \sim \frac{\gamma^2}{2} t^2$, $t \to 0$, avec $\gamma > 0$.

(2.2.5) $r(t) = O(1/\log t)$, $t \to \infty$.

Notons alors pour $\sigma_1 > 0$ et $\sigma_2 > 0$, $u(t) = \sqrt{2\log\frac{t\gamma}{2\pi\sigma_1}}$,

$$v(t) = \sqrt{2\log\frac{t\gamma}{2\pi\sigma_2}} ,$$

et $M(t)$ (resp. $N(t)$) le nombre de montées (resp. descentes) par rapport à $u(t)$ (resp. $-v(t)$) de $X(s)$ dans l'intervalle $[0,t]$).

Dans ces conditions, $(M(t),N(t))$ converge en loi quand $t \to \infty$ vers un couple de lois de Poisson indépendantes d'espérance σ_1 et σ_2 respectivement.

nb : La conclusion du théorème reste vraie si (2.2.5) est remplacé par

(2.2.6.) $(\int_0^\infty r^2(t)\,dt < \infty)$.

2.2.1. Lemmes préliminaires. La démonstration du théorème 2.2.2. repose avant tout sur le lemme de comparaison suivant.

LEMME 2.2.1. Soient A un ensemble fini et X un vecteur gaussien sur A centré de covariance $r(\alpha,\beta)$ avec $r(\alpha,\alpha) = 1$ pour tout α.

Soient de plus $\mathfrak{S}$ une partition de A de terme générique σ et (x), (y) deux suites de réels indexées sur A et vérifiant pour tout σ, $-\infty \le x_\alpha \le y_\alpha \le +\infty$. On leur associe

$$C_\alpha =]\![\, x_\alpha , y_\alpha \,]\![,$$

$$V_\sigma = \prod_{\alpha\in\sigma} C_\alpha , \qquad V = \prod_{\sigma\in\mathfrak{S}} V_\sigma ,$$

$$X_\sigma = \{X(\alpha),\ \alpha \in \sigma\} .$$

Dans ces conditions, on a :

(2.2.7) $$|P\{X \in V\} - \prod_{\sigma\in\mathfrak{S}} P\{X_\sigma \in V_\sigma\}| \le \sum_{\sigma\neq\sigma'} \sum_{\alpha\in\sigma} \sum_{\beta\in\sigma'} k(\alpha,\beta) ,$$

où

$$k(\alpha,\beta) = \int_{-|r(\alpha,\beta)|}^{|r(\alpha,\beta)|} [\Phi(x_\alpha,x_\beta,y) + \Phi(x_\alpha,y_\beta,y) + \Phi(y_\alpha,y_\beta,y) + \Phi(y_\alpha,x_\beta,y)]dy ,$$

et

$$\Phi(x,y,\rho) = (2\pi)^{-1} (1-\rho^2)^{-\frac{1}{2}} \exp\left(- \frac{x^2 + y^2 - 2\rho xy}{2(1 - \rho^2)}\right) .$$

Démonstration : On la fait pour $A = [1,n]$.

a) $\Gamma_1 = \mathrm{Cov}(X)$ est inversible : Soit pour chaque σ appartenant à $\mathfrak{S}$, X'_σ un vecteur gaussien sur σ de même loi que X_σ et tel que les X'_σ soient mutuellement indépendants ; nous notons alors :

(1) $\Lambda = (X'_\sigma ,\ \sigma \in \mathfrak{S})$, $\Gamma_0 = \mathrm{Cov}(\Lambda)$,

et pour tout λ compris entre 0 et 1 ,

$$\Gamma(\lambda) = \lambda\, \Gamma_1 + (1 - \lambda)\Gamma_0 .$$

Posons aussi

(2) $$F(\lambda) = \int_{\mathbb{R}^n} I_V(u)\, g_\lambda(u)\, du$$

où

$$g_\lambda(u) = K_n \int_{\mathbb{R}^n} \exp(i < u,y >) \exp(-\frac{1}{2}{}^t y\, \Gamma(\lambda) y) dy$$

$$= \frac{1}{(2\pi)^{\frac{n}{2}} \sqrt{\det \Gamma(\lambda)}} \exp(-\frac{1}{2}{}^t u\, \Gamma^{-1}(\lambda) u) .$$

On constate (cf par exemple lemme 2.1.4. p. 19, [5]) que $F(\lambda)$ est dérivable et que sa dérivée $F'(\lambda)$ peut être évaluée sous la forme

(3) $$F'(\lambda) = \int_{\mathbb{R}^n} I_V(u) \frac{\partial}{\partial\lambda} (g_\lambda(u))\, du$$

et

$$\frac{\partial}{\partial\lambda}(g_\lambda(u)) = \frac{1}{2} \operatorname{Tr}[\frac{d\Gamma(\lambda)}{d\lambda} \cdot \frac{d^2}{du^2} (g_\lambda(u))] .$$

Mais $$(\frac{d\Gamma(\lambda)}{d\lambda})\alpha,\beta = \begin{cases} r(\alpha,\beta) & \text{si } \alpha \in \sigma,\ \beta \in \sigma',\ \sigma \neq \sigma' \\ 0 & \text{sinon .} \end{cases}$$

Par conséquent,

(4) $$\frac{\partial}{\partial\lambda} (g_\lambda(u)) = \frac{1}{2} \sum_{\sigma\neq\sigma'} \sum_{\alpha\in\sigma} \sum_{\beta\in\sigma'} r(\alpha,\beta) \frac{\partial^2}{\partial_{u_\alpha} \partial_{u_\beta}} (g_\lambda(u)) ,$$

Ainsi

(5) $$F'(\lambda) = \frac{1}{2} \sum_{\sigma\neq\sigma'} \sum_{\alpha\in\sigma} \sum_{\beta\in\sigma'} r(\alpha,\beta) \int_{\mathbb{R}^n} I_V(u) \frac{\partial^2}{\partial_{u_\alpha} \partial_{u_\beta}} (g_\lambda(u))\, du .$$

Or

$$\int_{\mathbb{R}^n} I_V(u) \frac{\partial^2}{\partial u_\alpha \partial u_\beta} (g_\lambda(u))\, du = \int_{x_1}^{y_1} du_1 \int_{x_2}^{y_2} du_2 \cdots \int_{x_\alpha}^{y_\alpha} du_\alpha \int_{x_\beta}^{y_\beta} \frac{\partial^2}{\partial_{u_\alpha} \partial_{u_\beta}} (g_\lambda(u)) du_\beta$$

$$= \int_{\mathbb{R}^{n-2}} I_{\prod_{\substack{r\neq\alpha \\ r\neq\beta}} C_r} (v) \,\square_{\alpha,\beta}(g_\lambda(v)) dv ,$$

où on a noté

$$\square_{\alpha,\beta}(g_\lambda(v)) = \begin{aligned}[t] & \quad (\alpha \text{ ième}) \qquad (\beta \text{ ième}) \\ & g(v_1,v_2,\ldots, y_\alpha ,\ldots, y_\beta ,\ldots) \\ & -g(v_1,v_2,\ldots, x_\alpha ,\ldots, y_\beta ,\ldots) \\ & -g(v_1,v_2,\ldots, y_\alpha ,\ldots, x_\beta ,\ldots) \\ & +g(v_1,v_2,\ldots, y_\alpha ,\ldots, y_\beta ,\ldots) . \end{aligned}$$

Il suit que,

$$(6) \qquad \left|\int_{\mathbb{R}^{n-2}} I_{\prod_{\substack{r\neq\alpha\\ r\neq\beta}} C_r}(v) \square_{\alpha,\beta}(g_\lambda(v))dv\right| \leq \sum_{(s,t)}^{*} \Phi(s,t,\lambda\ r(\alpha,\beta)) ,$$

où la sommation Σ^* est indexée par

$$\{(y_\alpha,y_\beta) , (x_\alpha,x_\beta) (x_\alpha,y_\beta) , (y_\alpha,x_\beta)\} .$$

Par conséquent,

$$|P\{X \in V\} - \prod_{\sigma\in\mathcal{O}} P\{X_\sigma \in V_\sigma\}| = |F(1) - F(0)| = \left|\int_0^1 F'(\lambda)d\lambda\right| ,$$

$$= \frac{1}{2} \left| \sum_{\sigma\neq\sigma'} \sum_{\alpha\in\sigma} \sum_{\beta\in\sigma'} r(\alpha,\beta) \int_0^1 d\lambda \left(\int_{\mathbb{R}^{n-2}} I_{\prod_{\substack{r\neq\alpha\\ r\neq\beta}} C_r}(v) \square_{\alpha,\beta}(g_\lambda(v))dv\right)\right|$$

$$(7) \qquad \leq \frac{1}{2} \sum_{\sigma\neq\sigma'} \sum_{\alpha\in\sigma} \sum_{\beta\in\sigma'} k(\alpha,\beta) .$$

b) $\Gamma_1 = \text{Cov}(X)$ <u>n'est pas inversible</u>. Soit N un vecteur gaussien normal à valeurs dans $\mathbb{R}^n$ indépendant de X ; pour tout réel u, posons

$$X_u = X + u N \qquad\qquad \Lambda_u = \Lambda + u N .$$

Si u est non nul, les covariances Γ_{X_u} et Γ_{N_u} sont inversibles. La première étape de la démonstration montre que X_u vérifie les conclusions du lemme ; de plus $\Gamma_{X_u}(\alpha,\beta) = r(\alpha,\beta) + u^2$. Il suffit par conséquent de faire tendre u vers zéro pour conclure. $\square$

Dans la suite, on utilisera souvent le cas particulier suivant du lemme 2.2.1.

Pour tout couple de v.a. gaussiennes normalisées (X,Y) avec $E\,X\,Y = \rho$, x et y réels,

(2.2.8) $$P\{X < x, Y < y\} = \int_0^\rho \Phi(a,b,\lambda)\,d\lambda + P\{X < x\}\,P\{Y < y\}\,,$$

$$= \int_1^\rho \Phi(a,b,\lambda)\,d\lambda + P\{X < x_\wedge y\}\,.$$

LEMME 2.2.2. <u>Pour tout entier positif</u> n, <u>soit</u> $M'_n(t)$ <u>(resp.</u> $D'_n(t)$<u>) le nombre de montées (resp. descentes) par rapport à</u> $u(t)$ <u>(resp.</u> $-v(t)$<u>) du processus interpolé de</u> X <u>aux points</u> $\{\frac{jt}{n}, j \le n\}$. <u>Soit</u> g $\mathbb{R}_+ \longrightarrow \mathbb{R}_+$ <u>une application croissante non bornée.</u>

<u>Dans ces conditions, si</u> $n(t) = \mathrm{Ent}[t\,u(t)\,g(t)]$.

(2.2.9) $$\lim_{t\to\infty} E\,|M(t) - M'_{n(t)}(t)| = 0 \quad \underline{\text{et}} \quad \lim_{t\to\infty} E\,|D(t) - D'_{n(t)}(t)| = 0\,.$$

<u>Démonstration</u> : Compte tenu de la symétrie de la loi de X il suffit de faire le calcul pour M. D'après le théorème 2.1.1. et puisque X est réversible, on a

(1) $$E[M(t)] = \frac{1}{2}\,E[N(t)] = \frac{t\,\gamma}{2\pi}\,\exp\left(-\frac{u^2}{2}\right) = \sigma_1\,.$$

D'autre part, par stationnarité,

$$E\,M'_{n(t)}(t) = n(t)\,P\{X(0) \le u(t) < X(\tfrac{t}{n(t)})\}\,,$$

et par (2.2.8),

$$P\{X(0) \le u(t) < X(\tfrac{t}{n(t)})\} = \frac{1}{2\pi}\int_{r(\frac{t}{n(t)})}^{1} e^{-\frac{u^2(t)}{1+y}}\,\frac{dy}{\sqrt{1-y^2}}$$

donc

(2) $$E[M'_{n(t)}(t)] \sim \frac{n(t)}{2\pi}\int_{r(\frac{t}{n(t)})}^{\frac{\gamma t}{n(t)}} e^{-\frac{u^2\theta^2}{4(2-\frac{\theta^2}{2})}}\,d\theta \sim \sigma_1\,,$$

pourvu que $\frac{u(t)t}{n(t)} = \frac{1}{g(t)} \to 0$ quand $t \to \infty$ ce qui est par hypothèse.

Comme $M'_{n(t)}(t) \le M(t)$, la conclusion est immédiate. □

Dans le lemme suivant, $n(t)$ est précisé par le lemme 2.2.2.; on pose aussi

$$G_t = \{ \frac{jt}{n(t)} , j \le n(t)\} .$$

On considère dans $[0,t]$ m intervalles ε-distants ; m et $\varepsilon > 0$ sont fixes, on leur associe les v.a. suivantes

$$U_j = \sup_{I_j \cap G_t} X \qquad Vj = \inf_{I_j \cap G_t} X .$$

<u>LEMME 2.2.3. Sous les hypothèses du théorème</u> 2.2.2. ; <u>si de plus</u> $g(t)$ <u>vérifie</u>

(2.2.10) $\forall\, p > 0 , \quad \lim_{t\to\infty} g(t)\, t^{-p} = 0 ,$

(2.2.11) $\forall\, \beta > 0 , \quad \lim_{t\to\infty} g^2(t) \sup_{s\ge t^\beta} |r(s)| \log s = 0$

<u>alors pour tout m-uplet de variables</u> $(u_1,\dots,u_m)$, (<u>resp.</u> $(v_1,\dots,v_m)$) , <u>égales</u> <u>à</u> $u(t)$ <u>ou</u> $+\infty$, (<u>resp.</u> $v(t)$ ou $+\infty$)

$$\lim_{t\to\infty} | P\{\forall\ j = 1,\dots,m , -v_j \le V_j \le U_j \le u_j\} - \prod_{j=1}^{m} P\{-v_j \le V_j \le U_j \le u_j\}| = 0$$

<u>Démonstration</u> : On déduit du lemme 2.2.1. :

(1) $$|P\{\forall\ j = 1,\dots,m , -v_j \le V_j \le U_j \le u_j\} - \prod_{j=1}^{m} P\{-v_j \le V_j \le U_j \le u_j\}|$$

$$\le n(t) \sum_{\frac{n(t)\varepsilon}{t}\le j\le n(t)} \int_{-|r(\frac{jt}{n(t)})|}^{|r(\frac{jt}{n(t)})|} [\Phi(u(t),u(t),y) + \Phi(v(t),v(t),y) + \Phi(u(t),v(t),y)]dy .$$

Comme les zéros de $1 - r(t)$ forment un sous-groupe additif, la condition (2.2.5) nécessite que $1 - r(t)$ ne s'annule qu'en 0 ; par continuité, pour tout $\varepsilon > 0$ il existe $0 < \delta < 1$ tel que

(2) $\sup \{|r(s)| \mid s > \varepsilon\} < \delta$.

Nous établissons tout d'abord

$$(3)\qquad \lim_{t\to\infty} n(t) \sum_{\frac{n(t)\varepsilon}{t}\le j\le n(t)} \int_{-|r(\frac{jt}{n(t)})|}^{|r(\frac{jt}{n(t)})|} \Phi(u(t),u(t),\lambda)d\lambda = 0$$

ce qui ramène à montrer que

$$(4)\qquad n(t) \sum_{\frac{n(t)\varepsilon}{t}\le j\le n(t)} |r(\tfrac{jt}{n(t)})|\,(1 - r^2(\tfrac{jt}{n(t)}))^{-\frac{1}{2}} \exp(- \frac{u^2(t)}{1+r(\frac{jt}{n(t)})})$$

tend vers zéro quand $t \to \infty$.

On partage la sommation (4) en deux sommes partielles, la première sur les indices $j < n^\beta$ la seconde sur $j \geqslant n^\beta$ pour β vérifiant $0 < \beta < \frac{1-\delta}{1+\delta}$ et $\delta = \delta(\varepsilon)$ ($\varepsilon > 0$ est fixé dans la construction des I_j).

La première somme partielle x se majore à partir de (2) et (4) par

$$(5)\qquad O(1)\,(n(t))^{\beta+1} \exp(- \frac{u^2(t)}{1+\delta}) = O(1)\,(t^{1-\frac{2}{(1+\delta)(1+\beta)}} u(t)g(t))^{\beta+1} \to 0,\quad t\to\infty ,$$

par 2.2.

La seconde somme partielle s'évalue sous la forme

$$(6)\qquad n(t)\, e^{-u^2(t)} \sum_{n^\beta(t)<j<n(t)} |r(\tfrac{jt}{n(t)})|\; e^{r(\frac{jt}{n(t)})u^2} .$$

Comme pour tout t assez grand $t(n(t))^{\beta-1} > t^{\beta/2}$, il vient

$$(7)\qquad \begin{aligned} |r(\tfrac{tj}{n(t)})|\, u^2(t) &\le O(1)(\sup_{s\ge t^{\beta/2}} |r(s)|\log s)\ \frac{\log t}{\log t\; n(t)^{\beta-1}} \\ &\le O(1)\ (\sup_{s\ge t^{\beta/2}} |r(s)|\log s) \ ; \end{aligned}$$

(6) est donc majoré par

$$(8)\qquad O(1)\; g^2(t)\ (\sup_{s\ge t^{\beta/2}} |r(s)|\ \log s) ,$$

expression qui tend vers zéro quand t tend vers l'infini.

On obtiendrait de la même façon

$$\text{(9)} \qquad \lim_{t\to\infty} n(t) \sum_{\frac{\varepsilon n(t)}{t}\le j\le n(t)} \int_{-|r(\frac{jt}{n(t)})|}^{|r(\frac{jt}{n(t)})|} \Phi(v(t),v(t),y)\,dy = 0 \quad .$$

Pour la contribution du terme $\Phi(u(t),v(t),y)$, on remarque que

$$\Phi(u,v,\lambda) = \Phi(u,v,\lambda) + \exp\left[\frac{u(u-v)}{1+\lambda} - \frac{(u-v)^2}{2(1-\lambda)^2}\right] \quad ,$$

$$u(t) = \sqrt{2\log t} + \frac{\log \gamma/2\pi\sigma_1}{\sqrt{2\log t}} + O(1/\log t) \ , \quad t\to\infty \quad .$$

Par conséquent, $\lim_{t\to\infty} u(t)(v(t)-u(t)) = \log \frac{\sigma_2}{\sigma_1}$ et $\lim_{t\to\infty} (u(t)-v(t))^2 = 0$.

On est ramené aux cas précédents. □

LEMME 2.2.4. <u>Soient</u> Y_1 , Y_2 <u>deux v.a. gaussiennes centrées réduites indépendantes.</u>

<u>Posons</u>

$$\forall\, t \ge 0 \ , \quad Y(t) = Y_1 \cos t + Y_1 \sin t$$

$$\forall\, x \in \mathbb{R} \ , \quad \Phi(x) = (2\pi)^{-\frac{1}{2}} \int_{-\infty}^{x} \exp(-u^2/2)\,du \quad .$$

Alors pour tout $u > 0$, $0 < t < \frac{\pi}{2}$

$$\text{(2.2.12)} \qquad P\{\sup\{Y(s),\ 0 \le s \le t\} \le u\} = \Phi(u) - \left(\frac{t}{2\pi}\right) e^{-u^2/2} \quad .$$

<u>Démonstration</u> : On décompose suivant le schéma ci-contre la région $A = \{(x,y) : \forall\, 0 \le s \le t \quad x \cos s + y \sin s \le u\}$ en cinq domaines A_i , $i = 1,\dots,5$. Il s'agit donc de calculer

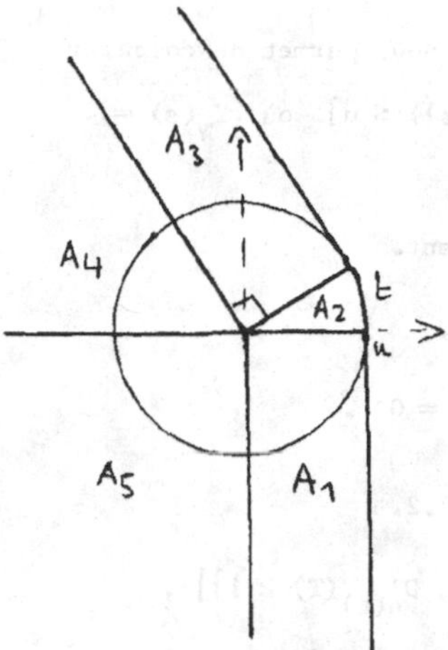

$$\sum_{i=1}^{5} \iint_{A_i} d\Phi(x)\, d\Phi(y) .$$

Pour $A_1 = \{0 \le x \le u,\ y \le 0\}$, on obtient $\frac{1}{2}[\Phi(u) - \frac{1}{2}]$. Ensuite

$$\iint_{A_2} d\Phi(x) d\Phi(y) = \int_0^t \int_0^u \frac{e^{-\rho^2/2}}{2\pi}\, \rho\, d\rho\, d\theta$$

$$= \frac{t}{2\pi}\,[1 - \exp(-u^2/2)] .$$

Par invariance , par rotation, l'intégrale correspondant à A_3 égale celle pour A_1 . Enfin

$$\iint_{A_4} d\Phi(x)\, d\Phi(y) = \int_0^\infty \int_{t+\frac{\pi}{2}}^{\pi} \frac{e^{-\rho^2/2}}{2\pi}\, \rho\, d\rho\, d\theta = (\tfrac{1}{4}) - (t/2\pi)$$

et

$$\iint_{A_5} d\Phi(x)\, d\Phi(y) = \frac{1}{4}$$, d'où le lemme.

COROLLAIRE 2.2.1. <u>Soit</u> $X(t)$ <u>un processus gaussien stationnaire vérifiant les conditions du théorème 2.2.2. Alors pour tout</u> $\varepsilon > 0$ <u>, il existe</u> $T > 0$ <u>, tel que</u>

(2.2.13) <u>Quels que soient</u> $u \ge 0$, $0 \le t \le T$,

$$\Phi(u) - \frac{\gamma t(1+\varepsilon)}{2\pi} \exp(-u^2/2) \le P\{\max\{X(s), 0\le s\le t\} \le u\} \le \Phi(u) - \frac{\gamma t(1-\varepsilon)}{2\pi} \exp(-\frac{u^2}{2}) .$$

Démonstration : Pour tout $\varepsilon > 0$, on peut trouver $T > 0$ tel que

(1) $\forall\ 0 \le t \le T$, $\cos[(1+\varepsilon)\gamma t] \le r(t) \le \cos[(1-\varepsilon)\gamma t]$.

En vertu du lemme de Slépian cette relation nous permet de comparer $P\{\max(0 \le s \le t\ ,\ X(s)) \le u\}$ et $P\{\max(0 \le s \le t,\ Y_\gamma(s)) \le u\}$ où $Y_\gamma(s) = Y_1 \cos[(1\pm\varepsilon)\gamma s] + Y_2 \cos[(1\pm\varepsilon)\gamma s]$.

Il suffit par conséquent d'appliquer le lemme précédent.

LEMME 2.2.5. Sous les hypothèses du théorème 2.2.2. ,

(2.2.14) $\forall\ T > 0$, $\lim\limits_{t\to\infty} t\, P\{M(T) \ge 1\ ,\ D(T) \ge 1\} = 0$.

Démonstration (pour $T = 1$) . En vertu du lemme 2.2.2.

$$t\ |\ P\{M(T) \ge 1,\ D(T) \ge 1\} - P\{M'_{n(t)}(T) \ge 1\ ,\ D'_{n(t)}(T) \ge 1\}|\ ,$$
$$\le t[P\{|M(T) - M'_{n(t)}(T)| \ge 1\} + P\{|D(T) - D'_{n(t)}(T)| \ge 1\}]\ ,$$
$$\le t[E\ |\ M(T) - M'_{n(t)}(T)| + E\ |\ D(T) - D'_{n(t)}(T)|]\ ,$$

et cette dernière expression tend vers zéro quand t tend vers l'infini. Il suffit donc de montrer

(1) $\lim\limits_{t\to\infty} t\, P\{M'_{n(t)}(T) \ge 1\ ,\ D'_{n(t)}(T) \ge 1\} = 0$.

On remarque

(2) $t\, P\{M'_{n(t)}(T) \ge 1,\ D'_{n(t)}(T) \ge 1\} \le 2n(t) \sum\limits_{j=1}^{[\frac{n(t)}{t}]} P\{X(0) < v(t), X(\frac{jt}{n}) > u(t)\}$.

Sans restreindre la généralité, nous pouvons supposer $\sigma_1 \le \sigma_2$. Soit aussi $\varepsilon > 0$ tel que $\inf\limits_{|s|\le\varepsilon} r(s) \ge \frac{1}{2}$.

Pour tout j tel que $j \le [\frac{n(t)\varepsilon}{t}]$, on a , à partir de

$$\forall\ x \ge 0\ ,\quad 1 - \Phi(x) \le \frac{\sqrt{2\pi}}{x}\ \exp(-\frac{x}{2})^2\ ,$$
$$P\{X(0) < -v(t),\ X(\frac{jt}{n(t)}) > u(t)\} \le P\{X(\frac{jt}{n(t)}) - X(0) > 2u(t)\}$$

$$\leq O(1)\, u(t)^{-1} \exp\left(-\frac{u^2}{1-r(\frac{jt}{n(t)})}\right) \leq O(1)\, u(t)^{-1} \exp(-2u^2(t))$$

$$\leq O(1)\, u(t)\, t^{-4} \quad ,$$

d'où,

$$(3) \qquad n(t) \sum_{j=1}^{[\frac{n(t)\epsilon}{t}]} P\{X(0) < -u(t), X(\tfrac{jt}{n(t)}) > u(t)\} \leq O(1)\, n^2(t) u(t) t^{-5}$$

qui tend vers zéro pourvu que (2.2.10) soit réalisé.

Notons maintenant $\delta = \sup_{|s|\geq\epsilon} |r(s)|$; on a $\delta < 1$.

A l'aide de (2.2.8) on obtient

$$(4) \qquad P\{X(0) < -u(t), X(\tfrac{jt}{n(t)}) > u(t)\} = [1 - \Phi(u(t))]^2 + \int_0^{-r(\frac{jt}{n(t)})} \Phi(u(t),u(t),y)dy ,$$

or si $j \geq [\frac{n(t)\epsilon}{t}]$,

$$(5) \qquad \left|\int_0^{-r(\frac{jt}{n(t)})} \Phi(u(t),u(t),y)dy\right| \leq \int_{-\delta}^{\delta} \Phi(u(t),u(t),y)dy \leq O(1) \exp\left(-\frac{u^2}{1+\delta}\right)$$

$$\leq O(1)\, t^{-\frac{2}{1+\delta}} \quad ;$$

(4) et (5) montrent donc

$$n(t) \sum_{j=[\frac{n(t)\epsilon}{t}]}^{[\frac{n(t)}{t}]} P\{X(0) < -u(t), X(\tfrac{jt}{n(t)}) > u(t)\}$$

$$= O(n^2(t)\, u(t)^{-2}\, t^{-3}) + O(n^2(t)\, t^{-(3+\delta)/(1+\delta)})$$

$$= O(1) , \qquad t \to \infty$$

d'où le lemme. □

Le lemme suivant constitue le deuxième outil principal de la démonstration du théorème de S.M. Berman. Il énonce une propriété limite comparable à celle décrite en (2.2.4) .

LEMME 2.2.6. <u>Sous les hypothèses du théorème 2.2.2.</u> , <u>on a</u>

$$(2.2.15) \qquad \lim_{t\to\infty} P\{\sup \{X(s),\ 0 \leq s \leq t\} \leq u(t)\} = e^{-\sigma_1}$$

$$(2.2.16) \qquad \lim_{t\to\infty} P\{\inf\{X(s),\ 0 \le s \le t\} > -u(t)\} = e^{-\sigma_1}$$

$$(2.2.17) \qquad \lim_{t\to\infty} P\{\sup\{|X(s)|,\ 0 \le s \le t\} \le u(t)\} = e^{-2\sigma_1} .$$

Démonstration : Nous utilisons le découpage suivant $(0 < \beta, \tau < 1)$

$$\forall\, k \ge 0 , \quad I_k = [k\tau,\ (k+1-\beta)\tau] ,$$

$$J_k = [(k+1-\beta)\tau,\ (k+1)\tau] .$$

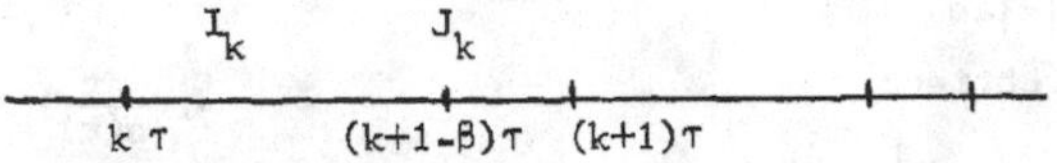

On note $\quad I = \Sigma_{k\ge 0} I_k \qquad J = \Sigma_{k\ge 0} J_k$.

Remarquons en premier lieu (en utilisant le corollaire 2.2.1) ,

$$(1) \qquad |P\{\sup_{I\cap[0,t]} X \le u(t)\} - P\{\sup_{[0,t]} X \le u(t)\}|$$

$$\le P\{\sup_{J\cap[0,t]} X > u(t)\} \le [\tfrac{t}{\tau}]\, P\{\sup_{[0,\tau\beta]} X > u(t)\}$$

$$\le [\tfrac{t}{\tau}]\, [1 - \Phi(u(t)) + \frac{\beta\gamma\tau(1+\varepsilon)}{2\pi}\, e^{-\frac{u^2(t)}{2}}]$$

d'où,

$$(2) \qquad \overline{\lim_{t\to\infty}}\, |P\{\sup_{I\cap[0,t]} X \le u(t)\} - P\{\sup_{[0,t]} X \le u(t)\}| \le \beta(1+\varepsilon)\,\sigma_1 .$$

D'autre part,

$$|P\{\sup_{I\cap[0,t]} X \le u(t)\} - P\{\sup_{I\cap[0,t]\cap G_t} X \le u(t)\}|$$

$$(3)$$

$$\le P\{|M(t) - M'_{n(t)}(t)| > 0\} \le E|M(t) - M'_{n(t)}(t)| .$$

Le lemme 2.2.2. montre donc

$$(4) \qquad \lim_{t\to\infty} [P\{\sup_{I\cap[0,t]} X \le u(t)\} - P\{\sup_{I\cap[0,t]\cap G_t} X \le u(t)\}] = 0 .$$

En outre $P\{\sup_{I\cap[0,t]\cap G_t} X \le u(t)\}$ est équivalent à $P\{\bigcap_{j=0}^{[\frac{t}{\tau}]} (\sup_{I_j\cap G_t} X(s) \le u(t))\}$

quand t est grand.

Les lemmes 2.1.2. et 2.1.3. impliquent

(5) $$\left|P\left\{\bigcap_{j=1}^{[\frac{t}{\tau}]} \sup_{I_j \cap G_t} X(s) \le u(t)\right\} - \left(P\left\{\sup_{[0,\beta\tau]\cap G_t} X(s) \le u(t)\right\}\right)^{[\frac{t}{\tau}]}\right|$$

$$\le n(t) \sum_{[\frac{n(t)\beta\tau}{t}] \le j \le n(t)} \int_{-|r(\frac{jt}{n(t)})|}^{|r(\frac{jt}{n(t)})|} \Phi(u(t),u(t),y)\,dy = O(1)\ , \quad t\to\infty\ .$$

Par conséquent $P\{\sup_{[0,t]} X \le u(t)\}$ se comporte à l'infini comme

$P\{\sup_{[0,\beta\tau]\cap G_t} X(s) \le u(t)\}^{[\frac{t}{\tau}]}$ ou encore, en réutilisant (3), comme

$P\{\sup_{[0,\beta\tau]} X(s) \le u(t)\}^{[\frac{t}{\tau}]}$, c'est-à-dire, grâce au lemme (2.2.5) (si $\tau < T = T(\varepsilon)$),

(6) $$\left[\Phi(u(t) - \frac{\tau\gamma(1-\beta)(1+\varepsilon)}{2\pi} \exp(-\frac{u^2(t)}{2}\right]^{[\frac{t}{\tau}]}$$

ce qui nous permet d'établir (2.2.15) et (2.2.16) après simplification. Nous établissons (2.2.17) maintenant. Posons $T = \tau(1-\beta)$; par symétrie

$$P\{\sup_T |X| > u(t)\} = 2\,P\{\sup_T X > u(t)\} - P\{\sup_T X > u,\ \inf_T X < -u\}$$

or

$$P\{\sup_T X > u(t),\ \inf_T X < -u(t)\} \le P\{\sup_T X > u(t),\ \inf_T X < -u(t), |X(s)| \le u(t)\}$$
$$+ P\{|X(o)| > u(t)\}$$

$$\le P\{M(T) \ge 1\ ,\ D(T) \ge 1\} + P\{|X(0) \ge U(t)\}$$

$$= O(\tfrac{1}{t})\ . \quad \text{(lemme 2.2.5.)}$$

(7) donc $$t\,P\{\sup_{[0,T]} |X| > u(t)\} \sim 2\,t\,P\{\sup_T X > u(t)\}\ .$$

Cette quantité s'évalue à partir de

$$-\,2\log P\{\sup_{[0,\tau(1-\beta)]} X(s) \le u(t)\}^{[t]}$$

sous la forme

(8) $\quad - 2\sigma_1(1 - \beta)\, \tau\, (1 \pm \epsilon)$

quand t est grand, ceci grâce au corollaire 2.2.1.

Il en résulte que

$$\log P\{\sup_{[0,\tau]} |X| \le u(t)\}^{\frac{t}{\tau}}$$ est voisin quand t est grand dans

(9) $\quad 2(1 - \beta)(1 + \epsilon)\sigma_1$.

Ce point étant établi, on traite (2.2.17) de la même façon que 2.2.15 par approximation sur $I \cap [0,t]$ puis sur $I \cap [0,t] \cap G_t$.

COROLLAIRE 2.2.2. Sous les hypothèses du théorème 2.2.2.

(2.2.18) $\quad \forall\, \theta > 0, \quad \lim_{t\to\infty} P\{\sup_{[0,\theta t]} X \le u(t)\} = e^{-\theta\sigma_1}$

(2.2.19) $\quad \lim_{t\to\infty} P\{\sup_{[0,\theta t]} X \le u(t)\} = e^{-2\theta\sigma_1}$.

Démonstration : par changement de temps.

Dans les deux derniers lemmes qui suivent ainsi que pour la démonstration du théorème 2.2.2., nous ferons intervenir le découpage suivant : m est un entier positif fixé, β est compris entre zéro et un strictement et t est un paramètre positif muet. Notons pour tout $j = 1,\dots,m$.

$$I_j = [\frac{(j-1)t}{m} \quad \frac{(j-\beta)t}{m}] \ .$$

On lui associe les v.a. suivantes :

$$\xi_j = \chi_{\ \sup_{I_j \cap G_t} X > u(t)}$$

$$\eta_j = \chi_{\ \inf_{I_j \cap G_t} X \le -v(t)}$$

$$\xi'_j \begin{cases} 1 & \text{si } \{X(s),\ s \in I_j\} \text{ a au moins une montée par rapport à } u(t) \\ 0 & \text{sinon .} \end{cases}$$

$$\eta'_j = \begin{cases} 1 & \text{si } X(s) \text{ a au moins une descente par rapport à } -v(t) \text{ dans } I_j \\ 0 & \text{sinon} \end{cases}$$

LEMME 2.2.7. Les hypothèses du théorème 2.2.2. étant supposées vérifiées

(2.2.20) $\lim_{t\to\infty} P\{\forall\, j = 1,\dots,m \,,\ \xi'_j = \xi_j \,,\ \eta'_j = \eta_j\} = 1$.

Démonstration : Compte tenu du lemme 2.2.2., il suffit d'établir (2.2.20) en remplaçant ξ'_j par $\xi''_j = \chi_{M'_{n(t)}[I_j] \leq 1}$ et η'_j par $\eta''_j = \chi_{D'_{n(t)}[I_j] \leq 1}$,

car $\lim_{t\to\infty} P\{\exists\, j=1,\dots,m : |M'_{n(t)}[I_j] - M[I_j]| > 0$ ou

$|D'_{n(t)}(I_j) - D(I_j)| > 0\} = 0$.

Or $\xi''_j \leq \xi_j$ et $\{\xi''_j < \xi_j\} \subset \{\sup_{I_j \cap G_t} X > u(t)$ et $M'_{n(t)}[I_j] = 0\}$

$\subset \{\inf_{I_j \cap G_t} X > u(t)\}$

donc $P\{\exists\, j = 1,\dots,m : \xi''_j \neq \xi_j\} \leq m(1 - \Phi(u(t))) = o(1) \quad t \to \infty$.

En tenant compte aussi de la symétrie de X, on obtient la même chose pour η_j, d'où (2.2.20).

Notons enfin pour tout entier $j \leq m$

M_j = nombre de montées de X par rapport à $u(t)$ dans I_j,

D_j = nombre de descentes de X par rapport à $-v(t)$ dans I_j.

LEMME 2.2.8. Avec les hypothèses du théorème 2.2.2.

(2.2.21) $\lim_{t\to\infty} P\{\sup\{M_j,\ j \leq m\} \geq 2\} \leq (1-\beta)\sigma_1 - m(1-\exp(-\sigma_1(1-\beta)(m))$,

(2.2.22) $\lim_{t\to\infty} P\{\sup\{D_j,\ j \leq m\} \geq 2\} \leq (1-\beta)\sigma_2 - m(1-\exp(-\sigma_2(1-\beta)(m))$,

(2.2.23) $\lim_{t\to\infty} P\{\xi_1 = \eta_1 = 1\} \leq [\frac{(\sigma_1 \vee \sigma_2)(1-\beta)}{m}]^2$.

Démonstration : On utilise l'inégalité suivante, valable pour toute v.a. $L \geq 0$ à valeurs entières

$$P\{ L \geq 2 \} \leq EL - P\{ L \geq 1 \} ,$$

laquelle fournit la majoration suivante

$$P\{\sup\{ M_j , j \leq m\} \geq 2\} \leq \sum_{j=1}^{m} (E[M_j] - P\{\xi_j' = 1\}) ,$$

car $\xi_j' \leq M_j$.

Or, d'une part, $\Sigma_{j=1}^{m} E(M_j) = (1-\beta)\sigma_1$ et, d'autre part, par stationnarité et par suite du lemme 2.2.7 et du corollaire 2.2.2.,

$$\lim_{t\to\infty} \Sigma_{j=1}^{m} P\{\xi_j' = 1\} = \lim_{t\to\infty} \Sigma_{j=1}^{m} P\{\xi_j = 1\}$$

$$= m \lim_{t\to\infty} P\{ \sup_{[0,\frac{(1-\beta)}{m}t]} X > u(t)\}$$

$$= m(1 - e^{-\frac{(1-\beta)}{m}\sigma_1})$$

d'où (2.2.21) . On obtient de la même façon (2.2.22) .

Posons $u(t) \wedge v(t) = w(t)$; alors

$$P\{\xi_1 = \eta_1 = 1\} \leq P\{ \sup_{[0,\frac{t(1-\beta)}{m}]} X \}> \omega(t) \;\; \inf_{[0,\frac{(1-\beta)t}{m}]} X < -\omega(t)\}$$

$$= 1 - 2P\{ \sup_{[0,\frac{t(1-\beta)}{m}]} X \leq \omega(t)\} + P\{ \operatorname{Sup}_{[0,\frac{t(1-\beta)}{m}]} |X| \leq \omega(t)\} .$$

Cette dernière expression converge vers $[1-\exp(-\sigma_1 \vee \sigma_2(1-\beta)/m]^2$ d'après le corollaire 2.2.2., d'où le lemme .

$\square$

Démonstration du théorème 2.2.2. Soient $m \in \mathbb{N}_*$, $0 < \beta < 1$, $t > 0$. Les lemmes précédents ont utilisé le découpage suivant

$$\forall\, j = 1,\dots,m, \qquad I_j = [\frac{(j-1)t}{m} , \frac{(j-\beta)t}{m}]$$

$$G_t = \{\frac{jt}{n(t)} , j \leq n(t)\}$$

$\frac{(j-1)t}{m}$ $\frac{(j-\beta)t}{m}$ $\frac{jt}{m}$

et les v.a. associées

$$\xi_j(t) = \chi_{\{\sup_{I_j \cap G_t} X > u(t)\}}$$

$$\eta_j(t) = \chi_{\{\inf_{I_j \cap G_t} X < -v(t)\}}$$

$M_j(t)$ = nombre de montées de X par rapport à $u(t)$ dans $I_j \cap G_t$,

$D_j(t)$ = nombre de descentes de X par rapport à $-u(t)$ dans $I_j \cap G_t$,

$$\xi'_j(t) = M_j(t) \wedge 1$$

$$\eta'_j(t) = D_j(t) \wedge 1 \ .$$

L'idée directrice consiste à montrer que les fonctions génératrices des couples $(M(t),D(t))$ et $(\Sigma_{j=1}^{m} \xi_j(t), \Sigma_{j=1}^{m} \eta_j(t))$ sont voisines pour les grandes valeurs de t , puis de conclure en calculant celle du second couple.

Comme $M(t) \geq M_1(t) + \dots + M_m(t)$, on établit directement, au moyen du théorème 2.1.1.,

(1) $$E\left|M(t) - \sum_{j=1}^{m} M_j(t)\right| = E\,M(t) - \sum_{j=1}^{m} E\,M_j(t) = \beta\,\sigma_1,$$

de même

$$E\left|D(t) - \sum_{j=1}^{m} D_j(t)\right| = \beta\,\sigma_2$$

soient $0 < \omega, z < 1$; on en déduit

(2) $$E\left|\omega^{M(t)} z^{D(t)} - \omega^{\Sigma_{j=1}^{m} M_j(t)} z^{\Sigma_{j=1}^{m} D_j(t)}\right|$$

$$\leq P\{|M(t) - \Sigma_{j=1}^{m} M_j(t)| > 0 \text{ ou } |D(t) - \Sigma_{j=1}^{m} D_j(t)| > 0\}$$

$$\leq \beta(\sigma_1 + \sigma_2) \ .$$

D'autre part, le lemme 2.2.8 implique

$$(3) \qquad \overline{\lim_{t\to\infty}}\ P\{\exists\, j \le m : \xi'_j \ne M_j(t) \ \text{ ou } \ \eta'_j \ne D_j(t)\}$$

$$= \overline{\lim_{t\to\infty}}\ P\{\exists\, j \le m : M_j(t) \ge 2 \ \text{ ou } \ D_j(t) \ge 2\}$$

$$\le (1-\beta)(\sigma_1+\sigma_2) - m[2 - \exp(-\sigma_1(1-\beta)/m) - \exp(-\sigma_2(1-\beta)/m)] \,.$$

Mais, en vertu du lemme 2.2.7., on peut remplacer dans (3) $\xi'_j(t)$ par $\xi_j(t)$ et $\eta'_j(t)$ par $\eta_j(t)$; donc

$$(4) \qquad \overline{\lim_{t\to\infty}}\ P\{\ \Sigma_{j=1}^m |(M_j(t) - \xi_j(t)| > 0 \ \text{ ou } \ \sum_{j=1}^m |D_j(t) - \eta_j(t)| > 0\}$$

$$\le (1-\beta)(\sigma_1+\sigma_2) - m[2 - \exp(-\sigma_1(1-\beta)/m) - \exp|-\sigma_1(1-\beta)/m] \,.$$

On en déduit comme pour (2)

$$(5) \qquad \overline{\lim_{t\to\infty}}\ E\left|w^{\Sigma_{j=1}^m M_j(t)} z^{\Sigma_{j=1}^m D_j(t)} - w^{\Sigma_{j=1}^m \xi_j(t)} z^{\Sigma_{j=1}^m \eta_j(t)}\right|$$

$$\le (1-\beta)(\sigma_1+\sigma_2) - m(2 - \exp(-\sigma_1(1-\beta)/m) - \exp(-\sigma_2(1-\beta)/m)) \,.$$

Evaluons $\lim_{t\to\infty} E[w^{\Sigma_{j=1}^m \xi_j(t)} z^{\Sigma_{j=1}^m \eta_j(t)}]$.

$$(6) \qquad E[w^{\Sigma_{j=1}^m \xi_j(t)} z^{\Sigma_{j=1}^m \eta_j(t)}] =$$

$$= \sum_{\underset{\sim}{x},\underset{\sim}{y}\in\{0,1\}^m} w^{\Sigma_{j=1}^m x_j} z^{\Sigma_{j=1}^m y_j}\ P\{\forall\ j \le m,\ \xi_j(t) = x_j\,,\ \eta_j(t) = y_j\} \,.$$

Mais pour tout $\underset{\sim}{x}\,,\ \underset{\sim}{y} \in \{0,1\}^m$

$$(7) \qquad \lim_{t\to\infty} |P\{\forall j\le m,\ \xi_j(t)=x_j\,,\ \eta_j(t)=y_j\} - \prod_{j=1}^m P\{\xi_j(t)=x_j\,,\ \eta_j(t)=y_j\}| = 0$$

En effet, notons $\delta = \{j \le m : \xi_j = 0 = \eta_j\}$, $\delta' = \{j_1,\dots,j_p\}$ et, pour tout $1 \le i \le p$, $\mathcal{E}_i = \{\{\xi_{j_i}(t) = 0\}\,,\ \{\eta_{j_i}(t) = 0\}\,,\ \{\xi_{j_i}(t) = \eta_{j_i}(t) = 0\}\}$;
enfin, $E_0 = \bigcap_{j\in\delta} \{\xi_j(t) = \eta_j(t) = 0\}$.

Dans ces conditions, on voit assez facilement que

$$(8)\qquad \Big|P\Big[\bigcap_{j=1}^{m}\{\xi_j(t)=x_j\ ,\ \eta_j(t)=y_j\}\Big] - \prod_{j=1}^{m} P\{\xi_j(t)=x_j\ ,\ \eta_j(t)=y_j\}\Big|$$

$$\leq \Sigma_{l=1}^{p} \sum_{\substack{0=i_0<i_1<\ldots<i_l\leq p \\ \forall k\leq l\ ,\ E_{i_k}\in \mathcal{E}_{i_k}}} \Big|P\Big\{\bigcap_{k=0}^{\ell} E_{i_k}\Big\} - \prod_{k=0}^{\ell} P\{E_{i_k}\}\Big|$$

le majorant ayant moins de 6^{n-1} termes. Le lemme 2.3.3. appliqué à chacun de ceux-ci montre donc (7), et :

$$(9)\qquad \lim_{t\to\infty} E\{w^{\Sigma_{j=1}^{m}\xi_j(t)}\ z^{\Sigma_{j=1}^{m}\eta_j(t)}\}$$

$$= \lim_{t\to\infty}\Big[\sum_{x,y\in\{0,1\}^m} w^{\Sigma_{j=1}^{m}x_j}\ z^{\Sigma_{j=1}^{m}y_j}\prod_{j=1}^{m}P\{\xi_j(t)=x_j\ ,\ \eta_j(t)=y_j\}\Big]$$

$$= \lim_{t\to\infty}\ (E[w^{\xi_1(t)}\ z^{\eta_1(t)}])^m\ ,$$

par stationarité de X.

Mais,

$$(10)\qquad E[w^{\xi_1(t)}\ z^{\eta_1(t)}] = 1-(1-w)\,P\{\xi_1(t)=1\} - (1-z)\,P\{\eta_1(t)=1\} + (1-w)(1-z)\,P\{\eta_1(t)=\xi_1(t)=1\}\ .$$

Le lemme 2.2.6. montre

$$(11)\qquad \lim_{t\to\infty}\ 1-(1-w)\,P\{\xi_1(t)=1\} - (1-z)\,P\{\eta_1(t)=1\}$$

$$= 1-(1-w)\,(1-\exp(-\sigma_1(1-\beta)/m)) - (1-z)(1-\exp(-\sigma_2(1-\beta)/m))$$

$$= A\ .$$

Par ailleurs, on a aussi (cf lemme 2.2.8)

$$(12)\qquad \overline{\lim_{t\to\infty}}\ P\{\xi_1(t)=\eta_1(t)=1\} \leq B = \left(\frac{\max(\sigma_1,\sigma_2)(1-\beta)}{m}\right)^2,$$

si bien que combinant (9), (10), (11) et (12)

$$(13)\qquad [A-B(1-z)(1-w)]^m \leq \lim_{t\to\infty} E\{w^{\Sigma_{j=1}^{m}\xi_j(t)}\ z^{\Sigma_{j=1}^{m}\eta_j(t)}\} \leq [A+B((1-w)(1-z)]^m\ .$$

Dans (2), (5) et (13), $\beta > 0$ et $m > 0$ sont arbitraires. Faisons tendre β vers zéro et m vers l'infini, dans ces expressions on obtient

$$\lim_{t\to\infty} E[w^{M(t)} z^{D(t)}]$$

$$= \lim_{t\to\infty} [1 - (1-w)(1- \exp(\frac{-\sigma_1(1-\beta))}{m}) - (1-z)(1 - \exp(-\sigma_2(1-\beta))$$

$$+ (1-w)(1-z)(\frac{\max(\sigma_1,\sigma_2)(1-\beta)}{m})^2]^m$$

$$= \exp[-\sigma_1(1-w) - \sigma_2(1-z)] \; .$$

□

2.3. Les instants de grande amplitude de certains processus gaussiens stationnaires.

2.3.0. Considérons un processus gaussien $X(t)$, t réel stationnaire normalisé ($E[X(0)] = 1$, $E[X^2(0)] = 1$) à trajectoires continues. Nous savons que l'amplitude maximum de X sur $[0,T]$ se comporte au plus comme $\sqrt{2\log T}$ quand T devient grand (en fait, comme la médiane de $\sup_{[0,T]} X$, voir [6]) . Nous voudrions, dans un premier temps, préciser la répartition des grandes valeurs de X en fonction du temps. Soient $\varphi : \mathbb{R}_+ \to \mathbb{R}_+$ une application continue croissante non bornée, et λ la mesure de Lebesgue. On note $A_\varphi(\omega) = \{t \geq 0 : X(\omega,t) > \varphi(t)\}$, et on étudie $\lambda\{[0,T] \cap A_\varphi(\omega)\}$ pour les grandes valeurs de T .

La seule définition 0.3. (chap. I) des classes à l'infini implique

1) si $\varphi \in \mathcal{U}_\infty(X)$,

$$A^c_\varphi(\omega) \supset [t(\omega),\infty) \text{ et } t_\omega < \infty \text{ p.s.}$$

et, a fortiori, $$P\{ \lim_{t\to\infty} \frac{\lambda\{[0,T] \cap A^c_\varphi(\omega)\}}{T} = 1\} = 1\} \; .$$

2) si $\varphi \in \mathcal{L}_\infty(X)$, on peut seulement affirmer

$$P\{\forall n \, , \, \#(A_\varphi(\omega) \cap [n,\infty)) = \infty\} = 1 \; ,$$

ce qui est peu. En fait, on montre dans [10] qu'il existe une partie dénombrable $\mathcal{E}(X)$ de $\mathbb{R}_+$, s'accumulant à l'infini telle que

a) $P\{ \#(A_\varphi(\omega) \cap [n,\infty)) = \infty\} = 1$,

b) les classes de $\{X(s),\ s \in \mathcal{E}(X)\}$ et $\{X(s),\ s \in \mathbb{R}\}$ sont dans un certain sens proches.

Soit φ un élément de $\mathcal{L}_{\infty}(X)$. Comme $A_{\varphi}(\omega)$ est presque sûrement un ouvert non vide, on peut donc naturellement poser les questions suivantes :

A) $P\{\lambda(A_{\varphi}(\omega)) = \infty\} = 1$?

B) Si oui, quel est l'ordre de grandeur de $\lambda(A_{\varphi}(\omega) \cap [0,T])$?

C) Soit $\rho > 0$, $A_{\varphi}(\omega)$ contient-il presque sûrement des intervalles de longueur ρ ?

Le théorème qui suit apporte une réponse positive à la question A) lorsque la covariance $r(t)$ tend suffisamment vite vers zéro à l'infini. Il fournit aussi des éléments de réponse à la deuxième question. Enfin C) sera l'objet du théorème 2.3.3.

Dans les deux premiers théorèmes, les propriétés ergodiques de X interviennent. Rappelons celles-ci.

2.3.1. Soit $X(t)$, $t \in \mathbb{R}$ un processus strictement stationnaire. Notons $E = \{f : \mathbb{R} \to \mathbb{R}\}$, $\mathcal{B} = \bigotimes_{t \in \mathbb{R}} \mathcal{B}(\mathbb{R}_t)$, $\tilde{P} = X(P)$ et $S(X) = (E,\mathcal{B},\tilde{P})$ l'espace de probabilité canonique associé à X. Notons $\tilde{X}$ la représentation canonique X, i.e. $\tilde{X}(f,t) = f(t)$; X et $\tilde{X}$ ont les mêmes marges et pour tout $A \in \mathcal{B}$, $P\{X \in A\} = \tilde{P}(A)$. Soient $T_u : E \to E$ la transformation mesurable définie par $T^u(f) = f(\cdot+u)$, G le groupe multiplicatif engendré. Puisque X est strictement stationnaire, G préserve $\tilde{P}$, $(\tilde{P}(T^{-u}A) = \tilde{P}(A))$, de sorte que $(S(X),G)$ est un système dynamique mesuré [3]. Les invariants de $S(X)$ sont des éléments A de B tels que $A \Delta TA$ ou $A \Delta T^{-1}A$ soient $\tilde{P}$-négligeables. Le système $(S(X),G)$ est ergodique lorsque les seuls invariants sont de probabilité zéro ou un ; par extension, la f.a. $\tilde{X}$ est dite ergodique et,par abus de language,la f.a. X est aussi dite ergodique.

Notons $\mathcal{B}_X = \bigvee_{t \in \mathbb{R}} X_t^{-1}(\mathcal{B}(\mathbb{R}))$ la tribu engendrée par X. Supposons que

$X : (\Omega, \mathcal{B}_X) \times \mathbb{R}, \mathcal{B}(\mathbb{R})) \to (\mathbb{R}, \mathcal{B}(\mathbb{R}))$ soit mesurable, alors $\lambda(0 \leq t \leq T : X(t) > \varphi(t))$ est $\mathcal{B}_X$-mesurable (théorème de Fubini). Il existe donc une partie dénombrable S de $\mathbb{R}$ telle que si on note $\mathcal{B}_X(S)$, la tribu $\bigvee_{s \in S} X_s^{-1}(\mathcal{B}(\mathbb{R}))$, alors la v.a. $\lambda(0 \leq t \leq T : X(t) > \varphi(t))$ est $\mathcal{B}(S)$-mesurable. Par conséquent, elle a même loi que la v.a. du même type liée à $\tilde{X}$. On a en particulier le

LEMME 2.3.1. <u>Soient</u> $X : (\Omega, \mathcal{B}_X, P) \times (\mathbb{R}, \mathcal{B}(\mathbb{R})) \to (\mathbb{R}, \mathcal{B}(\mathbb{R}))$ <u>un processus strictement stationnaire mesurable ergodique et</u> $g, \varphi : \mathbb{R}_+ \to \mathbb{R}_+^*$ <u>deux applications croissantes. Alors les v.a.</u>

$$\overline{\lim_{t\to\infty}} \frac{\lambda(A_\varphi(\omega) \cap [0,t])}{g(t)}$$

<u>et</u>

$$\underline{\lim_{t\to\infty}} \frac{\lambda(A_\varphi(\omega) \cap [0,t])}{g(t)} ,$$

<u>ont des lois dégénérées.</u>

2.3.2. Dans le théorème suivant on a posé pour tout réel x

$$\Psi(x) = \int_x^\infty e^{-u^2/2} \frac{du}{\sqrt{2u}} .$$

THEOREME 2.3.1. <u>Soit</u> $\{X(\omega,t) , \omega \in \Omega , t \in \mathbb{R}\}$ <u>une f.a. gaussienne stationnaire,</u> $B_X \otimes B(\mathbb{R})$ <u>mesurable, normalisée, de covariance</u> $r(t)$. <u>On suppose que</u>

(h_1) $r(t) = O(1/\log t)$, $t \to \infty$

<u>Soit</u> $\varphi : \mathbb{R}_+ \to \mathbb{R}_+$ <u>continue croissante telle que</u>

(h_2) $J(\varphi) = \int^{+\infty} \Psi \circ \varphi(t)\, dt = \infty$

(h_3) $\varphi(t) \asymp^{(*)} (\log t)^{\frac{1}{2}}$, $t \to \infty$.

<u>Dans ces conditions, on a aussi</u>

$$(2.3.1.) \quad P\{\omega : \overline{\lim_{T\to\infty}} \frac{\lambda([0,T] \cap A_\varphi(\omega))}{\int_0^T \Psi_o \varphi(t) dt} \geq 1\} = 1 ,$$

<u>et</u>

<u>(2.3.2.) $P\{\omega : \lambda(A_\varphi(\omega)) = \infty\} = 1$.</u>

(*) On écrit $f(t) \asymp g(t)$, $t \to \infty$ lorsque $f = O(g)$ et $g = O(f)$, $t \to \infty$.

La démonstration utilise le petit lemme technique suivant :

LEMME 2.3.2. Soient (η_1,η_2) un couple de v.a. gaussiennes centrées réduites, $\rho = E[\eta_1,\eta_2]$ et $\epsilon > 0$.

a) Pour tous $x > 0$, $y > 0$ tels que $\rho\, xy < \epsilon$, on a

$$P\{\eta_1 > x, \eta_2 > y\} \le c(\epsilon)\, P\{\eta_1 > x\}\, P\{\eta_2 > y\}.$$

b) Il existe un nombre $C_1 > 0$ tel que pour tous $0 \le a \le b$ et $0 \le \rho \le 1$ on ait

$$P\{\eta_1 > a, \eta_2 > b\} \le O(1) \exp\left(-\frac{1-\rho}{4} b^2\right) . P\{\eta_1 > a\}.$$

Démonstration du théorème 2.3.1. Il est clair que 2.3.1. implique 2.3.2. Nous montrons (2.3.1). L'inégalité classique de Paley-Zygmund pour des v.a. $X \ge 0$ de carré intégrable

$$\forall\, 0 < \delta < 1 \quad , \quad P\{X \ge \lambda EX\} \ge (1-\delta)^2 \frac{(EX)^2}{(EX^2)} ,$$

montre ici pour tous $S > 0$ et $0 < \delta < 1$,

$$P\{ \overline{\lim_{T\to\infty}} \frac{\lambda(A_\varphi(\omega) \cap [0,T])}{\int_0^T \Psi_o\varphi(t)dt} > \delta\} \ge P\{ \overline{\lim_{\substack{T\to\infty \\ T>\delta}}} \frac{\lambda(A_\varphi(\omega) \cap [S,T])}{\int_S^T \Psi o\varphi(u)du} \}$$

$$\ge \overline{\lim_{T\to\infty}}\; P\{\frac{\lambda(A_\varphi(\omega) \cap [S,T])}{\int_S^T \Psi o\varphi(u)du} > \delta\} \ge (1-\delta)^2\; \overline{\lim_{T\to\infty}}\; Q_\varphi([S,T])$$

où on a posé
$$Q_\varphi([S,T]) = \frac{(\int_S^T \Psi o\varphi(t)dt)^2}{\int_S^T \int_S^T P\{X(s)>\varphi(s), X(t)>\varphi(t)\}dsdt}$$

Le théorème 2.3.1. sera établi grâce au lemme 2.3.1. si l'on montre que $\overline{\lim_{T\to\infty}}\; Q_\varphi([S,T]) > 0$. Or nous avons le

LEMME 2.3.3. Les conditions du théorème 2.3.1. impliquent qu'il existe $S_o > 1$, $\tau_o > 1$, et $k_o > 0$, $k_1 > 0$ indépendants de S_o, tels que pour tout $S > S_o$ et $T > \tau_o S$,

$$(2.3.3.)\qquad Q_\varphi([S,T]) \geq \left(k_1 + \frac{k_o}{\int_S^T \Psi\circ\varphi(s)}\right)^{-1}$$

On a aussi, par conséquent : $\overline{\lim_{T\to\infty}}\ Q_\varphi([S,T]) \geq k_1^{-1}$.

Démonstration du lemme 2.3.3. Les hypothèses h_1 et h_2 impliquent qu'il existe $0 < a \leq b < \infty$, $0 < M_o < \infty$ et $S_o > 1$ tels que pour tout $t > S_o$

$$(1)\qquad a \leq \frac{\varphi(t)}{\sqrt{2\log t}} \leq b ,$$

$$(2)\qquad |r(t)| \leq \frac{M_o}{\log t} .$$

Nous avons aussi, par symétrie de $P(s,t) = P\{X(s) > \varphi(s)\ , X(t) > \varphi(t)\}$

$$(3)\qquad \int_S^T \int_S^T P(s,t)\, ds\, dt = 2\int_S^T ds \left(\int_S^T P(s,t)\, dt\right)$$

soient alors $\alpha > 0$ et $A > 0$ tels que

$$(4)\qquad \forall\, t \geq A ,\ |r(t)| \leq \frac{1}{2} \quad \text{et} \quad \alpha < \frac{a^2}{4} .$$

Nous effectuons le découpage suivant

$$(5)\qquad \Lambda(s) = \int_S^T P(s,t)\, dt = \Lambda_1(s) + \Lambda_2(s) + \Lambda_3(s) ,$$

où

$$\Lambda_1(s) = \int_s^{s+A} P(s,t)\, dt ,$$

$$\Lambda_2(s) = \int_{s+A}^{s+A+s^\alpha} P(s,t)\, dt ,$$

$$\Lambda_3(s) = \int_{s+A+s^\alpha}^{T} P(s,t)\, dt .$$

Cette décomposition peut s'expliquer par le schéma suivant,

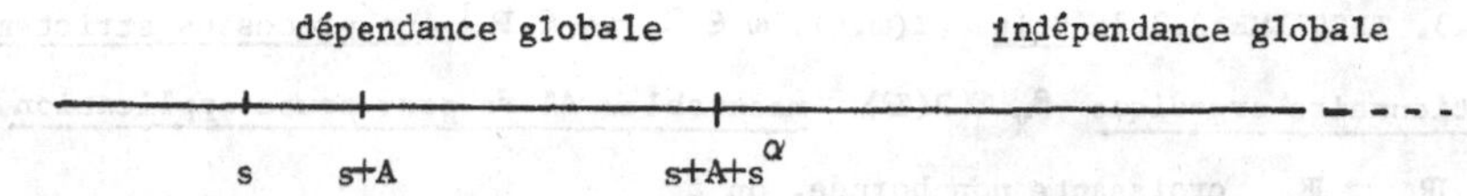

Nous majorons $\Lambda(s)$; tout d'abord, on voit facilement que

(6) $$\Lambda_1(s) \leq A\, \Psi o\varphi(s) \ .$$

Par l'intermédiaire du lemme 2.3.2., on établit successivement

$$\Lambda_2(s) \leq O(1)\, \Psi o\varphi(s) \int_{s+A}^{s+A+s^\alpha} \exp\left(-\frac{1-r(t-s)}{4}\cdot\varphi^2(t)\right) dt,$$

(7) $$\leq O(1)\, \Psi o\varphi(s)\, s^{\alpha - a^2/4} \ ,$$

et

(8) $$\Lambda_3(s) \leq k_1\, \Psi o\varphi(s) \int_{s+A+s^\alpha}^{T} \Psi o\varphi(t)\, dt \ ,$$

où $k_1 > 1$ ne dépend que du choix de S_o .

On déduit de (6) , (7) et (8)

$$2\int_S^T \Lambda(s)\, ds \leq 2(A + O(s^{\alpha - \frac{a^2}{4}})) \int_S^T \Psi o\varphi(u)\, du + k_1\left(\int_S^T \Psi o\varphi(u)\, du\right)^2 \ .$$

Finalement pour tout $S > S_o$ et $T > \tau_o S$, on a montré

$$Q_\varphi([S,T]) \geq \left(k_1 + k_O\left(\int_S^T \Psi o\varphi(u)\, du\right)^{-1}\right)^{-1} ,$$

où $k_O > 0$ ne dépend que de S_o ; d'où le lemme 2.3.3. □

On montrerait aussi, à l'aide du lemme 2.3.3., que les conditions (h_1) et (h_2) impliquent

(2.3.4.) $$P\{\overline{\lim_{T\to\infty}}\ \frac{\lambda(A_\varphi^c(\omega) \cap [0,T])}{T} = 1\} = 1 \ .$$

(indication : traiter séparément les cas $J(\varphi) < \infty$ et $J(\varphi) = \infty$) .

Cependant (2.3.4.) peut-être très nettement renforcée. C'est l'objet du théorème suivant.

2.3.3. THEOREME 2.3.2. Soit $\{X(\omega,t),\ \omega \in \Omega\ ,\ t \in \mathbb{R}\}$ un processus strictement stationnaire ergodique $\mathcal{B}_X \otimes B(\mathbb{R})$ mesurable. Alors pour toute application $g : \mathbb{R}_+ \to \mathbb{R}_+$ croissante non bornée, on a

$$(2.3.5.)\qquad P\{\lim_{t\to\infty} \frac{\lambda(0 \le u \le t : |X(t)| \le g(t))}{t} = 1\} = 1 \ .$$

Nous démontrons d'abord deux lemmes.

LEMME 2.3.4. Les hypothèses du théorème 2.3.2. impliquent pour tout élément A de la tribu produit $\mathcal{B}$, si

$$T_o(\omega) = \chi_A(X(\omega))$$

$$\forall\, n \ge 1,\ T_n(\omega) = \inf\{k > T_{n-1}(\omega) : T^k(X(\omega)) \in A\}$$

et $\quad P\{T_o > 0\} > 0$, alors

$$P\{\lim_{j\to\infty} \frac{T_j}{j} = \frac{1}{P\{T_o > 0\}} \mid T_o > 0\} = 1 \ .$$

Démonstration : Il revient au même d'établir le lemme pour $\tilde{X}$. Notons alors $\mathcal{B}_A = \mathcal{B} \cap A$, $\mu_A = \mu(\cdot|A)$ et pour tout élément f de E , $T_A(f) = T^{T_1(f)}(f)$. Par hypothèse le système $(S(X),G)$ est ergodique. On sait (cf [3] p. 130) que cela implique que le système induit $(A,\mathcal{B}_A,\mu_A,T_A)$ l'est aussi.

Considérons l'application mesurable $F = T_1 : (A,\mathcal{B}_A) \to \mathbb{N}$. On constate que F est intégrable et

$$E_A(F) = \int_A F\, d\mu_A = \sum_{\ell=1}^{\infty} \mu_A(F \ge \ell) = \frac{\mu(\cup_{k=0}^{\infty} T^k A)}{\mu(A)} = \frac{1}{\mu(A)} \ .$$

Le lemme (2.3.4.) apparaît donc comme une conséquence facile du théorème de Birkhoff.

LEMME 2.3.5. Les hypothèses du théorème 2.3.2. impliquent aussi pour tout $x > 0$,

$$(2.3.6.)\qquad P\{\liminf_{U\to\infty} \frac{\lambda(0 \le u \le U : |X(\omega,u)| \le x)}{U} \ge (1 - \sqrt{G(x)})^2\} = 1 \ ,$$

où on a posé $\quad G(x) = P\{|X(0)| > x\}$.

Démonstration : Il suffit de traiter le cas $G(x) < 1$.

Posons pour tout $x > 0$ et $0 < \gamma < 1$,

$$\alpha(\gamma,x) = \{\omega : \lambda(0 \le t \le 1 : |X(\omega,t)| \le x\} > \gamma\}$$

$$\gamma_x = 1 - \sqrt{G(x)} .$$

On obtient facilement pour tout $0 < \gamma < \gamma_x$

$$P\{\alpha(\gamma,x)\} = 1 - P\{\lambda(0 \le t \le 1 : |X(\omega,t)| > x) > 1 - \gamma\}$$

$$(1) \qquad \ge 1 - \frac{E[\lambda(0 \le t \le 1 : |X(\omega,t)| > x)]}{1 - \gamma}$$

$$= 1 - \frac{G(x)}{1-\gamma} > 0 .$$

Fixons $x > 0$ et $0 < \gamma < \gamma_x$ de façon à ce que $P\{\alpha(\gamma,x)\} > 0$. On constate à l'aide de (1),

$$(2) \qquad \lim_{x \to \infty} P\{\alpha(\gamma,x)\} = 1 .$$

Définissons la suite des temps de retour du processus dans $\alpha(\gamma,x)$

$$R_o(\omega) = \chi_{\alpha(\gamma,x)}(\omega)$$

$$\forall\, j > 0, \quad R_j(\omega) = \inf\{k > R_{j-1}(\omega) : T^k(X(\omega)) \in \alpha(\gamma,x)\} .$$

Les lemmes 2.3.1. et 2.3.4. impliquent

$$(3) \qquad P\{ \lim_{j\to\infty} \frac{R_j}{j} = \frac{1}{P\{\alpha(\gamma,x)\}} \mid R_o > 0\} = 1 .$$

De plus, pour tout n entier et tout ω dans $\alpha(\gamma,x)$, nous avons

$$(4) \qquad \begin{aligned} &\lambda(0 \le t \le 1 : |X(\omega, R_n(\omega) + t)| \le x) \\ &= \lambda(R_n(\omega) \le t \le R_n(\omega) + 1 : |X(\omega,t)| \le x\} \ge \gamma . \end{aligned}$$

Soit maintenant ω dans $\alpha(\gamma,x) \cap \{ \lim_{j\to\infty} \frac{R_j}{j} = \frac{1}{P\{\alpha(\gamma,x)\}} \}$. Pour tout $\varepsilon > 0$, il existe $j(\varepsilon,\omega) < \infty$ tel que

$$(5) \qquad \forall\, j > j(\varepsilon,\omega) , \quad j(\frac{1}{P\{\alpha(\gamma,x)\}} - \varepsilon) \le R_j < j(\frac{1}{P\{\alpha(\gamma,x)\}} + \varepsilon) .$$

Soit alors $t > R_{j(\varepsilon,\omega)}$ et j l'entier tel que

$$(6) \qquad R_j + 1 \leq t < R_{j+1} + 1 .$$

On a par inclusion

$$(7) \qquad \frac{\lambda(0 \leq u \leq t : |X(\omega,u)| \leq x)}{t} \geq \frac{\lambda(0 \leq u \leq R_{j+1} : |X(\omega,u)| \leq x)}{R_{j+1} + 1}$$

$$\geq \frac{\gamma_j}{(j+1)(\epsilon + \frac{1}{P(\alpha(\gamma,x))}) + 1} .$$

Cette dernière inégalité montre

$$(8) \qquad \liminf_{t \to \infty} \frac{\lambda\{0 \leq u \leq t : |X(\omega,u)| \leq x\}}{t} \geq \frac{\gamma}{\epsilon + (P\{\alpha(\gamma,x)\})^{-1}} .$$

Mais comme $\epsilon > 0$ est arbitraire, on obtient donc pour tout ω dans

$\alpha(\gamma,x) \cap \{ \lim_{j \to \infty} \frac{R_j}{j} = \frac{1}{P\{\alpha(\gamma,x)\}}\}$, événement de probabilité strictement positive,

$$(9) \qquad L(\omega) = \liminf_{t \to \infty} \frac{\lambda(0 \leq u \leq t : |X(\omega,u)| \leq x\}}{t} \geq \gamma(1 - \frac{G(x)}{1-\gamma}) .$$

Mais puisque X est ergodique, la v.a. L a une loi dégénérée. On a donc, pour tout $x > 0$, en faisant tendre γ vers γ_x

$$(10) \qquad P\{ \liminf_{t \to \infty} \frac{\lambda(0 \leq u \leq t : |X(\omega,u)| \leq x\}}{t} \geq (1 - \sqrt{G(x)})^2\} = 1$$

ce qui établit le lemme 2.3.5. □

<u>Remarques</u> : On établit de la même façon

$$(2.3.7.) \qquad P\{ \liminf_{t \to \infty} \frac{\lambda(0 \leq u \leq t : |X(\omega,u)| \geq x\}}{t} \geq (1 - \sqrt{1 - G(x)})^2\} = 1 .$$

La conjugaison des résultats (2.3.6) et (2.3.6) a pour conséquence le corollaire suivant, dans lequel on a posé

$$\forall U > 0 , \forall x > 0 , \forall \omega \in \Omega ,$$

$$L_x(\omega,U) = \frac{\lambda(0 \leq t \leq U : |X(\omega,t)| \leq x\}}{U} .$$

COROLLAIRE 2.3.1. Les hypothèses du théorème de 2.3.2. impliquent que l'oscillation à l'infini de la v.a. $L_x(\omega,U)$ est presque sûrement majorée par $2(\sqrt{1-G(x)} + \sqrt{G(x)}-1)$. En d'autres termes,

$$(2.3.8)\quad \text{p.s.}\quad W_\infty(L_x(\omega,.)) = \lim_{U\to\infty} \sup_{t,t'>U} |L_x(\omega,t) - L_x(\omega,t')|$$
$$\leq 2(\sqrt{1-G(x)} + \sqrt{G(x)-1}) .$$

On en déduit, lorsque $P\{X(0) = 0\} = 0$

$$(2.3.9)\text{p.s.}\quad \lim_{x\to 0} W_\infty(L_x(\omega)) = \lim_{x\to\infty} W_\infty(L_x(\omega)) = 0 .$$

Démonstration du théorème 2.3.2. Soient g et $x > 0$ fixés ; notons $T = T(g,x)$ un entier tel que

$$(1)\qquad \forall\, t > T ,\quad g(t) > x .$$

On a, par inclusion, pour tout $t > 0$,

$$(2)\qquad \begin{aligned} &\lambda(0 \leq u \leq t+T : |X(\omega,t)| \leq g(u)\} \\ &\geq \frac{\lambda\{T \leq u \leq T+t : |X(\omega,u)| \leq x\}}{t+T} - \frac{\lambda(0 \leq u \leq T : |X(\omega,u)| \leq x\}}{t+T} . \end{aligned}$$

On applique le lemme (2.3.5) à (2) et on obtient

$$(3)\ \text{p.s.}\ \liminf_{U=t+T\to\infty} \frac{\lambda(0 \leq u \leq U : |X(\omega,u)| \leq g(u))}{U} \geq (1 - \sqrt{G(x)})^2 .$$

Le théorème se déduit de (3) en faisant tendre x vers l'infini. □

Remarques : 1) Soit $h : \mathbb{R}_+ \to \mathbb{R}_+$ décroissante telle que $\lim_{t\to\infty} h(t) = 0$.

On montre de la même façon, en utilisant (2.3.7) cette fois,

$$(2.3.10)\qquad P\{\omega : \lim_{t\to\infty} \frac{\lambda(0 \leq u \leq t : |X(\omega,u)| \geq h(u))}{t} = 1\} = 1$$

2) Supposons que le processus étudié soit de plus partiellement majoré au sens suivant

(C) il existe $a > 0$ tel que notant $\|X\|_a = \sup\{|X(t)|\ 0 \leq t \leq a\}$, on ait : $P\{\|X\|_a < \infty\} > 0$.

Notons alors $\mathcal{S}(a)$ l'ensemble des familles dénombrables d'intervalles non empiétants $I = \{I_n , n \geq 1\}$, de longueur commune a, et, $\hat{I} = \bigcup_{n\geq 1} I_n$. Notons aussi pour tout $0 \leq \rho \leq 1$,

$$\mathcal{S}_\rho(a) = \{I \in \mathcal{S}(a) : D_\infty(\hat{I}) = \liminf_{t \to \infty} \frac{\lambda(\hat{I} \cap [0,t])}{t} \geq \rho\}$$

et posons pour tout $x > 0$, $a > 0$

$$A(a,x) = [\ \|X\|_a \leq x\}\ .$$

Pour tout x tel que $P\{A(a,x)\} > 0$, on a alors

(2.3.11) $\quad P\{\omega : \exists\, I \in \mathcal{S}(a, P\{A(a,x)\} : \forall\, t \in I\ ,\ |X(\omega,t)| \leq x\} = 1$.

Ceci montre que $|X|$ est presque sûrement inférieur à x sur une suite d'intervalles dont la répartition est justement précisée par $P\{\|X\|_a \leq x\}$.

Démonstration de (2.3.11) (esquisse) : Soit T_a la transformation $X(t) \to X(t+a)$ où $a > 0$ est déterminé par (C) ; notons $\{R_n , n \geq 0\}$ la suite des temps de retour de X dans $A(a,x)$. On calque sur la démonstration du lemme (2.3.5) puisqu'ici aussi, on a

$$\text{p.s.} \quad \lim_{j\to\infty} \frac{R_j}{j} = \frac{1}{P\{A(a,x)\}} \ .$$

3) Supposons que X soit presque sûrement majoré sur tout intervalle borné. On déduit de (2.3.11) que toute application $g : \mathbb{R}_+ \to \mathbb{R}_+$ croissante non bornée majore presque sûrement $|X|$ sur des intervalles de longueur arbitraires.

Dans le même ordre d'idées, on peut s'interroger sur l'existence d'intervalles de longueur $\alpha > 0$ fixée sur lesquels $|X|$ est minoré par $g(t)$. C'est l'objet des paragraphes suivants.

2.3.4. Dans le paragraphe $X(t)$, $t \in \mathbb{R}$ est un processus gaussien stationnaire, normalisé à trajectoires continues et tel que $\lim_{t\to\infty} E[X(0).X(t)] = 0$.

On rappelle que ces hypothèses entraînent ([8]) ,

$$(2.3.12) \qquad P\{\limsup_{t\to\infty} \frac{|X(\omega,t)|}{\sqrt{2\log t}} = 1\} = 1 \ .$$

Nous étudions les instants durant lesquels l'amplitude de la trajectoire $X(\omega)$ est maximale. Nous commençons par poser

$$\mu(X,\varphi) = \sup\{\alpha \geq 0 : \forall\, T,\ P\{\exists\, t > T : \forall\, u \in\,]t,t+\alpha[,\ |X(u)| > \varphi(u)\} = 1\} \ .$$

Nous montrons que nos hypothèses entraînent aussi lorsque

$\varphi(t) = \varphi_\epsilon(t) = (1-\epsilon)\sqrt{2\log t}$, $0 < \epsilon \leq 1$, alors ,

$$(2.3.13) \qquad 0 < \mu(X,\varphi) < \infty \ .$$

Nous montrons aussi qu'il existe une classe de processus gaussiens telle que $\mu(X,\varphi)$ soit arbitrairement grand dans cette classe. Celle-ci est constituée des processus gaussiens stationnaires tels que $r(t)$ soit deux fois dérivable en 0. Posons alors

$$\mu(X,\epsilon) = \mu(X\ ,\ \varphi_\epsilon) \ .$$

Nous établirons

$$(2.3.14) \qquad \mu(X,\epsilon) \geq \frac{\epsilon}{C_o\sqrt{-r''(0)\log\frac{C_o}{\epsilon}}} \ ,$$

où $C_o > 0$ est une constante absolue.

Ces résultats reposent avant tout sur une estimation de la loi de $\inf_{0\leq t\leq 1} |X(t)|$ qui fait l'objet des lemmes suivants

2.3.5. LEMMES PRELIMINAIRES. Considérons un processus gaussien $\{X(t),\ t \in T\}$ normalisé, où T est un ensemble arbitraire. Notons d l'écart sur T associé à X , et, pour tout $u > 0$, $N_d(T,u)$ le cardinal minimal des suites de centres de d-boules ouverte de rayon u suffisant pour recouvrir T .

Lorsque $N_d(T,u)$ est fini, on note $S_d(T,u)$ une des suites minimales.

LEMME 2.3.6. <u>Soient</u> $\{X(t),\ t \in T\}$ <u>un processus gaussien normalisé,</u> d <u>l'écart sur</u> T <u>associé à</u> X . <u>On suppose l'intégrale</u> $\int_0 \sqrt{\log N_d(T,u)}\, du$ <u>convergente. Alors pour tout</u> ϵ <u>compris entre</u> 0 <u>et</u> 1 <u>strictement, on peut trouver deux</u>

nombres $0 < B(\varepsilon) < \infty$ et $0 < z(\varepsilon) < \infty$ tels que quels que soient $\lambda > \sqrt{2}$, $0 \le z \le z(\varepsilon)$ et $\sigma \in T$, on ait :

$$(2.3.15)\quad P\{\inf\{|X(s)|,\ s \in B_d(\sigma,z) < \lambda(1-\varepsilon),\ |X(\sigma)| > \lambda\} \le B(\varepsilon)\, \Psi(\lambda(1+\varepsilon)) .$$

Démonstration : Pour tout ε,ρ compris entre 0 et 1 strictement tout σ élément de T et $z > 0$, on commence par poser

$$Y(\rho,\sigma,z) = \Sigma_{k=1}^{\infty}\ (\rho^{k-1}z)^2 \vee (2z\rho^{k-1} \sqrt{\log \frac{N_d(B_d(\sigma,z),\rho^k z)}{\rho^k z}} ,$$

$$H(\rho,\sigma,z) = \frac{2}{\rho^2(1-\rho^2)} \int_0^{\rho z} [u \vee \sqrt{\log \frac{N_d(B_d(\sigma,z),\ u)}{u}}]\, du \ ,$$

$$H(\rho,T,z) = \frac{2}{\rho^2(1-\rho^2)} \int_0^{\rho z} [u \vee \log \frac{N_d(T,u)}{u}]du \ .$$

On constate aisément que $Y(\rho,\sigma,z) \le H(\rho,\sigma,z) \le H(\rho,T,z) < \infty$, et $H(\rho,T,.)$ est croissante et s'annule en zéro. Nous fixons maintenant ρ dans l'intervalle $]0, \exp(-4\varepsilon)[$ par exemple $\rho = e^{-4}$, et nous posons

$$\forall\, 0 < \varepsilon < 1,\ \ z(\varepsilon) = \sup\{0 < z < \sqrt{2} :\ H(\rho,T,z) \le \varepsilon\}$$

$$\forall\, z > 0,\ \forall\, k \ge 0 \quad A_k = \{\inf\{|X(s)|\ s \in S_d(B(\sigma,z),z\rho^k\} < \lambda(1-\sum_{j=1}^{k} y_j)\}$$

où $y_j = (\rho^{j-1}z)^2 \vee 2(z\rho^{j-1}) \sqrt{\log \frac{N_d(B_d(\sigma,z),\rho^j z)}{\rho^j z}}$.

Alors pour tout $0 < z < z(\varepsilon)$ et tout élément σ de T , $Y(\rho,\sigma,z)$ est majoré par ε ; par conséquent

$$E = P\{\inf\{|X(s)|,\ s \in B(\sigma,z) < \lambda(1-\varepsilon)\ ,\ |X(\sigma)| > \lambda\}$$

$$\le P\{\inf\{|X(s),\ s \in B(\sigma,z)\} < \lambda(1 - Y(\rho,\sigma,z)),\ |X(\sigma)| > \lambda\}$$

$$\le P\{\bigcup_{k=1}^{\infty} A_k \cap A_o^c\} \le \sum_{\substack{k=1 \\ s \in S_d(B(\sigma,z),z\rho^k)}}^{\infty} \alpha_k(s),$$

où $\alpha_k(s) = P\{|X(s)| < \lambda(1-\sum_{j=1}^{k} y_j),\ |X(\tau_k(s))| \ge \lambda(1-\sum_{j=1}^{k-1} y_j)\}$

et $\tau_k : B(\sigma,z) \to S_d(B(\sigma,z),z\rho^{k-1})$ est une des applications définies par $d(s,\tau_k(s)) < z\rho^{k-1}$.

Décomposons $X(s)$ suivant $X(\tau_k(s))$,

$$X(s) = \eta\, X(\tau_k(s)) + \sqrt{1-\eta^2}.Z \ ,$$

où $(Z,X(\tau_k(s))$ est un couple de v.a. gaussiennes indépendantes centrées réduites et

$$\eta = E\{X(s)\ .\ X(\tau_k(s))\} = 1 - \frac{1}{2}\, d^2(s,\tau_k(s)) > 1 - \frac{z^2}{2} \geq 0 \ .$$

A partir de

$$|Z\sqrt{1-\eta^2}| \geq \eta|X(\tau_k(s)| - |X(s)| \ ,$$

on obtient la majoration suivante

$$\alpha_k(s) \leq P\{|X(\tau_k(s))| > \lambda(1-\Sigma_{b=1}^{j} y_b),\ |Z\sqrt{1-\eta^2}| > \lambda(y_k - \tfrac{1}{2}(z\rho^{k-1})^2)\} \ ,$$

dans laquelle $y_k - \frac{1}{2}(z\rho^{k-1})^2 > 0$.

D'où,

$$(2) \qquad \alpha_k(s) \leq \Psi(\lambda(1-\epsilon))\ \Psi(\lambda\sqrt{\log \frac{N(B(\sigma,z),z\rho^k)}{z\rho^k}}\ ,$$

$$\leq O(1)\ \Psi(\lambda(1-\epsilon))(N(B(\sigma,z),z\rho^k))^{-\frac{\lambda^2}{2}}\ (\rho^k z)^{\frac{\lambda^2}{2}} \ .$$

On en déduit en reportant (2) dans (1) :

$$E \leq O(1)\ \Psi(\lambda(1-\epsilon))\ \frac{(z\rho)^{\frac{\lambda^2}{2}}}{1-\rho^{\frac{\lambda^2}{2}}} \leq O(1)\ \Psi(\lambda(1+\epsilon))(1+2\epsilon)(1-\rho^{\frac{\lambda^2}{2}})^{-1}$$

$$\leq O(1)\ (1+2\epsilon)\ \Psi(\lambda(1+\epsilon)) \ ,$$

car $z(\epsilon) \leq \sqrt{\epsilon(1-\rho^2)} < 1$; ce qui établit le lemme 2.3.6.

THEOREME 2.3.3. <u>Soit</u> $\{X(t),\ t \in \mathbb{R}\}$ <u>un processus gaussien stationnaire normalisé à trajectoires continues</u> $r(t)$ <u>et tel que</u> $\lim_{t\to\infty} r(t) = 0$. <u>Alors on a</u>

a) $\forall\ 0 < \alpha \leq 1\ ,\quad 0 < \mu(X,\alpha) < \infty$

b) <u>Si, de plus,</u> $r(t)$ <u>a une dérivée seconde en zéro,</u> a) <u>est renforcée par</u>

$$\forall\, 0 < \alpha \le 1\ ,\quad \mu(X,\alpha) \ge \frac{\alpha}{C_o \sqrt{-r''(0)\log \frac{C_o}{\alpha}}}\ ,$$

où $C_o > 0$ est une constante indépendante de X.

<u>Démonstration</u> : a) Puisque $\lim_{t\to\infty} r(t) = 0$, on a, a fortiori,

$\lim_{\beta\to\infty} \frac{1}{\beta^2} \int_0^\beta (\beta-u)\, r(u)\, du = 0$. Soit $0 < \alpha \le 1$ fixé et $\beta > 0$ lié à α par

la relation

(1) $\quad \beta^2(1-\alpha)^2 \left(\int_0^\beta (\beta-s)\, r(s)\, ds\right)^{-1} > 1$.

Notons pour tout entier $k \ge 1$, $A_k = \{\forall\, s \in [n\beta,(n+1)\beta]\ |X(s)| \ge \varphi_\alpha(s)\}$.

On obtient facilement

(2) $\quad \forall\, k \ge 1\ ,\quad P(A_k) \le \Psi\left(\frac{\beta\ \varphi_\alpha(n\beta)}{\sqrt{\int_0^\beta (\beta-s) r(s) ds}}\right)$

ce qui, par (1), montre que la série de terme général $P(A_k)$ est convergente.

Le lemme de Borel-Cantelli montre que presque sûrement pour tout n suffisamment grand, il existe $t \in [n\beta\ , (n+1)\beta]$ tel que $|X(\omega,t)| < \varphi_\alpha(t)$. Ceci implique par conséquent

(3) $\quad \mu(X,\alpha) \le 2\beta < \infty$.

Nous établissons $\mu(X,\alpha) > 0$. Fixons α et σ dans $]0,1[$ et posons

$\epsilon = \epsilon(\alpha) = \frac{\alpha(3-\alpha)}{2(2-\alpha)}$.

Soit $z(\epsilon)$ défini dans la démonstration du lemme 2.3.6. et notons τ_x la translation par x.

Nous posons pour tout entier $n \ge 1$

(4)
$$\Omega'_n = \{\inf\{|X(s)|,\ s \in \tau_{n-1}(B(\sigma,z(\epsilon))) \cap\,]n-1,n[\} < \varphi_\alpha(n)\}$$
$$\Omega''_n = \{|X(\tau_{n-1}(\sigma))| > \frac{\varphi_\alpha(n)}{1-\epsilon}\}$$

La continuité de $r(t)$ en zéro montre que $B(\sigma,z(\varepsilon)) \cap]0,1]$ contient un petit intervalle $[0,\eta]$, où $\eta > 0$ dépend de α ; de même pour la suite des translatés $\tau_{n-1}(B(\sigma,z(\varepsilon))) \cap]n-1,n[$.

Le lemme 2.3.6. montre que pour tout n suffisamment grand

$$(5) \qquad P\{\Omega'_n \cap \Omega''_n\} \leq C(\varepsilon)\,\Psi\Big(\frac{1+\varepsilon}{1-\varepsilon}\,\varphi_\alpha(n)\Big)$$

où $0 < C(\varepsilon) < \infty$ ne dépend que de ε .

La majoration précédente ainsi que le choix de ε en fonction de α permettent de conclure à $\sum_{n=1}^{\infty} P\{\Omega'_n \cap \Omega''_n\} < \infty$.

D'où en vertu du lemme de Borel-Cantelli

$$1 = P\{\liminf_{n\to\infty} ((\Omega'_n)^c \cup (\Omega''_n)^c)\}$$

$$\leq P\{\limsup_{n\to\infty} (\Omega'_n)^c\} + 1 - P\{\limsup_{n\to\infty} \Omega''_n\}$$

$$= P\{\limsup_{n\to\infty} (\Omega'_n)^c\} ,$$

car d'après (2.3.12), $(1-\varepsilon)^{-1}\varphi_\alpha$ appartient à la classe $\mathcal{L}_\infty(\{X(\tau_n(\sigma)),\ n \geq 1\})$ de sorte que l'événement $\limsup_{n\to\infty} \Omega''_n$ a pour probabilité 1 . Cette argumentation permet donc d'établir

$$P\{\limsup_{n\to\infty} (\Omega'_n)^c\} = 1 ,$$

et, a fortiori,

$$\mu(X,\alpha) \geq \eta > 0 .$$

Enfin b) s'obtient en explicitant les calculs effectués dans la démonstration du lemme 2.3.6.

□

REFERENCES DU CHAPITRE II

[1] S.M. BERMAN : Asymptotic independance of the numbers of high and low level crossings stationary gaussian processes. Ann. of Math. and Statis. (1971). V. 42, N° 3, 927-945.

[2] S.M. BERMAN : A compound poisson limit for stationary sums and sojourns of gaussian processes. Ann. of Prob. (1980), Vol. 8, n°3,511-538.

[3] J.P. CONZE : Systèmes topologiques et métriques en théorie ergodique. Ecole d'Eté de probabilité de Saint-Flour (1974), Lect. Notes Math. 480 (1975), p. 100-187.

[4] H. CRAMER, M.R. LEADBETTER : Stationary and Related stochastic processes (Wiley).

[5] X. FERNIQUE : Régularité des trajectoires de fonctions aléatoires gaussiennes. Ecole d'Eté de probabilité de Saint-Flour. Lect. Notes Math. 480 (1975), 2-95.

[6] X. FERNIQUE : L'ordre de grandeur à l'infini de certaines fonctions aléatoires, préprint.

[7] K. ITO : The expected number of zeros of continuous stationary gaussian processes. J. Math. Kyoto Univ. 3-2, (1964) 207-216.

[8] G. MARUYAMA : The harmonic analysis of stationary gaussian processes. Memoirs of the faculty of Sci. Kyusyu Univ. (1949) Ser. A, V4, 45-106.

[9] Y.A. ROSANOV, V.A. VOLKONSKI : Some limit theorems for randoms functions I and III. Teoriya Veroyatnostei i ee Primaneniya 186-206 et 6 (1961), 202-215.

[10] M. WEBER : Analyse asymptotique de processus gaussiens stationnaires. Ann. Inst. H. Poincaré Vol. XVI, n°2 (1980), 117-176.

[11] M. WEBER : Sur les D-modules asymptotiques de processus strictement stationnaires ergodiques. Z. Wahr. (1980) 53, 231-246.

[12] M. WEBER : Sur les instants de grande amplitude des trajectoires de processus gaussiens stationnaires, Z. Wahr.(1980) 53,221-229.

[13] D. YLVISAKER : On the absence of tangencies of gaussian paths. Ann. of Math. Statis. 36 (1965) 1043-1046.

Vol. 817: L. Gerritzen, M. van der Put, Schottky Groups and Mumford Curves. VIII, 317 pages. 1980.

Vol. 818: S. Montgomery, Fixed Rings of Finite Automorphism Groups of Associative Rings. VII, 126 pages. 1980.

Vol. 819: Global Theory of Dynamical Systems. Proceedings, 1979. Edited by Z. Nitecki and C. Robinson. IX, 499 pages. 1980.

Vol. 820: W. Abikoff, The Real Analytic Theory of Teichmüller Space. VII, 144 pages. 1980.

Vol. 821: Statistique non Paramétrique Asymptotique. Proceedings, 1979. Edited by J.-P. Raoult. VII, 175 pages. 1980.

Vol. 822: Séminaire Pierre Lelong-Henri Skoda, (Analyse) Années 1978/79. Proceedings. Edited by P. Lelong et H. Skoda. VIII, 356 pages, 1980.

Vol. 823: J. Král, Integral Operators in Potential Theory. III, 171 pages. 1980.

Vol. 824: D. Frank Hsu, Cyclic Neofields and Combinatorial Designs. VI, 230 pages. 1980.

Vol. 825: Ring Theory, Antwerp 1980. Proceedings. Edited by F. van Oystaeyen. VII, 209 pages. 1980.

Vol. 826: Ph. G. Ciarlet et P. Rabier, Les Equations de von Kármán. VI, 181 pages. 1980.

Vol. 827: Ordinary and Partial Differential Equations. Proceedings, 1978. Edited by W. N. Everitt. XVI, 271 pages. 1980.

Vol. 828: Probability Theory on Vector Spaces II. Proceedings, 1979. Edited by A. Weron. XIII, 324 pages. 1980.

Vol. 829: Combinatorial Mathematics VII. Proceedings, 1979. Edited by R. W. Robinson et al.. X, 256 pages. 1980.

Vol. 830: J. A. Green, Polynomial Representations of GL_n. VI, 118 pages. 1980.

Vol. 831: Representation Theory I. Proceedings, 1979. Edited by V. Dlab and P. Gabriel. XIV, 373 pages. 1980.

Vol. 832: Representation Theory II. Proceedings, 1979. Edited by V. Dlab and P. Gabriel. XIV, 673 pages. 1980.

Vol. 833: Th. Jeulin, Semi-Martingales et Grossissement d'une Filtration. IX, 142 Seiten. 1980.

Vol. 834: Model Theory of Algebra and Arithmetic. Proceedings, 1979. Edited by L. Pacholski, J. Wierzejewski, and A. J. Wilkie. VI, 410 pages. 1980.

Vol. 835: H. Zieschang, E. Vogt and H.-D. Coldewey, Surfaces and Planar Discontinuous Groups. X, 334 pages. 1980.

Vol. 836: Differential Geometrical Methods in Mathematical Physics. Proceedings, 1979. Edited by P. L. García, A. Pérez-Rendón, and J. M. Souriau. XII, 538 pages. 1980.

Vol. 837: J. Meixner, F. W. Schäfke and G. Wolf, Mathieu Functions and Spheroidal Functions and their Mathematical Foundations Further Studies. VII, 126 pages. 1980.

Vol. 838: Global Differential Geometry and Global Analysis. Proceedings 1979. Edited by D. Ferus et al. XI, 299 pages. 1981.

Vol. 839: Cabal Seminar 77 – 79. Proceedings. Edited by A. S. Kechris, D. A. Martin and Y. N. Moschovakis. V, 274 pages. 1981.

Vol. 840: D. Henry, Geometric Theory of Semilinear Parabolic Equations. IV, 348 pages. 1981.

Vol. 841: A. Haraux, Nonlinear Evolution Equations- Global Behaviour of Solutions. XII, 313 pages. 1981.

Vol. 842: Séminaire Bourbaki vol. 1979/80. Exposés 543–560. IV, 317 pages. 1981.

Vol. 843: Functional Analysis, Holomorphy, and Approximation Theory. Proceedings. Edited by S. Machado. VI, 636 pages. 1981.

Vol. 844: Groupe de Brauer. Proceedings. Edited by M. Kervaire and M. Ojanguren. VII, 274 pages. 1981.

Vol. 845: A. Tannenbaum, Invariance and System Theory: Algebraic and Geometric Aspects. X, 161 pages. 1981.

Vol. 846: Ordinary and Partial Differential Equations, Proceedings. Edited by W. N. Everitt and B. D. Sleeman. XIV, 384 pages. 1981.

Vol. 847: U. Koschorke, Vector Fields and Other Vector Bundle Morphisms – A Singularity Approach. IV, 304 pages. 1981.

Vol. 848: Algebra, Carbondale 1980. Proceedings. Ed. by R. K. Amayo. VI, 298 pages. 1981.

Vol. 849: P. Major, Multiple Wiener-Itô Integrals. VII, 127 pages. 1981.

Vol. 850: Séminaire de Probabilités XV. 1979/80. Avec table générale des exposés de 1966/67 à 1978/79. Edited by J. Azéma and M. Yor. IV, 704 pages. 1981.

Vol. 851: Stochastic Integrals. Proceedings, 1980. Edited by D. Williams. IX, 540 pages. 1981.

Vol. 852: L. Schwartz, Geometry and Probability in Banach Spaces. X, 101 pages. 1981.

Vol. 853: N. Boboc, G. Bucur, A. Cornea, Order and Convexity in Potential Theory: H-Cones. IV, 286 pages. 1981.

Vol. 854: Algebraic K-Theory. Evanston 1980. Proceedings. Edited by E. M. Friedlander and M. R. Stein. V, 517 pages. 1981.

Vol. 855: Semigroups. Proceedings 1978. Edited by H. Jürgensen, M. Petrich and H. J. Weinert. V, 221 pages. 1981.

Vol. 856: R. Lascar, Propagation des Singularités des Solutions d'Equations Pseudo-Différentielles à Caractéristiques de Multiplicités Variables. VIII, 237 pages. 1981.

Vol. 857: M. Miyanishi. Non-complete Algebraic Surfaces. XVIII, 244 pages. 1981.

Vol. 858: E. A. Coddington, H. S. V. de Snoo: Regular Boundary Value Problems Associated with Pairs of Ordinary Differential Expressions. V, 225 pages. 1981.

Vol. 859: Logic Year 1979–80. Proceedings. Edited by M. Lerman, J. Schmerl and R. Soare. VIII, 326 pages. 1981.

Vol. 860: Probability in Banach Spaces III. Proceedings, 1980. Edited by A. Beck. VI, 329 pages. 1981.

Vol. 861: Analytical Methods in Probability Theory. Proceedings 1980. Edited by D. Dugué, E. Lukacs, V. K. Rohatgi. X, 183 pages. 1981.

Vol. 862: Algebraic Geometry. Proceedings 1980. Edited by A. Libgober and P. Wagreich. V, 281 pages. 1981.

Vol. 863: Processus Aléatoires à Deux Indices. Proceedings, 1980. Edited by H. Korezlioglu, G. Mazziotto and J. Szpirglas. V, 274 pages. 1981.

Vol. 864: Complex Analysis and Spectral Theory. Proceedings, 1979/80. Edited by V. P. Havin and N. K. Nikol'skii. VI, 480 pages. 1981.

Vol. 865: R. W. Bruggeman, Fourier Coefficients of Automorphic Forms. III, 201 pages. 1981.

Vol. 866: J.-M. Bismut, Mécanique Aléatoire. XVI, 563 pages. 1981.

Vol. 867: Séminaire d'Algèbre Paul Dubreil et Marie-Paule Malliavin. Proceedings, 1980. Edited by M.-P. Malliavin. V, 476 pages. 1981.

Vol. 868: Surfaces Algébriques. Proceedings 1976–78. Edited by J. Giraud, L. Illusie et M. Raynaud. V, 314 pages. 1981.

Vol. 869: A. V. Zelevinsky, Representations of Finite Classical Groups. IV, 184 pages. 1981.

Vol. 870: Shape Theory and Geometric Topology. Proceedings, 1981. Edited by S. Mardešić and J. Segal. V, 265 pages. 1981.

Vol. 871: Continuous Lattices. Proceedings, 1979. Edited by B. Banaschewski and R.-E. Hoffmann. X, 413 pages. 1981.

Vol. 872: Set Theory and Model Theory. Proceedings, 1979. Edited by R. B. Jensen and A. Prestel. V, 174 pages. 1981.

Vol. 873: Constructive Mathematics, Proceedings, 1980. Edited by F. Richman. VII, 347 pages. 1981.

Vol. 874: Abelian Group Theory. Proceedings, 1981. Edited by R. Göbel and E. Walker. XXI, 447 pages. 1981.

Vol. 875: H. Zieschang, Finite Groups of Mapping Classes of Surfaces. VIII, 340 pages. 1981.

Vol. 876: J. P. Bickel, N. El Karoui and M. Yor. Ecole d'Eté de Probabilités de Saint-Flour IX – 1979. Edited by P. L. Hennequin. XI, 280 pages. 1981.

Vol. 877: J. Erven, B.-J. Falkowski, Low Order Cohomology and Applications. VI, 126 pages. 1981.

Vol. 878: Numerical Solution of Nonlinear Equations. Proceedings, 1980. Edited by E. L. Allgower, K. Glashoff, and H.-O. Peitgen. XIV, 440 pages. 1981.

Vol. 879: V. V. Sazonov, Normal Approximation – Some Recent Advances. VII, 105 pages. 1981.

Vol. 880: Non Commutative Harmonic Analysis and Lie Groups. Proceedings, 1980. Edited by J. Carmona and M. Vergne. IV, 553 pages. 1981.

Vol. 881: R. Lutz, M. Goze, Nonstandard Analysis. XIV, 261 pages. 1981.

Vol. 882: Integral Representations and Applications. Proceedings, 1980. Edited by K. Roggenkamp. XII, 479 pages. 1981.

Vol. 883: Cylindric Set Algebras. By L. Henkin, J. D. Monk, A. Tarski, H. Andréka, and I. Németi. VII, 323 pages. 1981.

Vol. 884: Combinatorial Mathematics VIII. Proceedings, 1980. Edited by K. L. McAvaney. XIII, 359 pages. 1981.

Vol. 885: Combinatorics and Graph Theory. Edited by S. B. Rao. Proceedings, 1980. VII, 500 pages. 1981.

Vol. 886: Fixed Point Theory. Proceedings, 1980. Edited by E. Fadell and G. Fournier. XII, 511 pages. 1981.

Vol. 887: F. van Oystaeyen, A. Verschoren, Non-commutative Algebraic Geometry, VI, 404 pages. 1981.

Vol. 888: Padé Approximation and its Applications. Proceedings, 1980. Edited by M. G. de Bruin and H. van Rossum. VI, 383 pages. 1981.

Vol. 889: J. Bourgain, New Classes of $\mathcal{L}^p$-Spaces. V, 143 pages. 1981.

Vol. 890: Model Theory and Arithmetic. Proceedings, 1979/80. Edited by C. Berline, K. McAloon, and J.-P. Ressayre. VI, 306 pages. 1981.

Vol. 891: Logic Symposia, Hakone, 1979, 1980. Proceedings, 1979, 1980. Edited by G. H. Müller, G. Takeuti, and T. Tugué. XI, 394 pages. 1981.

Vol. 892: H. Cajar, Billingsley Dimension in Probability Spaces. III, 106 pages. 1981.

Vol. 893: Geometries and Groups. Proceedings. Edited by M. Aigner and D. Jungnickel. X, 250 pages. 1981.

Vol. 894: Geometry Symposium. Utrecht 1980, Proceedings. Edited by E. Looijenga, D. Siersma, and F. Takens. V, 153 pages. 1981.

Vol. 895: J.A. Hillman, Alexander Ideals of Links. V, 178 pages. 1981.

Vol. 896: B. Angéniol, Familles de Cycles Algébriques – Schéma de Chow. VI, 140 pages. 1981.

Vol. 897: W. Buchholz, S. Feferman, W. Pohlers, W. Sieg, Iterated Inductive Definitions and Subsystems of Analysis: Recent Proof-Theoretical Studies. V, 383 pages. 1981.

Vol. 898: Dynamical Systems and Turbulence, Warwick, 1980. Proceedings. Edited by D. Rand and L.-S. Young. VI, 390 pages. 1981.

Vol. 899: Analytic Number Theory. Proceedings, 1980. Edited by M.I. Knopp. X, 478 pages. 1981.

Vol. 900: P. Deligne, J. S. Milne, A. Ogus, and K.-Y. Shih, Hodge Cycles, Motives, and Shimura Varieties. V, 414 pages. 1982.

Vol. 901: Séminaire Bourbaki vol. 1980/81 Exposés 561–578. III, 299 pages. 1981.

Vol. 902: F. Dumortier, P.R. Rodrigues, and R. Roussarie, Germs of Diffeomorphisms in the Plane. IV, 197 pages. 1981.

Vol. 903: Representations of Algebras. Proceedings, 1980. Edited by M. Auslander and E. Lluis. XV, 371 pages. 1981.

Vol. 904: K. Donner, Extension of Positive Operators and Korovkin Theorems. XII, 182 pages. 1982.

Vol. 905: Differential Geometric Methods in Mathematical Physics. Proceedings, 1980. Edited by H.-D. Doebner, S.J. Andersson, and H.R. Petry. VI, 309 pages. 1982.

Vol. 906: Séminaire de Théorie du Potentiel, Paris, No. 6. Proceedings. Edité par F. Hirsch et G. Mokobodzki. IV, 328 pages. 1982.

Vol. 907: P. Schenzel, Dualisierende Komplexe in der lokalen Algebra und Buchsbaum-Ringe. VII, 161 Seiten. 1982.

Vol. 908: Harmonic Analysis. Proceedings, 1981. Edited by F. Ricci and G. Weiss. V, 325 pages. 1982.

Vol. 909: Numerical Analysis. Proceedings, 1981. Edited by J.P. Hennart. VII, 247 pages. 1982.

Vol. 910: S.S. Abhyankar, Weighted Expansions for Canonical Desingularization. VII, 236 pages. 1982.

Vol. 911: O.G. Jørsboe, L. Mejlbro, The Carleson-Hunt Theorem on Fourier Series. IV, 123 pages. 1982.

Vol. 912: Numerical Analysis. Proceedings, 1981. Edited by G. A. Watson. XIII, 245 pages. 1982.

Vol. 913: O. Tammi, Extremum Problems for Bounded Univalent Functions II. VI, 168 pages. 1982.

Vol. 914: M. L. Warshauer, The Witt Group of Degree k Maps and Asymmetric Inner Product Spaces. IV, 269 pages. 1982.

Vol. 915: Categorical Aspects of Topology and Analysis. Proceedings, 1981. Edited by B. Banaschewski. XI, 385 pages. 1982.

Vol. 916: K.-U. Grusa, Zweidimensionale, interpolierende Lg-Splines und ihre Anwendungen. VIII, 238 Seiten. 1982.

Vol. 917: Brauer Groups in Ring Theory and Algebraic Geometry. Proceedings, 1981. Edited by F. van Oystaeyen and A. Verschoren. VIII, 300 pages. 1982.

Vol. 918: Z. Semadeni, Schauder Bases in Banach Spaces of Continuous Functions. V, 136 pages. 1982.

Vol. 919: Séminaire Pierre Lelong – Henri Skoda (Analyse) Années 1980/81 et Colloque de Wimereux, Mai 1981. Proceedings. Edité par P. Lelong et H. Skoda. VII, 383 pages. 1982.

Vol. 920: Séminaire de Probabilités XVI, 1980/81. Proceedings. Edité par J. Azéma et M. Yor. V, 622 pages. 1982.

Vol. 921: Séminaire de Probabilités XVI, 1980/81. Supplément Géométrie Différentielle Stochastique. Proceedings. Edité par J. Azéma et M. Yor. III, 285 pages. 1982.

Vol. 922: B. Dacorogna, Weak Continuity and Weak Lower Semicontinuity of Non-Linear Functionals. V, 120 pages. 1982.

Vol. 923: Functional Analysis in Markov Processes. Proceedings, 1981. Edited by M. Fukushima. V, 307 pages. 1982.

Vol. 924: Séminaire d'Algèbre Paul Dubreil et Marie-Paule Malliavin. Proceedings, 1981. Edité par M.-P. Malliavin. V, 461 pages. 1982.

Vol. 925: The Riemann Problem, Complete Integrability and Arithmetic Applications. Proceedings, 1979-1980. Edited by D. Chudnovsky and G. Chudnovsky. VI, 373 pages. 1982.

Vol. 926: Geometric Techniques in Gauge Theories. Proceedings, 1981. Edited by R. Martini and E.M.de Jager. IX, 219 pages. 1982.